Middle Eastern Oil and the Western World

PROSPECTS and PROBLEMS

THE MIDDLE EAST

Economic and Political Problems and Prospects

Studies from a research program of
The Rand Corporation
and
Resources for the Future, Inc.
Sidney S. Alexander, *Program Director*

Published

Marion Clawson, Hans H. Landsberg,
and Lyle T. Alexander
The Agricultural Potential of the Middle East

Sam H. Schurr *and* Paul T. Homan
with Joel Darmstadter, Helmut Frank, John J. Schanz, Jr.,
Thomas R. Stauffer, *and* Henry Steele
Middle Eastern Oil and the Western World
PROSPECTS AND PROBLEMS

In Preparation

Sidney S. Alexander
The Economics and Politics of the Middle East

Charles A. Cooper *and* Sidney S. Alexander *(eds.)*
Economic Development and
Population Growth in the Middle East

Paul Y. Hammond *and* Sidney S. Alexander *(eds.)*
Political Dynamics in the Middle East

Middle Eastern Oil and the Western World

PROSPECTS and PROBLEMS

Sam H. Schurr

Director, Energy and Minerals Studies
Resources for the Future, Inc.
Washington, D.C.

Paul T. Homan

Consultant, Resources for the Future, Inc.
Washington, D.C.

**with Joel Darmstadter, Helmut Frank
John J. Schanz, Jr., Thomas R. Stauffer
and Henry Steele**

American Elsevier Publishing Company, Inc.

NEW YORK · 1971

AMERICAN ELSEVIER PUBLISHING COMPANY, INC.
52 Vanderbilt Avenue, New York, N.Y. 10017

ELSEVIER PUBLISHING COMPANY, LTD.
Barking, Essex, England

ELSEVIER PUBLISHING COMPANY
335 Jan Van Galenstraat, P.O. Box 211
Amsterdam, The Netherlands

International Standard Book Number 0-444-00094-1

Library of Congress Card Number 72-135059

Manufactured in the United States of America

CONTENTS

ILLUSTRATIONS

TABLES

Foreword

NUMEROUS studies of the economic and related development problems of the Middle Eastern countries have accumulated over the years as individual scholars and various institutes have undertaken to probe the history and future prospects of particular countries or particular industries or natural resources. But these separate studies, important though each one may be, do not add up to an orderly and comprehensive view of the economic development potentialities and problems of that great region of the world. Without such a view, as up-to-date as possible, it is difficult for those in the Middle East and elsewhere concerned with policies and decisions relating to Middle Eastern economic development to discharge their responsibilities effectively.

With this situation in mind, The RAND Corporation and Resources for the Future, Inc., responded to a request from the Ford Foundation to cosponsor, with Ford Foundation support, a series of studies which would aim to delineate the main economic development and related potentialities and problems of the Middle East region, stretching—for some of the studies—from Iran on the east through Morocco on the west. In addition to its role as cosponsor, RFF took particular responsibility for two of the principal component studies: *The Agricultural Potential of the Middle East* by Marion Clawson, Hans Landsberg, and Lyle Alexander; and the study presented here, *Middle Eastern Oil and the Western World: Prospects and Problems.*

Oil has been and for some time will continue to be a leading element in the economies of many Middle Eastern countries. As Mr. Schurr and his associates point out, for the five major oil-exporting countries considered in this study—Iran, Iraq, Kuwait Saudi Arabia, and Libya—revenues directly associated with oil have accounted for between 70 and 90 percent of foreign-exchange earnings during recent years and between 60 and 90 percent of government income, with variations from country to country. Imports of oil from these countries are also of tremendous importance to the countries of Western Europe, to Japan, and to a number of other consuming countries in various parts of the world. Oil is by far the leading commodity in international trade; and the commercial link between those countries that sell and those that buy oil is provided by great industrial companies involved in exploration, discovery, development, production, transportation, refining, and marketing. The web of policies and arrangements, both public and private, by means of which Middle Eastern oil is produced and brought to market is extremely complex and responds to a variety of forces—economic, political, and technological.

The present study describes important aspects of this total activity, delineates certain main trends and problems, and seeks to lay a basis for an understanding of future prospects. It aims also to bring out the symbiotic relationship—at times imperiled by temporary disturbances and their aftermaths—between the Middle Eastern oil-producing countries and the consuming countries of the Western world.

This study, made under the direction of its principal author, Sam H. Schurr, brings together in an organized way the contributions of a number of scholars experienced in research on this subject. Paul T. Homan joined Mr. Schurr as coauthor; together they are responsible for the overall analysis and conclusions of the study. Helmut Frank, Professor of Economics at the University of Arizona, and John J. Schanz, Jr., Professor of Natural Resources at the University of Denver, prepared the background paper on which the analysis of prospects for unconventional sources of liquid fuels was based. Thomas R. Stauffer of the Harvard Center for Middle Eastern Studies wrote the background papers for the analysis of the dependence of the exporting countries on oil earnings; he also contributed other materials used in Part II. Henry Steele, Professor of Economics at the University of Houston, developed background studies used in the analysis of oil prospects in the North American Arctic and in the estimation of excess productive capacity in the Western Hemisphere. Joel Darmstadter, a member of the RFF staff, prepared the Statistical Appendix, which served as the quantitative basis for the projections of consumption, production, and trade flows used throughout the study.

Two additional points should be noted: (1) Research on many specific matters covered in this study has been done by others and the authors have drawn heavily on such analyses, as well as on the critical comments of specialists who reviewed the draft manuscript. (2) Although this is essentially an economic study, a realistic treatment of the subject requires that noneconomic factors be taken into account. An effort has been made to go no further than necessary in the treatment of political and strategic questions, so that even the analysis of supply interruptions concentrates on the economic dimensions of this problem and the methods for dealing with it.

This study stands on its own feet as a treatment of the prospects and problems of Middle Eastern oil in relation to the Western importing countries, but at the same time forms a unit in the larger RAND-RFF program of Middle East studies, being prepared under the overall direction of Sidney Alexander, consultant to RAND and Professor of Economics at MIT. It is hoped that readers involved in oil matters will find in this book some clarification of the major issues and a better basis for understanding both the present situation and the future possibilities.

Washington, D.C.
January 1970

Joseph L. Fisher, President
Resources for the Future, Inc.

Acknowledgments

THIS book covers so wide a range of subject matter that its preparation would not have been possible without the wholehearted cooperation of all those listed as authors. Certain of the aspects covered would, in fact, have been outside our reach without the specialized competence brought to their subjects by Joel Darmstadter, Helmut J. Frank, John J. Schanz, Jr., Thomas R. Stauffer, and Henry Steele. To each of them I express my deep appreciation.

The final responsibility for the organization and content of the study and its conclusions is that of the two senior authors. Paul T. Homan, who died shortly after the first draft of the manuscript was completed, was a pillar of strength throughout the one-year period during which the basic research, analysis, and writing were done; whatever merit the study possesses is due in no small measure to him.

We profited greatly from the review of draft manuscript by many individuals whose generous assistance is gratefully acknowledged: M. A. Adelman, Georges Brondel, Paul H. Frankel, Mac Goodman, Brandon H. Grove, J. E. Hartshorn, Joseph Lerner, Jonathan Levin, V. E. McKelvey; Edith Penrose, Harry Perry, E. F. Schumacher, John J. Simpson, Edward Symonds, A. Tsumura, and J. A. van den Heuvel. The penetrating comments of Sidney S. Alexander, the director of the broad RAND-RFF project of which this book is one part, and of Walter J. Levy, acting as a member of an advisory committee to the project, were of great value. A great amount of technical review was received also from members of several oil companies who read the manuscript in draft.

The burdens of the authors were lightened considerably by the participation within RFF of Sally S. Nishiyama, who was research assistant to the group, and of Helen-Marie Streich who attended to our secretarial needs.

Sam H. Schurr

Summary and Conclusions

AN outstanding feature of the international oil trade is the economic interdependence it has created between two groups of countries—the large producing countries of the Persian Gulf region and North Africa and the large consuming countries of Western Europe and Japan. The present study is an examination of the prospects and problems presented by this interdependence, currently and as projected into the future.

For brevity and convenience, the "Middle East" is commonly referred to as the locus of the principal Eastern Hemisphere producers for international trade and this elliptical usage is adopted in the title of this study. However, North Africa must also be included because of its great and growing importance.[1] Similarly, our title uses the short term "Western World" for the principal consuming countries. But these include not just the United States and the importing countries of Western Europe, but also Japan, the largest single oil-importing country in the world.

The United States, the world's largest oil market, is a relatively unimportant participant in the oil trade with the Middle East and North Africa, but figures prominently in the study for three main reasons. (1) It is the world's largest oil consumer, as well as the largest producer, providing about 80 percent of its consumption from domestic sources, with the remainder originating mainly elsewhere in the Western Hemisphere. However, the present relative level of imports of the United States cannot be taken for granted, and if it were to change significantly, in whatever direction, there would be an enormous impact on world oil flows. (2) Excess productive capacity to produce liquid fuels in the United States has, in the past, been a factor in meeting supply emergencies in Western Europe, and it is often thought of as serving the same function in the future, if needed. (3) Five of the seven major international oil companies, which constitute the main commercial link between the oil exporting and importing countries are U.S. corporations. These companies explore, produce, transport, refine and market the major fraction of the world's oil supply moving in international trade.

The book contains two parts. Part I measures the interdependence between countries of the Western world and Japan as oil importers, and countries of the Middle East and North Africa as oil exporters, and assesses possible developments over the next decade. Because of the prospect of great and growing interdependence, the problem of how to reduce the potential importance of supply interruptions is analyzed. Various possibilities for doing so are examined for the period running to 1980, within the context of international oil flows as presently projected. In the last chapter of Part I, the potentials and implications of the North American Arctic, as they can be perceived at this early date, are examined; however, the possible impact of developments, particularly on the North Slope of Alaska, is also considered in other chapters, when judged to be especially relevant.

Part II deals with the economic importance of oil operations to the exporting countries. Two broad matters are examined. One is the extent to which these countries depend upon the earnings from oil exports, and upon other economic effects flowing from oil industry operations. The other is the matter of changing relationships between the host countries in which the oil is produced and the major oil companies which conduct most oil industry operations. Not only have these relationships undergone change in the past, they appear to be headed for additional, perhaps more fundamental, changes in the future. The nature of the changes and their implications are examined because of the challenges they present to, and the important impact they may have on, the producing and consuming countries, the companies, and the international oil trade.

The remainder of the present chapter presents the broad summary and conclusions of the entire study. The major findings are summarized first. This is followed by a concluding section in which the central problems, as we see them, are identified and discussed: the problems presented (1) by the possibility of future supply interruptions, and (2) by the future course of the relationships between the producing countries and the large international oil companies.

Summary of Major Findings

The Dependence of the Western World on Middle Eastern Oil Supplies

Attention in this study is centered on the United States, Western Europe, and Japan as the major consumer areas. In 1968, the world's total consumption of oil came to approximately 39 million barrels/day or 1.9 billion metric tons for the year.[2] Within this total, the 13 million barrels/day (636 million tons/year) consumed by the United States was far and away the largest of any country, accounting for about one-third. The countries of Western

[1] The following observation appears in a publication of the Middle East Institute: "There are no precise boundaries of the Middle East, but the term has long been considered to embrace the states from Egypt to the Persian Gulf and from Turkey to Iran to the Indian Ocean. Recently it has become general practice, adopted by the MEI, to group the countries of North Africa with those of the Middle East when considering the strategic, economic and cultural aspects of American interests and concerns in the area." (*American Interests in the Middle East,* [Washington: The Middle East Institute, 1969], p. 1).

[2] On the basis of average specific gravity of crude oil around the world, 1 barrel/day equals approximately 50 metric tons/year; or, 1 metric ton equals approximately 7.2 barrels. Text references to both tons and barrels usually reflect this approximate conversion, although in some instances the underlying sources have used more precise figures. Such variations will result in slightly different conversion factors for different regions.

Europe, taken as a group, were the second largest consuming area with 10 million barrels/day (500 million tons/year), a share of the world's total of approximately 25 percent. Japan's oil consumption, 2.8 million barrels/day (nearly 140 million tons/year), was equal to about 7 percent of the world's total; its consumption, although well below that of the United States, ranked second among the individual non-Communist countries of the world. Together, then, the United States, Western Europe, and Japan accounted for about two-thirds of the world's oil consumption.

The Communist countries, taken together, now account for about 15 percent of the total world consumption. However, they are essentially self-sufficient. Out of a total production of about 7 million barrels/day in 1968, they were able to export about 1 million barrels/day to the rest of the world. According to recent Western appraisals, necessarily conjectural, their net export position is likely to decline. In this study it has been assumed that the Communist countries, as a group, will neither be drawing upon nor competing with Middle Eastern and North African supplies to an extent sufficient to invalidate the broad picture here described.

1. *United States*

The 13 million barrels/day of liquid fuels consumption accounted for by the United States in 1968, is projected in this study to reach a level of about 16 million barrels/day in 1975 and 18½ million barrels/day in 1980. U.S. consumption would thus decline from its current 33 percent share of world consumption to about 27 percent in 1975 and to less than 25 percent in 1980. However, throughout this period the United States will continue to rank first among the oil-consuming countries of the world. The importation of oil is currently subject to severe governmental restriction, with the net result that only about 20 percent of total liquid fuels consumption consists of imports. Of these imports, only a small portion (about 18 percent, or about 3½ percent of total oil consumption) originated in the Middle East and North Africa in the mid-1960's.

A major question posed in Chapter 2 is whether the assumption employed in this study's projections—that U.S. imports can continue to be held at no more than about 20 percent of total oil consumption over the next decade or so at a cost penalty to the economy comparable to that now being incurred—is a realistic one.

The conclusion reached is that the comprehensive analysis of U.S. oil prospects published by the U.S. Department of the Interior in mid-1968 cannot support an unqualified "yes" in reply to this crucial question. However, in light of the potentials of Northern Alaska, as inferred from recent discoveries, and also taking into account the fallback position provided by liquid fuels from unconventional sources in North America (as described in Chapter 4), we conclude that the United States would face no serious problem in holding the import line up to 1980 if it should choose to do so.[3]

For unconventional sources of liquid fuels (shale oil, tar sands, and coal), more than one technology appears to be on the borderline of economic feasibility. On this basis alone it would be possible to conclude that if a judgment were to be reached at an early date that crude oil from conventional sources threatened to be inadequate, accelerated research and development on the unconventional sources of liquid fuels could result in commercial facilities to provide supplies—at cost levels comparable to those of crude oils now produced in the United States—in sufficient amounts to sustain the level of liquid fuels self-sufficiency of the late 1960's. This is an important option for the United States (probably involving cooperation with Canada where tar sands are concerned) to have at its disposal.

However, the intense activity of oil companies in Northern Alaska, although of far more recent origin than the experimentation with unconventional sources, appears, as of 1969, to make the Arctic a more likely prospective source of sizeable liquid fuels supplies for the United States over the next decade than such substitutes as shale oil, tar sands, and coal. A new field described as "one of the largest petroleum accumulations known to the world today" was discovered at Prudhoe Bay in 1968; and a great volume of additional exploratory activity, which has resulted in further discoveries, is under way. Even if vast accumulations of oil should exist there, severe technical problems in production and transportation will need to be overcome. The oil industry has an outstanding record for overcoming natural obstacles; consequently, optimism runs high, and billions of dollars of oil company money have already been committed to lease purchases and to the effort to find oil and meet the associated problems of production and transport.

(Alaska may in fact pose an unexpectedly new problem in the United States at a relatively early date. If and when large Alaskan supplies become available at low delivered costs relative to much of the production in the "lower forty-eight," a major problem would be posed for domestic policy: Can the U.S. market absorb Alaskan supplies without significant changes in the regulatory systems in most of the major oil-producing states? These systems of regulation, among their other results, serve to limit statewide production to "market demand" and to allocate

---

[3] A major Cabinet-level review of U.S. oil import policy was under way as this study was drawing to a close. Our conclusion that the United States could hold the import line for another decade if it were to choose this course is not meant as a prediction of the policy that will result from the new government study, nor as a recommendation that this should be the policy conclusion. Moreover, it is important to bear in mind that our conclusion refers to the relative importance of total imports into the United States as a percentage of oil consumption. Even if this proportion were to rise, for whatever reason, this could happen without a significant increase in the amounts originating in the Middle East and North Africa—the producing regions upon which this study is focused. (See also Postscript to this chapter, pp. 18-19.)

shares of the total to all producers within the affected states.)

2. *Western Europe*

The 25 percent share of the world's oil consumed by Western Europe in the mid-1960's is projected to remain at about the same proportion in the decade of the 1970's. The absolute quantities consumed are estimated to rise from about 10 million barrels/day (500 million tons/year) in 1968 to about 14½ million barrels/day (730 million tons/year) in 1975, and to over 18 million barrels/day (more than 900 million tons/year) in 1980. Throughout this period Western Europe will remain the leading market for international oil, originating mainly in the Middle East and North Africa, with some increase in the absolute and relative amounts coming from West Africa.

Approaching 50 percent of total inland energy consumption in the mid-1960's, oil is projected to supply over 60 percent by 1980. The rapid transformation of Western Europe after 1955 from a coal-based to an oil-based energy economy, and the virtual absence of indigenous oil deposits, makes importation of oil supplies critical.*

The estimate that oil will supply about 60 percent of Western Europe's total energy in 1980 is, of course, related to collateral estimates for other sources of energy—coal, natural gas, and nuclear power. Coal has become an increasingly uneconomic source of energy, and is likely to supply no more than 15 percent in 1980 (compared to nearly 40 percent in recent years), even allowing for protective measures—designed partly to maintain some reliance on indigenous sources of energy, partly to soften the impact on mining employment and the local and regional repercussions that would follow the loss of jobs by miners.

The recent discovery of rich deposits of natural gas could ease the dependence on imported oil; but it remains to be seen whether the amounts discovered and the pace of development will place it in a more important position than the somewhat more than 10 percent of total energy in Western Europe estimated for 1980. As for nuclear energy as a source of electrical power, even under the most optimistic estimates it appears likely to play a relatively modest role in total energy supply—at least up to 1980, when it is carried in our estimates at slightly over 10 percent of the West European total.

Given the possibilities of these alternative sources of energy, there is a chance (albeit an unlikely one) of a lower reliance on oil than the 60 percent estimate of this study, but probably at a substantial economic cost to the European nations (particularly if this were to be achieved through greater reliance on coal), and hardly to a point lower than 50 percent for Western Europe as a whole.

3. *Japan*

Japan's share of world oil consumption is estimated to rise from 7 percent in 1968 to more than 10 percent by the end of the 1970's. In absolute terms consumption is projected to increase from 2.8 million barrels/day (140 million tons/year) in 1968 to almost 5 million barrels/day (240 million tons/year) in 1975 and over 7½ million barrels/day (385 million tons/year) in 1980. These supplies will continue to originate largely in the Middle East, although Southeast Asia (particularly Indonesia) will become an increasingly important source.

The Japanese energy economy, like that of Western Europe, has been undergoing a rapid transformation, moving from a heavy reliance on coal and hydroelectric power in 1955 to a dependence on oil of well over 50 percent in the mid-1960's and an estimated 80 percent in 1980. And, as in Europe, the virtual absence of indigenous oil deposits places the importation of oil in a position of crucial importance.

The possibilities of greatly modifying this dependence on imported oil appear to be slight in the period to 1980. Reserves of coal are limited; and the level of production is falling off in spite of subsidized assistance. Indigenous supplies of natural gas are slight; and while liquefied natural gas may be more extensively used, it is probable that supplies would in large measure have to be imported from the same places as oil. Conditions for expansion of hydropower are not favorable. And there is considerable uncertainty regarding the rate of expansion of nuclear power, estimated in this study to be less than 5 percent of total energy in Japan by 1980.

Taking full account of the possibilities of these alternative sources of energy, as indicated in the most favorable estimates for their use, there is a chance for a lower reliance on oil than the 80 percent estimate of this study, but probably not to a point lower than 75 percent.

Diversification in Sources of Liquid Fuels Supplies

For many years, and particularly since the Suez crisis of 1956/57, the governments of countries in which imports are of major importance have been acutely conscious of the problem posed by the vulnerability of their sources of supply of oil. For the Western European countries and Japan, the problem arises out of their almost complete reliance on imports. The United States, though barely affected directly, has been concerned because of its

*A major North Sea oil discovery was announced while this book was in press (see Chap. 3, footnote 21). In early reports a rate of output of about 1 million barrels/day from this oil field was indicated as a possibility.

It would obviously take many such fields to bring about a significant reduction in Western Europe's dependence on the Middle East and North Africa; and much time would be needed for oil field exploration and development and to construct the necessary production and transportation facilities.

As of mid-summer 1970, a letter to us from the OECD (on whose estimates we have placed heavy reliance in Chap. 3), stated that the Organisation did not have "any estimates as yet on the possible importance of the recent oil discoveries in the North Sea for OECD Europe over the next decade."

contingent liability as a supplier to Western Europe in times of emergency. And there has been the possibility that the United States, too, might find itself more heavily dependent on imports from the Middle East and North Africa in the future. In light of these factors—and also for logistical, tactical, and economic reasons of their own—the oil companies have long been interested in developing supplies in a number of different regions. One approach, which has to date met with only slight success, has been to try to achieve a commercial status for those processes that involve the derivation of liquid fuels from unconventional sources in place of crude oil. The other approach, pushed with far more vigor, has been to achieve greater geographic diversification in crude oil supplies.

1. *Geographic Diversification in the Period Between the Mideastern Crises of 1956/57 and 1967*

Geographic diversification in crude oil supplies is always a lively prospect because oil deposits, although far from ubiquitous, do exist in many regions of the world. Supply diversification did indeed take place in the decade between the Suez crisis and the Arab-Israeli war, but there was no decrease in the degree of dependence on Middle Eastern and North African sources. This is because the principal change was the rise of North Africa, particularly Libya, as a prolific source of oil. In 1956, Western Europe imported 82 percent of its crude oil from the Middle East. In 1966, this had fallen to about 60 percent (though the volume of imports from the Middle East had risen by about 150 percent), while North Africa, by 1966, was supplying 25 percent of West European crude oil imports compared to practically nothing in 1956. In the case of Japan, the dependence on Middle Eastern crude oil increased during the decade from 80 percent in 1956 to 90 percent in 1966.

The problem of supply for Western Europe during and after the Suez crisis of 1956/57 was created primarily by the closure of the Suez Canal and the necessity of the long sea haul around Africa. Consequently, the shift in source to west-of-Suez was a welcome development viewed in the light of the earlier crisis. The Arab-Israeli war of 1967, in addition to the closure of the Canal, introduced a hazard of a different sort: the stoppage of supplies to some countries for political reasons by concerted action of the Arab producing countries. Against this new type of danger the shift towards Libya was of no help: it responded to the same pressures as other Arab countries in interrupting supplies in 1967.

During the decade only West Africa, mainly Nigeria, emerged as a potentially large and "politically safe" source of supply. But supplies from Nigeria have also been interrupted by political disorders, although completely unrelated to the disturbances in the Middle East. In any case, West Africa's 4 percent share of European imports was even in 1966 (prior to the disruption of Nigeria's production) quite small; and in 1968, the share, reflecting continuing hostilities in the area, had dropped to 2 percent. Among the major Middle Eastern producing countries the Iranian share rose somewhat, and proved to be a dependable source during the 1967 supply interruption.

As of the end of 1966, the Middle East and North Africa accounted for 73 percent of the proved reserves of oil outside the Communist countries, compared to less than 70 percent for the Middle East alone as of year-end 1955. In 1966, the two regions accounted for 41 percent of non-Communist world production compared to 22 percent in 1956. North America and the Caribbean region accounted for 52 percent of production in 1966, most of it directed to North American markets.

Western Europe and Japan are thus dependent for their supplies predominantly on five countries—Iran, Iraq, Kuwait, Saudi Arabia, and Libya—which among them controlled 85 percent of all the oil produced in the Middle East and North Africa in 1966. (In 1968 the comparable figure had risen to 88 percent.) Without Iran, the other four countries—all Arab—controlled about 70 percent of production within the broad region; and it is explosive political forces affecting the Arab world that create most of the uneasiness about interruptions of supply from that source. It is, of course, possible to discount the hazards because the source countries are themselves so fully dependent upon their exports of oil. (See Chapter 8 and pp. 9ff., below.) But the experience of 1967, when the Arab nations did act in concert—albeit for a short period—to cut off supplies to specific countries, is still an active memory.

2. *Geographic Diversification in the Future: The North American Arctic*

Although numerous prospects for discovery and development exist that could result in a further regional diversification of world oil supplies (and some of the more likely ones, such as West Africa and Indonesia, are already reflected in our projections), the North American Arctic now appears to be the most probable region of development on a massive scale in the next decade. An enormously large field, discovered in Northern Alaska in 1968, is the only significant one whose estimated size had been publicly announced as of late 1969. However, estimates have been made regarding potential discoveries in other sections of this vast region which, if validated, indicate the existence of new oil provinces rivaling those in North Africa.

Geographic diversification, which produced negligible strategic benefits during the decade following the Suez Canal crisis may, therefore, hold significant promise for the years ahead. The region around which the speculation centers encompasses a vast, physically inhospitable area in Northern Alaska and in the Canadian Arctic (including the Arctic Islands). While no firm basis for prediction exists, geological inference and informed speculation appear to support the possibility that Northern Alaska *might* by 1975/80 have "proved-up" oil reserves of as much as 30 to

40 billion barrels, and a productive capacity of between 1 and 2½ million barrels/day by 1975, and 2½ to 5 million barrels/day by 1980. In the Canadian Arctic, recoverable reserves have been placed at anywhere from 43 to 85 billion barrels, and we have estimated that productive capacity there during the decade of the 1970's might match that in Northern Alaska. Most present publicly-stated industry opinion, it should be noted, would regard these estimates as being on the high side, particularly where the Canadian Arctic is concerned.

The only relatively firm information in all this is confined to several commercial strikes in the Prudhoe Bay region of Northern Alaska, placed in the public announcements at upwards of 10 billion barrels of recoverable reserves; and to the commitment to build a pipeline to Valdez, an all-weather port in southcentral Alaska, with the original hope of having it completed in 1972. The initial capacity of this line is placed at 500,000 barrels/day, with an ultimate capacity of about 2 million barrels/day. Nearly all other information publicly available is in the category of speculation: either regarding the possible outcome of exploratory ventures to discover other oil deposits (based essentially upon geological inference), or regarding the possible results of activities, aside from the trans-Alaskan pipeline, relating to the transportation of oil to market. In the latter category, the most likely possibility envisaged for movement through the icebound waters of the Arctic—the icebreaking tanker—received its first test during the late summer of 1969. Although this voyage by the S. S. Manhattan, the largest tanker flying the U.S. flag, which had been especially equipped and modified in various ways to break its way through the ice, demonstrated that an icebreaking tanker could negotiate the Northwest Passage, commercial feasibility has not yet been established. Additional testing is being planned. However, optimism regarding eventual commercial success has been expressed by Humble Oil, the chief financial backer of the test voyage.

What is certain is that the exploratory rush is under way, and that much money is being spent on the search for oil and the solution of the technical problems associated with its development, production, and transportation.

Assuming that possibilities within the range indicated above are realized, how would the North American Arctic reserves compare with those of the Middle East and North Africa? Official estimates of reserves in these latter two regions as of the beginning of 1969 were: Middle East, 270 billion barrels; North Africa, 40 billion barrels. These figures are undoubtedly conservative. Within industry circles underestimation of between 50 and 100 percent of the amounts that could be proved up is not considered unlikely. If a 50 percent underestimation is assumed, a more realistic figure for potential recoverable reserves in the Middle East would be of the order of 400 billion barrels, and for North Africa, 60 billion barrels.

In comparative terms, therefore, estimated recoverable reserves in the North American Arctic, Middle East and North Africa, could shape up as follows:

	Billion barrels
Middle East	270–400 (as of January 1, 1969)
North Africa	40–60 (as of January 1, 1969)
North American Arctic:	
Northern Alaska[a]	30–40 (as of 1975–80)
Canadian Arctic[b]	45–85 (as of 1975–80)

[a] Referred to in the underlying source as reserves estimated to be "proved up" by 1975–80.

[b] Referred to in the underlying sources as estimated recoverable reserves, with no time schedule provided as to the rate at which the reserves might be proved up. We venture the guess that some figure within the range shown, more likely toward the lower end, will be proved up within the decade of the 1970's.

Although not nearly as large as the Middle East, Northern Alaska alone might, according to these estimates, rival North Africa's 1969 proved reserves level by the middle or late 1970's; and, together with the Canadian Arctic, might substantially exceed North Africa's 1969 level by 1980. But this is of course counting unhatched chickens.

Even if it is granted that the North American Arctic might reach these levels, the question still remains: *Will* it? There is, to begin with, the matter of incentives for exploration and development of reserves. Activity is great in the Arctic, but to date, much of this activity is in *anticipation* of further favorable discovery experience; solutions to transportation problems; and profitable delivered prices in major markets, particularly the United States (where restrictive import policies, undergoing Cabinet-level review in 1969 by order of the President, have served to keep prices well above the world market). Activity is, no doubt, also motivated by the political-strategic considerations that favor the development of a major new producing province outside the troubled regions of the Middle East and North Africa. Whether activity will be sustained at a high-enough rate to achieve the reserve levels indicated above must remain a moot question.

An important fact to be borne in mind is that the average reserves to production ratio in the Middle East in 1967, on the basis of official reserves figures, was almost 70:1. The Middle East, therefore, contains enough reserves to satisfy major markets for a long period of years into the future. For those international companies which have a strong position in Middle Eastern reserves, the strongest economic incentive for investing the large sums of money required to establish productive capacity and deliverability in the remote reaches of the Arctic would probably arise out of the ability to reach the rich and relatively high-priced U.S. oil market with supplies that can be produced and delivered at a lesser cost than in the "lower forty-eight" and that are not subject to import restrictions.

In addition, there are the strategic-political factors already referred to; the fact that the various oil companies are differently situated so far as the size of their reserve holdings in the Middle East are concerned; that oil companies are being subjected to increasing economic pressure, resulting in profit erosion, by governments in the Middle East and North Africa (see Part II, Chapter 10); and that supplies from the North American Arctic are also well located to serve foreign markets, if tanker transport problems can be solved.

Assuming that the incentives exist to push ahead at high speed, there remains the question of whether the levels of productive capacity that have been estimated could be achieved within the indicated time periods. Within a period of seven years from the first major discovery, a range of productive capacity of 1 to 2½ million barrels/day has been projected as a possibility for Northern Alaska, and within a period of twelve years, the possible productive capacity has been placed at 2½ to 5 million barrels/day. Inclusion of the Canadian Arctic resulted in a doubling of these estimates.

Without pushing the comparison too far, it may be noted that the midpoints between the high and low estimates of productive capacity for Northern Alaska in 1975 and 1980 are roughly comparable to the levels achieved in Libya over similar periods of time. Adding the Canadian Arctic estimates into the picture implies the equivalent of the development of two "Libyas" over the next decade in the North American Arctic.

Although such an accomplishment is not out of the question, its attainment would require strong motivation. Strategic reasons for diversification away from the Middle East and North Africa probably are stronger now than after the first Suez Canal crisis. As for the North Slope of Alaska (the rough equivalent of one Libya in our estimates), there is, of course, the powerful attraction of the rich, protected U.S. market. If not for the latter factor, so rapid a rate of development in a region in which wholly novel problems will have to be overcome might be considered a completely unwarranted exercise in optimism.

3. *Technological Diversification: Liquid Fuels from Unconventional Sources*[4]

The search for technologies which would yield an alternative to crude oil as a source of liquid fuels has been pursued in the main in North America, which is abundantly supplied with raw materials suitable for this purpose—in the forms of oil shale in the Rocky Mountains region, of tar sands in Canada, and of coal capable of being liquefied. Basic technologies for conversion are known; but, except in the case of Canadian tar sands, they have not reached the stage of commercial exploitation. There is, therefore, uncertainty concerning the state of technology, how costly the fuels might turn out to be, and at what rate commercial production might be expected to develop.

Shale Oil

Oil-bearing shales of high quality are known to contain very large quantities of oil, many times the known reserves of crude oil in the United States, with still larger quantities of lower grades. Different reports exist concerning the nearness of proven technology to commercial exploitation, under existing crude oil prices in the United States. Even assuming an established economic basis for exploitation, there are a number of barriers to rapid development. The bulk of the deposits are on federally-owned lands, and federal leasing policy has yet to be established. In addition, there are other mining claims, with unclear legal status, that complicate the leasing process. There would be gigantic problems of waste disposal. Water supply could be a limiting factor in the extent of exploitation.

On the technical side, despite favorable claims that had already been made for an existing privately-sponsored process, a 1968 U.S. Department of the Interior report proposed a "two-generation" model of experimental plants, the "second generation" incorporating improvements discovered in the first. Under the circumstances envisaged in the Interior Department approach, it appears unlikely that the first full-scale commercial plant could be operative before the mid-1970's. Thereafter, and assuming favorable operating results, plants with a capacity of 50,000 to 100,000 barrels/day might be brought on stream at the rate of one a year, at best. The "most likely" expectation is that 1980 production would not exceed some 200,000 barrels/day, a relatively minor contribution to estimated U.S. liquid fuels consumption at that time.

A factor making for delay is that oil from shale will have to face competition from crude oil as well as from other unconventional oil sources. Business interests seem unlikely to invest heavily in shale oil development if the prospects of new oil supplies from the Arctic region appear promising. And even if the latter do not measure up to the more optimistic expectations, the risk would still be great unless it were clear that a restrictive oil import policy would continue to support domestic U.S. oil prices. Finally, if unconventional sources are deemed necessary, shale development will, of course, have to compete for funds with such alternatives as tar sands and liquefied coal.

---

[4] The prospects for the several sources, as here summarized, are of necessity based on evaluations made prior to a reliable economic assessment of the apparently rich oil strike in Northern Alaska. It is almost needless to point out that the motivation, both public and private, for pushing forward on technology for synthetics is particularly susceptible to the degree of success that will be experienced in the development of commercial oil supplies in Alaska and other parts of the North American Arctic. The pace of unconventional liquid fuels research and development has already slackened as a result of the Alaskan oil discovery. Additional successes in Alaska and in the Canadian Arctic would almost certainly further depress the outlook for unconventional sources. A situation in which the determinants are characterized by so much uncertainty is virtually impossible to pin down with any precision. The quantities here assigned to productive capacity from the various sources that might be achieved during the 1970's must, therefore, be viewed as highly conjectural.

We take it to be significant, though, that high officials of Atlantic Richfield, one of the most successful companies in the developments to date in Alaska, were stating, in late 1969, that the United States "will undoubtedly begin supplementing conventional oil with synthetic crude hydrocarbons within the next ten years." (T.F. Bradshaw, President of Atlantic Richfield Co., the New York *Times*, December 3, 1969, pp. 77, 85.) See also remarks by Robert O. Anderson, Chairman of the Board, Atlantic Richfield Co., before the Society of Petroleum Engineers, Denver, September 30, 1969: "Vast quantities of synthetic crude are available from three sources [tar sands, oil shale, and coal], and the day when they will be needed is fast approaching."

Coal Hydrogenation

Synthetic oil from coal, although successfully produced in Germany during World War II, faces the problem of developing a technology that will permit production within an economic cost margin. The United States has vast reserves of coal that could be converted to liquid form; but the research and development program supported by the government has been on a very modest scale, with a single pilot plant now on the verge of full operation. The technology for movement to large-scale production is less advanced than with shale oil.

Under highly favorable circumstances, developments might move from early pilot plant results to a prototype commercial plant yielding evidence of commercial feasibility by the mid-1970's. On that basis, one or two large commercial plants per year (100,000 barrels/day appears to be the minimum size worth considering) might come on stream in the years between 1977 and 1980. Such a rate of progress is highly problematical, partly from uncertainties in the technological sphere, partly from the same general market uncertainties that affect the prospects for shale oil. The "most likely" possibility would appear to be three plants with a combined capacity of about 300,000 barrels/day by 1980.

An advantage of coal, compared to other sources of synthetic oils, is that substantially less materials would need to be handled for a particular output of liquids so that smaller waste disposal problems would be encountered. In addition, coal is not beset by the leasing and other legal problems that affect shale, owing to the latter's predominant location in federally-owned lands. Many coal deposits are also better located in relation to markets than is shale, but there could be strict limits on the size of a liquefied coal industry based solely upon the availability of sufficiently large deposits of well-located coal reserves not already preempted for other uses.

Tar Sands

The oil-impregnated Athabasca tar sands of central Canada represent one of the greatest oil deposits in the world. Although the total content of the entire formation is probably larger, recoverable pipeline-grade synthetic crude from the tar sands has been estimated at 285 billion barrels, with 85 billion barrels considered currently minable. A privately financed commercial-scale operation to produce oil from these sands made its first delivery to a refinery in 1968. This event climaxed eight years of preliminary work involving engineering, financial, and policy problems. It now remains to be demonstrated whether the operation will be economically successful. The 45,000 barrels/day plant is based on surface strip mining and is probably the minimum size that can be built with a reasonable expectation of profitable operation. (Another firm has received approval for an 80,000 barrels/day plant, but with a start-up of operation delayed until 1976.)

The results to date have been disappointing: after the first six months the venture was reported to be operating at a substantial loss. However, even assuming economically successful experience, the rate of development will depend heavily upon Canadian policy in permitting exploitation. Two main complicating factors are involved. The older, to which an interim solution has been found, resulted from the fact that much of the crude oil-producing capacity in the same region was shut in for lack of a market. Hence there has been the problem of accommodating both the flow of crude oil and tar sands oil within a limited market for Albertan liquid fuels. A subsequent complicating factor, which has led to a postponement of consideration of application for new tar sands capacity, resulted from the inability to assess the impact of the new Northern Alaskan discoveries, and possible discoveries in the Canadian Arctic, on the prospects for Albertan oil. Also involved has been United States policy concerning the level of oil imports that will be permitted from Canada.

Despite the technical and commercial lead held by tar sands over the other unconventional sources, the policy problems, existing technical difficulties, and the Alaskan situation all point to a go-slow approach for a period of years. The current permissible capacity is 150,000 barrels/day and it appears unlikely that this level will be exceeded by 1980.

General Outlook for Unconventional Sources

While the resource base for oil from unconventional sources is immensely large in North America, the prospects for development of new supplies by 1980 appear to be severely limited for technical, commercial, and policy reasons and also to be marked by a high degree of uncertainty. On a "most likely" basis, as viewed under technical circumstances favorable to the development of supplies from these sources, but restrained by the economic uncertainties created by the Arctic developments (as they can presently be taken into account) the analysis in Chapter 4 indicates about 500,000 barrels/day as an estimate of production in the United States from unconventional sources in 1980.

However, if the matter were deemed of sufficient urgency and an early decision were to be made, a stepped-up program of research and development and subsidized operation at least of first-generation facilities probably could increase the rate of development of these new sources, assuming that the technological problems of large-scale operations are inherently soluble on an economically acceptable basis.

We conclude, therefore, that while liquid fuels from unconventional sources stand as a great potential future source of energy, very little is to be expected from them before the mid-1970's, and not much through the remainder of the decade, unless there should be a deliberate governmental decision to push ahead with their development on an accelerated basis. Viewed against the background of available resources and the possibilities for technological progress, a stepped-up program, if it were deemed desirable and decided upon at an early date, could result in perhaps up to 4–5 million barrels/day of capacity by 1980. However, as sketched out in the appendix to Chapter 4, such a program of development would have to cope with formidable problems, such as the processing of enormous amounts of materials and the disposing of unusually large quantities of waste; attracting the necessary labor force and developing entirely new communities; and obtaining sufficient quantities of water required in production. On top of all these, the investment requirements would be very high, and the economic risks great, especially for brand new industries. For these reasons, then, any such level is unlikely to be achieved without governmental stimulation; experience will probably indicate, however, that the need for stimulation to achieve such capacity levels does not exist by reason of the favorable outlook for conventional crude oil supplies, particularly if developments in the Arctic show good progress.

Coping with Supply Interruptions

One of the conclusions of this study, summarized earlier, is that if it should choose to do so the United States would be able, at least over the time period considered in this study—to 1980—to remain as relatively self-sufficient as at present in liquid fuels supplies. This implies either sufficient additions to domestic reserves in the United States, or improvements in the recovery of oil from underground reservoirs, or the creation of synthetic liquid fuels capacity, if required. In any event, U.S. dependence has been primarily on Western Hemisphere sources, so that interruptions in Middle Eastern supplies have not, even in the past, posed a direct threat to this country. The United States has, however, felt a contingent responsibility for helping to meet supply interruptions affecting other countries, particularly in Western Europe.

The situation is entirely different for Western Europe and Japan. For them, as summarized earlier, dependence on oil is projected to increase and with it an increased dependence upon imports, as almost all of the liquid fuels consumed in the two regions will continue to consist of imports. Oil from the Middle East and North Africa is expected to dominate their import totals, although West Africa should become more important as a supply source for Europe, and Indonesia and other Far Eastern producers for Japan.

In formulating the problem that may face Europe and Japan if future supply interruptions were to occur, we have assumed interruptions that might occur in what for them are generally peaceful circumstances. The interruptions in supply, it is assumed, would arise from disturbed conditions in the producing countries and/or as a result of their use of oil as a political or commercial bargaining weapon. War involving the consuming countries would, of course, introduce important factors not here considered—such as the possible destruction of energy-consuming facilities, and the immediate imposition of severe oil rationing. In looking at the problem, we must remember that West European imports from the Middle East and North Africa are projected from nearly 7 million barrels/day in 1967 (the year of the Arab-Israeli war) to over 15 million in 1980, and Japanese imports from the Middle East from 2 million barrels/day to nearly 5.5 million.

Two of the methods for meeting major supply interruptions that could be used, and have to a degree been used in the past, are (1) drawing on existing idle crude oil-producing capacity, including that portion located in the Western Hemisphere; and (2) stockpiling of oil in the importing countries. Rationing, especially in nonessential uses, is of course also available if it appears that supply shortages will not be fully overcome. It would, however, be an unpopular step particularly if adopted to meet a supply crisis brought about by distant disturbances not directly involving the importing country.

1. *Existing Idle Capacity in the Western Hemisphere*

Idle productive capacity in the United States was estimated at about 4 million barrels/day at the beginning of 1968 and projected to 1980 by the U.S. Department of the Interior at 3 million barrels/day (a highly dubious figure).[5] The first figure greatly overstates *deliverability*, which may be placed at around 1 million barrels/day within a 90-day period based upon the 1967 experience, with sustainability at this level and some measure of increase over a longer period of time dependent upon expenditure of funds on additional field and transport facilities. (On possibilities for increases, as of 1967, see the appendix to Chapter 2.) "Instant" Canadian excess deliverable capacity as of 1967 was perhaps less than 200,000 barrels/day but might possibly have been built up by an additional several hundred thousand barrels/day deliverability over a year's time. Venezuelan deliverable excess capacity may be placed at around 500,000 barrels/day.

It is difficult to estimate what the situation might be in the future. We judge the Interior Department's 1968 estimate of 3 million barrels/day for the United States in 1980 to be unrealistically high, and even if it were an acceptable estimate, deliverability within 60-90 days would undoubtedly be very much less. Heavily discounting this figure, and taking into account the excess capacity that might exist in the traditional producing areas of Canada and Venezuela, the amounts in any case would be a very minor fraction of levels of consumption in Western Europe and Japan at that date, though they might somewhat ease temporary shortages in one or two countries. This would constitute a significant change from the situation that has existed in the past, when excess capacity in the Western Hemisphere has served as an important element in meeting supply crises in Europe.

---

[5] A January 1, 1969 estimate comparable to that for 1968 puts excess productive capacity at about 3½ million barrels/day. (American Petroleum Institute, *Reserves of Crude Oil, Natural Gas Liquids, and Natural Gas in the United States and Canada, As of December 31, 1968* [Washington, 1969], pp. 11, 23, and 78.)

2. *Stockpiling of Crude Oil in Importing Countries*

On the assumption that the type of supply interruptions against which protection is required would be of relatively short duration, an approach which has been in effect for some time is the stockpiling of oil in the consuming countries. Recent stated policy of the European Economic Community is for stocks equal to 65-days consumption; Japan is reported to be planning 60-day stocks; Great Britain is said to have maintained stocks of 120 days as of 1966. Stocks equal to 6-months consumption, or higher, have been proposed as a security measure.

On assumptions stated in Chapter 6, we have estimated the investment required for stocks and storage facilities to satisfy 60 days of West European imports from the Middle East and North Africa to be $2.4 billion at projected 1975 levels and $3.0 billion at 1980 levels of consumption. Thus, the investment in building up a 6-month reserve, if that were judged the appropriate level (a matter to be discussed on p. 15ff., below), would reach $7.2 billion by 1975 and $9.0 billion by 1980. To this must be added the cost of land which we were unable to estimate owing to its wide variations, but which would, in any case, probably constitute a minor fraction. While the required capital outlay is large, stockpiling would appear to be the most practical method of insuring against interruptions of supply for limited periods. The "surcharge" or "premium" per barrel of actual oil consumption implied by these costs (about 16 cents for 180 days of stockpile protection, according to our estimates) would still leave imported oil as the cheapest source of energy supply. (All of the above in 1967 prices.)

3. *Implications of the North American Arctic*

The preceding discussion of supply interruptions in Western Europe and Japan and of methods of coping with them did not take the potentials of the North American Arctic into account. Production is not expected to begin to flow from Northern Alaska before late 1972 at the earliest. But by 1975 and 1980 sizeable amounts of oil might be available from the Alaskan and Canadian Arctic.

Compared to estimated Japanese imports from the Middle East, the possible 1975 flows from the North American Arctic are of significance. Looking ahead to 1980, the high estimate of productive capacity in the North American Arctic (its questionable basis must be kept in mind) would constitute a significant fraction of both Japanese and West European imports, equaling about half of their combined projected imports from the Middle East and North Africa. However, a large portion of the total would undoubtedly be consumed in North American markets, particularly in the United States; and this fact alone introduces a major uncertainty in estimating to what extent the Arctic might serve as a future source of imports for Western Europe and Japan.

Although all of this must be cast in hypothetical terms, it does appear that a major new element of flexibility may be in the making so far as European and Japanese supplies are concerned. Consequently, as prospects unfold in the North American Arctic (and in other new "secure" producing regions that may be discovered), they should be kept under study for the opportunities created for further diversification in low-cost sources of vital oil supplies to Europe and Japan. However, until ample, dependable supplies are forthcoming from new "secure" regions such as the Arctic, stockpiling of oil in amounts geared to the import levels from the Middle East and North Africa will continue to be the prudent course for Western Europe and Japan to follow to cope with possible supply interruptions.

When supplies do begin to flow in substantial amounts from the Arctic, it is altogether possible that various factors, both economic and political, will continue to favor the Middle East and North Africa as major sources of supply for Japan and/or Western Europe for a long period of time. Even so, when the necessary facts are in hand the possibility should be studied of creating standby capacity in the North American Arctic to determine whether it might be a less costly way for the European countries and Japan to maintain a security reserve (perhaps beginning in the mid-1970's) than further enlarging their stockpiles. (The appendix to Chapter 6 undertakes a highly conjectural and sketchy analysis of the possibilities offered by the purposeful creation of standby capacity in the North American Arctic, as well as in other "secure" regions of the world, as compared to the building of stocks in the importing countries.)[6]

The Dependence of Middle Eastern Exporting Countries on Oil Earnings

The degree of dependence of the principal consuming countries upon oil supplies from the Middle East and North Africa has its counterpart in the dependence of the oil-producing countries upon the markets of Western Europe and Japan. Out of total Middle Eastern exports of approximately 10½ million barrels/day in 1968, Western Europe took close to one-half. North Africa's exports of over 3½ million barrels/day were sold almost entirely to Western Europe. Out of the *combined* exports of the Middle East and North Africa, Western Europe's 1968 share averaged out to about 60 percent. By 1980, this high overall share is projected as remaining essentially unchanged. Japan took around 25 percent of the Middle East's exports in 1968, and is projected to take nearly 30 percent by 1980.

---

[6] Also, on a "forced draft" basis, it might, as indicated earlier, be possible to accelerate development of unconventional sources—shale oil, synthetic oil from coal, tar sands—up to possibly 4 million barrels/day or more, but at great costs, and only if an early decision were to be made. To hold such capacity on a standby basis might prove impossible because its operation requires a large pool of skilled mining and other labor. On the other hand, its existence might release some crude oil capacity for standby treatment, but to achieve extra capacity in this way would involve a deliberate policy of public investment of billions of dollars (see appendix to Chapter 4).

The economic significance of the quantities involved can be appreciated by examining the financial importance of oil earnings to the exporting countries. In 1967, company payments of oil royalties, taxes, and miscellaneous fees to Middle Eastern governments amounted to just under $3 billion; including payments to Libya the total is increased to about $3.5 billion.[7] For 1980, on the assumption that existing institutional and economic arrangements continue (a matter on which there is much uncertainty, as indicated in Chapters 10 and 11, and summarized later in this chapter), the comparable figure is projected at nearly $9 billion for the Middle East and Libya taken together. To this must be added company payments to local factors of production (wages, salaries, materials, and services purchased locally), which came to about $400 million in the five leading export countries in 1967, and may be projected at approximately the same level (say, $500 million) in 1980.

The payments to governments are collected, in the main, via a tax levied upon the income ascribed to the companies on their oil exports, a mechanism described in Chapter 10 and summarized in the next section. These payments are made almost without exception in foreign exchange, either in dollars or in pounds sterling, except for a very small fraction of the transfers conducted in other currencies. The payments to local factors are made in local currency which has been purchased by the companies through payments in foreign exchange.

Together the companies' payments to governments and to local factors of production add up to the foreign earnings from oil that are retained by the host country. In Chapter 8 this sum is designated as the "retained value" generated by the exports of oil from the foreign-owned oil sector. A comparison of retained value, thus defined, with gross foreign exchange earnings, indicates the crucial role of oil foreign exchange earnings for the five major oil-exporting countries of the region. In 1966, the relevant figures were:[8]

	Oil sector retained value	Gross foreign exchange earnings	Oil sector as a per cent of total foreign exchange earnings
	..........(million dollars)........		
Iran	$757	$986	77%
Iraq	430	601	71
Kuwait	871	1,053	83
Libya	504	593	85
Saudi Arabia	882	987	89

The largest portion of the retained value of the oil sector consists, as indicated, of the payments made directly by the oil companies to the host governments. Thus, the most immediate impact is upon the government's own budget. Government oil revenues in 1966 in the five major exporting countries were as follows, shown also as a percentage of the combined ordinary and development budgets:[9]

	Oil company payments to governments (million dollars)	Percent of total government revenues
Iran	$652	58%
Iraq	392	63
Kuwait	818	83
Libya	390	73
Saudi Arabia	790	89

The importance of oil revenues is very great in all cases. However, even where oil company payments are lowest in comparative terms (Iran and Iraq), the revenues earned on tariffs on imported goods, which constitute a large element in the revenues from other sources, are in some measure an indirect consequence of the economy's oil earnings.

In contrast with the importance of the oil sector in foreign exchange earnings and in government revenues, the direct employment of labor by foreign oil companies is relatively small. The total is in the neighborhood of 60,000 for the five major exporting countries, having fallen between 1961 and 1967 by some 25 percent while production about doubled. In addition, there are large numbers of third-country nationals employed, chiefly in Kuwait and Saudi Arabia. These are mainly Pakistanis, Indians, and Palestinians. In Kuwait, in particular, nationals make up only about one-quarter of the industry's direct work force. In all of the countries there are also secondary employment impacts induced by the effects of government expenditures and local outlays by the companies, but we attempted no measure of these.

All of the foregoing measures relate to the economic dependence of the oil exporting countries on the ongoing

[7] These payments figures are those published by *Petroleum Press Service*; see footnote 8, below, for the basis upon which they are calculated. Provisional estimates for 1968 are $3.4 billion for the Middle East and almost $1 billion for Libya. (*Petroleum Press Service,* August 1969, p. 297.)

[8] This set of figures needs to be distinguished from another that is sometimes used which measures the share of oil in total merchandise exports. The percentage dependence runs much higher in the latter comparison; the measure is, however, less relevant than the one here used primarily because the gross value of oil exports includes the oil receipts that are repatriated or otherwise disbursed abroad by the oil companies.

In addition, the figures on payments to governments included within these totals are on a cash basis—i.e., they credit the cash disbursements to the year in which they were actually received by the country. The *Petroleum Press Service* figures cited earlier refer to the accrued tax liabilities of the companies, which can be appreciably more than the cash tax payments where production has been growing rapidly or where tax settlements significantly lagged exports.

[9] Only governmental revenues on current account, as narrowly defined, have been included. Specifically, revenues from government enterprises have been excluded, together with any net borrowings from abroad. The items thus excluded were large in some of the official budget presentations.

operations of the foreign owned oil industries within their borders. No effort was made to reach beyond these first-order effects to assess the impact on economic development of the several countries' earnings from oil industry operations. By the same token, no attempt was made to deal with the comparative economic effectiveness with which the different countries utilized their oil revenues. To do either would have required intensive analyses of each of the domestic economies well beyond the scope of this study. Estimates were made, however, of the foreign exchange reserves accumulated by the various countries, as compared to their current import expenditures on nonoil account. These are covered in Chapter 8 but will not be summarized here, although reference will be made to them in that part of the concluding portion of this chapter dealing with the problem of possible interruptions in the flow of oil supplies.

Relationships Between Oil Companies and Host Governments

The oil resources of the Middle East were, until the late 1950's, almost wholly developed under exclusive concessions granted before World War II to companies controlled by eight foreign parent companies.[10] These concessions covered most or all of the territory of the countries and ran for long periods, usually between 60 and 75 years. The first, dating from 1901, was in Persia (Iran); it is now operated by a Consortium of companies, including the eight major companies and others. The next, organized in Iraq after World War I, is mainly owned by five parent companies. The other major concessions, dating from the 1930's, are in Kuwait, owned by two parent companies, and in Saudi Arabia, owned by four parent companies. The same group of companies have a few lesser concessions; and since World War II a number of other companies have obtained producing rights in various countries.

In contrast to the exclusive rights granted under the older concessions, Libyan oil development, starting in the late 1950's, has been carried out under a considerable number of separate agreements.

Numerous changes in the older concessions have been introduced as a result of pressures applied by the host governments. In Iran, the concession of British Petroleum (then Anglo-Iranian) was canceled in 1951. All the properties were taken into titular ownership by the government; but operations were later assigned to companies jointly owned by the Consortium of foreign parent companies. In Iraq over 99 percent of the territory covered by the concession was reclaimed by unilateral action of the government in 1961, but the territory left to the concessionaire by the governmental decree contained all the properties then in actual production.

[10] The parent companies, under their present names, are Standard Oil Company (New Jersey), Standard Oil Company of California, Mobil Oil Corporation, Gulf Oil Corporation, Texaco, Inc., British Petroleum Company, Ltd., The Royal Dutch/Shell Group, and Compagnie Francaise des Petroles.

1. *Changes in Concession Terms*

The principal changes that have occurred in the terms of the old major concessions concern the financial terms and the territorial coverage.

Financial Terms

The financial terms of the early concessions, apart from bonuses and deadrents, called for flat royalty payments, usually 4 shillings (gold) or 3 rupees per ton of production. Currency depreciations led to revisions of the nominal rates, usually after severe bargaining. But, beyond this, the highly profitable operations led to demands by governments for higher payments. This was at first effected by increases in the royalty rate. But in the early 1950's, demands were made for enlarged payments based on profits; and one country after another achieved a tax formula of 50 percent of net profits, profits being calculated on the basis of the posted price of oil (less some discounts). The new formula greatly enhanced the revenues of the host governments, but cost the companies little because they could, in accord with accepted practice, offset the increased payments against income-tax liability to the home country of the parent companies.

Disputes, however, arose concerning the pricing of oil for tax purposes. As oil-product prices declined in world markets in the late 1950's, companies began to reduce the posted price of crude oil, an action that directly reduced the tax revenues of governments. Resisting this tendency, governments have succeeded in preventing the further decline of posted prices. To the extent that the latter have, as a consequence, lost their relationship to declining market prices, they have become essentially nominal in character, serving solely as a base for calculating tax liabilities. Given this base, taxes became in effect a fixed charge per barrel which is, from the companies' point of view, an inescapable cost, running generally between 80 and 90 cents (see Chapter 10, p. 123).

The tax take has further been increased by an arrangement, strongly resisted by the companies, through which companies now pay royalty separately from tax instead of including it within the 50 percent profit payment to governments. The new system is based on what is called "the expensing of royalties." In this calculation the royalty (now normally 12.5 percent of posted price), is deducted as an expense before arriving at the net "profit" before tax, of which the government receives 50 percent. In addition, the government receives the royalty. To illustrate, the computation of revenue per barrel to the government might then run as follows, using hypothetical figures:

Posted price	$1.80
less cost	0.20
less royalty (12.5%)	0.225
Putative company profit before tax	1.375
Tax (50%)	0.6875
plus royalty	0.225
Government take	$0.9125

The effect of expensing royalties, in this example, is to raise government revenue per barrel from 80 cents (i.e., one-half of net profits of $1.60) to 91.25 cents. This overstates the present difference, since some discounts (to be phased out over the next few years) are allowed from posted prices.

Relinquishment of Territory

The other principal change in concession terms has been the relinquishment, mainly under pressure by governments, of large portions of the territory included in the original agreements. In Iraq, almost the whole has been taken away by unilateral action of the country (a matter which is still in dispute); in Iran and Saudi Arabia, a major fraction—with provision for further progressive reduction; in Kuwait, more or less one-third of the original territory. In Libya relinquishment by stages is written into the agreements, and this feature also appears in new agreements in Middle Eastern countries. This places in the hands of the countries large areas which they can undertake to explore and develop through their own agencies or assign under new agreements. None of the countries has, however, expropriated properties actually developed and in production by concessionaires (except temporarily in Iran), an action which would cut off the secure revenues received from the companies.[11]

2. *New Agreements and the Role of the National Oil Companies*

In addition to the limitations placed upon the old concessionaires, all the countries have adopted a policy of entering into agreements with foreign companies, other than the original eight, to explore for oil and develop producing properties. These agreements place upon the companies the risks of exploration, but provide for amortization of the costs if oil is found in commercial quantities. Usually they specify the right of the government to acquire an ownership share up to 50 percent in the operations of successful ventures. In a few cases, the foreign companies are brought in as contractors with a right to share in the oil discovered, titular ownership remaining in a national oil company. It is not yet clear to what extent the new types of agreements offer positive financial advantages, as compared with the terms under older concessions (see appendix to Chapter 10); but they are at least a step toward satisfying the desire of countries to be active participants in the exploitation of their oil resources.

Correlatively with the recovery of territory from old concessionaires and the entrance of new companies on a joint-venture or contract basis, the countries have all established national oil companies to care for an enlarged direct government responsibility in oil affairs. These companies have acquired roles of different degrees of present importance in the various countries. The Iranian company is the most advanced in scale of activities. Besides operating the internal distributive market and promoting oil-related industrial development, it has general responsibility for exploiting the oil and gas resources of the country. It is a participating owner under some new agreements, and contracts for the services of foreign companies under others. It also looks after Iran's interests in Consortium operations.

All the countries hope for a large degree of participation by national companies in the future control and development of their oil resources. Taking a very long view, they presumably are planning to create agencies that can take over a high degree of responsibility when the present concessions run out. Currently, they gain experience through participation in certain activities under the new agreements, where they are partners from the start. But the bulk of oil operations is still under the control of the major concessionaire companies.

The role to be played by the national companies in the future appears to present the most troublesome issues now arising in the relations between governments and major concessionaires. One developing bone of contention is the part they may play in the existing production activities of the major concessions. Looking further ahead, some of them appear to wish also to build up a position in "downstream" transport, refining, and marketing operations.

A major problem for the national companies will be how to expand their market role without disrupting the international price structure on which country oil revenues are so strongly dependent. As indicated above, the income tax paid by the major concessionaires, when based on fixed posted prices, has the effect of levying a tax of about 80 to 90 cents/barrel. This is from the viewpoint of the concessionaire a fixed unit cost item. The consequence is to establish a high tax-paid cost figure for the companies. However, the real cost of oil, ex tax, is very low. According to estimates of one close student of the subject (cited in Chapter 10, p. 123), the current production cost—by which he means current operating expense plus a 20 percent return on the necessary development investment—is about 12 cents/barrel in the most expensive of the five major producing countries of the Middle East and North Africa. And he estimates that, within the fields now exploited, there are adequate reserves to support much higher rates of production over the next fifteen years, at operating and development costs of 20 cents/barrel, more or less. However, so long as government taxes and royalties of around 85 cents/barrel must be added as a cost that the concessionaire must meet, a figure of upwards of $1.00/barrel defines the limit to which prices might be driven by competition. Prices are still well above this cost base; but they are estimated to have gone down by roughly 40 percent since early 1957, and are likely to continue downward. The per-barrel tax now stands as a barrier to a deep competitive erosion of prices. When producer-country national companies attempt to enter the international market, they create the danger to themselves of further eroding the price structure upon which country revenues depend.

[11] In Iraq, the North Rumaila field which was forcibly relinquished, although not actually in production, had been partially drilled by the concessionaire, and to that extent was in the process of development.

3. *Organized Action by the Exporting Countries*

On a different though related subject, a striking development is the growing effort of the countries to engage in concerted, or at least parallel, action in their moves to gain greater control over the disposition of their oil resources. The most tangible evidence of this effort was the establishment of the Organization of Petroleum Exporting Countries (OPEC) in 1960. Since that date, this agency has directed the campaign for revised financial terms of concession agreements, referred to above. It has considered plans for a joint production program amounting in effect to a producing countries' cartel, though as yet without acceptance by member countries. It has also formulated an elaborate set of principles to guide countries in their efforts to improve their position vis-à-vis the companies.

The principles as laid down in the OPEC Declaratory Statement of June 1968 express a set of aspirations and a long-run strategy to guide individual countries in the future development of their oil resources. Emphasis is placed upon a number of concepts and procedures, which we here quote or paraphrase from the document:

(1) the principle expressed in a United Nations resolution of "the inalienable right of all countries to exercise permanent sovereignty over their natural resources";
(2) the desirability of direct exploitation of resources by agencies of the producing countries, rather than by outside agencies;
(3) the doctrine of "changing circumstances" which is said to justify countries in demanding changes in the terms of existing concession agreements;
(4) under the principle of changing circumstances, the right to acquire a participating ownership share for governments under the existing concession agreements;
(5) the right of governments to alter the financial terms of agreements where companies are receiving "excessively high net earnings";
(6) the right of governments to determine posted or tax reference prices;
(7) the accelerated relinquishment of concession territory.

One thing that these "principles" almost say is that the countries have the right unilaterally to change the terms of concessions as they see fit in the light of "changing circumstances." This raises a fundamental question of the nature of national sovereignty in relation to the contractual character of the agreements with companies.

Another body set up following the 1967 Arab-Israeli war, the Organization of Arab Petroleum Exporting Countries (OAPEC), had an initial membership of Kuwait, Saudi Arabia, and Libya.* It has so far not formulated a program of action; but it appears to contemplate various forms of action in such fields as industrial development, pipelines and tanker transport, production control, and marketing procedures. Being confined to Arab countries which are large oil exporters, it presumably has some implicit political orientation. Although the exact character of its political purposes has not been avowed, it may have been originally intended in part as a protective device against the more extreme proposals or efforts of other Arab countries to use oil as a political weapon.

*Enlarged, in 1970, to include Algeria, Abu Dhabi, Bahrain, Dubai, and Qatar.

4. *The Principle of Participation*

Almost all the postwar agreements with newcomers have provided for a degree of ownership participation by national oil companies. The idea has been brought up from time to time that the same principle should apply to the companies operating under the old major concessions; but it was only in 1968 that this idea began to force its way to the center of attention in the councils of OPEC.

One important role that the national companies are being called upon to fill in the future is that of participation in the ownership, and to a degree the management, of the old concessionary companies. In what ways such a relationship may develop cannot now be foreseen, but a few of the possibilities can be indicated.

When a national company becomes a partner in a producing company, it acquires title to a share of the oil. It can, if it wishes, arrange for this oil to be marketed by the foreign partner, merely sharing in the money proceeds. On the other hand, it can undertake to market some or all of its own oil. Up to the present the national companies have not had much oil at their disposal under the new agreements; and they have mostly left the marketing to the other partner because of their lack of independent outlets, although the Iranian company has been extending its activities in this direction.

So far, the matter has been of relatively minor significance because of the small amounts of oil involved. But it would enter an entirely different order of magnitude if the national companies were to become substantial partners in the old major concessions, as is now being sought.

With large amounts of oil available to them, the national companies could create a situation in which their effects on prices were great enough to take the profitability out of private concessions. If so, this could bring about a drastic restructuring of international oil markets—eventually leaving national companies in control of producing operations, with the international companies in the position of buyers. As purchasers of the oil, the major companies might be able to drive very hard bargains with national companies which were competing for the sale of their oil.

Some thoughts on these matters will be summarized in the final section of this chapter. At this point, a basic difference, already alluded to, must be noted between the position of a national company as seller of its country's oil and the position of a private concessionaire. The concessionaire cannot profitably offtake oil except at a price above tax-paid cost, say upwards of $1.00/barrel. A national company, if free of tax and royalty, has no base to which competition among countries trying to sell their oil might drive prices higher than the real cost, say 20 cents/barrel. In other words, the tax-paid cost of concessionaire oil gives an assurance of country revenues which would not be assured under competitive sales by national companies. Thus, it may be assumed that, even though national companies acquire a participating share in the major concessions, a prudent regard for their own economic self-interest would prevent governments from radically disturbing market structures. Nevertheless, there is the possibility of an erosion of the price structure from this source.

The Central Problems

The international oil trade constitutes what is perhaps the outstanding case of international economic interdependence in the world today. Both partners benefit in this exchange: the importing countries of the Western world obtain vital energy supplies at a lower price than from alternative sources, while the exporting countries of the Middle East earn sums of money which are vast by almost any standard, and particularly so when matched against their potential foreign earnings from any other goods and services which they might sell abroad.

A few statistics, presented earlier in this chapter, bring the extent of interdependence into sharp focus. In 1968, Western Europe imported more than 8 million barrels/day of oil from the Middle East and North Africa and the projected volume for 1980 is more than 15 million barrels/day. Over the same time period Japanese imports from these producing regions are estimated to increase from over 2½ million barrels/day to a projected amount of nearly 5½ million barrels/day. As a percentage of *total* energy consumption in Western Europe, oil, consisting mainly of imports from the Middle East and North Africa, is estimated to grow from less than 50 percent in the mid-1960's to over 60 percent in 1980; in Japan the percentage share of oil, mainly Middle Eastern imports, in *total* energy consumption is estimated to grow from over 50 percent in the mid-1960's to about 80 percent in 1980. Turning to the other side of the coin, the oil earnings of Middle Eastern countries and Libya, cited earlier, came to about \$3.5 billion in 1967, and are estimated (on the basis of existing arrangements) to come to roughly \$9.5 billion in 1980. In the mid-1960's, these earnings accounted for between 70 percent and 90 percent of the total foreign exchange earnings of the five major exporting countries, and between 60 percent and 90 percent of their total government revenues.

Despite these mutual economic benefits, serious problems have arisen. Two that are of special importance are considered in the remainder of this chapter: (1) the problem of supply interruptions; and (2) country-company relationships.

The Problem of Supply Interruptions

The three main occasions when the flow of oil from the Middle East was interrupted were: (1) from the three-year shutdown of Iranian production beginning in 1951 as a result of company actions occasioned by a breakdown of relations between the government and the concessionaire, the Anglo-Iranian Oil Company; (2) from the closure of the Suez Canal between November 1956 and March 1957 following the British-French-Israeli attack on Egypt in 1956, which in turn followed the prior nationalization of the Canal by Egypt; and (3) by the Canal and pipeline closures and partial embargo on deliveries by all the Arab producing countries at the time of the Arab-Israeli war in 1967. All three of these crises were on the whole successfully weathered from the standpoint of the consuming countries—the latter two mainly by reason of redeployment of supply sources and transport routes accomplished by the international oil companies, with the guidance and help of governmental and intergovernmental agencies.[12] The situation was helped by the existence of excess productive capacity in various parts of the world and, in 1967, by idle tanker capacity.

An essentially new factor was introduced into the oil supply interruption problem by the Arab-Israeli war of June 1967: for the first time the Arab countries engaged in a concerted effort to turn oil into a political weapon. As the crisis developed, before the outbreak of hostilities, the main producing countries expressed their intention of cutting off oil supplies to any Western countries which aided the Israelis in case of war; and President Nasser announced that in such an event the Suez Canal would be closed. In some quarters there was also the suggestion that the host countries might end, or greatly modify, the concessions of companies whose home governments were involved in Israeli "aggression." The warnings were clearly directed toward Great Britain and the United States.

After the outbreak of hostilities, upon the charge by Egypt, Jordan, and Syria that the United States and Great Britain had given direct military assistance to Israel, Egypt closed the Canal as Nasser had warned in advance that he would do in such an event. Also, the principal producing countries almost completely halted production, tanker loadings, and pipeline throughput, while considering the elements of a common policy. There was an understanding that, when movement was resumed, shipment to Great Britain and the United States, and possibly West Germany, should be forbidden, and that shippers should be pledged not to evade this restriction by indirect or roundabout means. On this understanding, countries were to resume shipments according to their own judgment. By the middle of June, Kuwait and Saudi Arabia had authorized general resumption of operations, and by the end of the month,

[12] Prior steps had been taken to meet possible emergencies in the two most recent crises. After the nationalization of the Canal in July 1956, but two months prior to actual closure, planning measures for such an eventuality were undertaken by American companies through a Middle East Emergency Committee, set up by the Foreign Petroleum Supply Committee. A similar organization was set up by the Organisation for European Economic Co-operation (OEEC). (American cooperation, however, was withheld during the first month of the crisis, November, while the American and British governments were at loggerheads over British Suez policy.) Essentially the same mechanism was reactivated by the OECD (successor agency to the OEEC) during the 1967 crisis.

For a description and analysis of the first two supply interruptions, see Harold Lubell, *Middle East Oil Crises and Western Europe's Energy supplies*, A RAND Corporation Study (Baltimore: Johns Hopkins Press, 1962).

Iraq had selectively raised the embargo. Being deterred by a workers' strike, Libya was the last to resume—during July.

Fortunately, even the partial embargo on oil supplies by the producing countries, confined to supplies to Great Britain and the United States, and in part to Western Germany, was of relatively brief duration. What would have happened if the war had lasted for a long time is impossible to know. Active discussion among ministers of the Arab states during August preceded a summit meeting of heads of state at the end of the month. An extreme proposal by Iraq, originally supported by Egypt, to embargo all oil shipments for three months, to be followed by selective penalties thereafter, gained little support. The action taken was to lift the embargo completely, Egypt concurring. The supporting argument of the rulers of the principal producing countries was that oil revenues were necessary to repair the damage of war and to return the Arab countries to the path of economic development and military strength. As a *quid pro quo*, Kuwait, Saudi Arabia, and Libya by the Arab countries, and to mitigate their impact if they should occur.

The concerted use of oil as a political instrument, attempted once, may be tried again at some future date, when it might have a much more severe impact. But, as indicated in Chapter 6, this is by no means the only cause that could give rise to the interruption of supplies in the future. Though obviously of great importance to importing countries in judging the possible situations to be guarded against, we do not speculate on other types of events that might cause such interruptions. We deal only with the question of what might be done to reduce the likelihood that supply interruptions would be undertaken, or threatened, by the Arab countries, and to mitigate their impact if they should occur.

It should be plain that from an economic standpoint such a question needs to be dealt with in the interests of both the importing and exporting countries because supply interruptions are economically damaging to both. Not only do they have sharp short-run effects which are economically painful, but their longer-run consequences can also be damaging if channels of commerce are diverted into alternatives which impose a permanent economic penalty upon both those countries that sell oil and those that buy.

The most realistic way to keep supply interruptions from being employed as an instrument of policy is to render such interruptions useless. If this could be accomplished, the threat to use oil as a weapon would be an empty one, and its actual use, if attempted, futile.

How can this objective best be achieved? The conclusion that emerges from our analysis is that the most practical approach now available is substantially to strengthen the method presently being employed—that is, the maintenance of emergency stocks of oil in the importing countries of Western Europe and Japan. (The situation should, however, be under constant appraisal for the possible existence of acceptable alternatives that are less costly, including the deliberate creation of standby capacity in regions judged to be secure. See appendix to Chapter 6.) However, although we find it to be the best course of action presently available, the total investment involved in creating a strategic reserve of oil stocks is by no means small. For example, according to our estimates a stockpile of oil, equivalent to 60 days of projected 1975 import levels in Western Europe from the Middle East and North Africa, would represent an investment of $2.4 billion in oil and storage facilities, increased for longer periods by multiples of that amount. (See p. 81.)

How many days of stocks should the countries of Western Europe and Japan have in their security reserve? Policy on this matter, as noted, has not been uniform: the targets have varied from about two months in Japan and the European Economic Community to a reported four months in Great Britain in 1966. Although never specifically acknowledged, that we are aware of, it is a reasonable speculation that these levels presuppose the existence of a substantial amount of idle capacity in the United States and other parts of the Western Hemisphere. Looking to the future, it is necessary to realize that, unless deliberate measures are taken, such excess capacity (except for the situation, presently unpredictable, that may evolve in the North American Arctic) probably will decline sharply in relative terms, and also in absolute amounts, and that stocks in the importing countries will need to carry practically the full burden of meeting supply interruptions, except for excess capacity in other non-Arab producing countries.

From a short-run economic standpoint, the analysis in Chapter 8 indicates that the Arab countries would have been able in 1966 to cover their current import levels for periods ranging from 1½ months for Iraq to 31 months for Kuwait through the use of their cumulated foreign exchange reserves. Libya's and Saudi Arabia's staying ability, calculated on the same basis, would have been 5 and 8 months, respectively. These are, however, conservative calculations in that they bring into play only the disposable foreign exchange reserves (i.e., foreign-exchange holdings less those foreign assets statutorily designated as required for currency cover), and also in that they do not distinguish between essential and nonessential imports. In a crunch, the staying power of the exporting countries, calculated even in strictly short-run economic terms, would be substantially greater. With such political factors as patriotism and nationalistic fervor brought into the equation, and with recognition that by far the major portion of the population is rural and largely self-sufficient, the ability of the countries to get by without oil earnings could be extended over even longer periods of time.[13]

Against all this must be put the awareness on the part of

[13]The gist of much of the argument in this paragraph was cogently put by Walter Levy, a leading international oil authority, in an article written around the time of the Suez Canal crisis: "Nor can the West rely on the importance of uninterrupted oil operations and oil revenues to Middle East governments as a deterrent to hostile actions. Economic considerations, important as they are to the relatively impoverished countries of the area, become insignificant when confronted with political necessities or political pretensions." And, in the same article: "At present, continued availability of Middle Eastern oil [to the West] is so urgent that no interruption can be borne. In contrast, the dependence of some Middle East

many who are now in positions of leadership in the producing countries of the long-run importance of oil industry operations as the foundation upon which the economic development of these countries must be built. We will have more to say about the views of members of this group, which includes those well-versed in the intricacies of the international oil industry, in the following section.

On balance, especially in view of the imponderables of the political and emotional elements in the situation, including the possible harassment by hostile forces, of those countries that may not want to participate in an oil embargo, we believe that it would be prudent to build up stockpiles beyond the official targets that now exist. *Stock levels equal to oil imports for six months or even longer do not seem an unreasonable planning target.*[14] No proof can be offered in support of this belief, except for two factors: (1) the relatively comfortable foreign-exchange reserve position of most of the Arab exporting countries involved; and (2) that stocks will increasingly need to act as a substitute for dwindling excess capacity in the Western Hemisphere relative to imports in Europe and Japan. Even at the six-month level the "insurance premium," as calculated per barrel of oil consumption in Chapter 6, does not result in a high price for Middle Eastern and North African oil when compared with other energy sources. The assignment of costs as between the governments and the companies within the consuming countries should reflect the fact that the insurance provided by stocks is needed for the protection of the economy as a whole, and not merely to enable the oil companies to continue to sell petroleum products.

Country-Company Relations

During recent years there have developed two persistent drives on the part of the oil-producing countries: one is to increase the revenues secured from concessionaire companies; the other is to secure an enlarged role for the countries in controlling the exploitation of their oil resources. Observers in the West have engaged in a considerable analysis of the forces at work and the effects they may have upon the future state of the relations between oil companies and the governments of oil-producing countries. While there is no precise consensus concerning what the future holds, there tends to be a similarity in their views concerning the directions of institutional change. In its extreme form, the conclusion some have reached is that the demands of countries upon companies, for larger revenues and participation in control, in association with declining prices, will reach a point so unacceptable that companies will withdraw from their concessionary status, to the serious financial disadvantage of the countries. A more moderate view is that, short of that outcome, the countries will have to moderate their demands in order to perpetuate a modus vivendi mutually advantageous to both parties. There is no basic difference of view concerning the character of the evolutionary forces at work.

The problem starts from the basic fact of very large reserves of oil producible at very low real cost. Under these conditions, prices have been moving downward, and may be expected to continue to do so. If producer countries attempt to enter markets through their national companies they will accelerate this trend. At the same time, as these countries attempt to increase their tax take, some concessionaires may be ground between higher taxes and lower prices to the point where they can no longer secure an acceptable rate of return on development investment. They might then terminate their concessionaire status and simply become contractors or purchasing offtakers of oil produced by national companies. Or, short of that, countries would be under pressure to reduce taxes in order to maintain the offtake of their oil.

A series of crises could occur from one country and company to another—striking the higher-cost sources first—leading progressively to the breakdown of concession agreements. Once companies were no longer producers and taxpayers, countries would, according to this view, have to look to the international companies as purchasers of their oil since only the latter have the necessary marketing and transportation network; and these companies would be in a position to strike hard bargains. With countries competing in the sale of oil, prices would be driven downward, with no bottom limit except the low real cost of oil production, ex tax.

Some observers appear to think that this is the likely course of events. Others, accepting it as a credible progression, consider that those now in positions of responsibility in the countries have a sense of economic realities, and the companies an adaptability to changing circumstances, sufficient to avoid a breakdown in company-country relations. It is a point upon which no one can entertain firm expectations.

If the countries were able to cooperate in the operation of a strong producers' cartel, thereby supporting the price structure, through putting an end to competitive efforts to dispose of their oil, they would be able to press their demands upon companies further, or even end the major concessions, without disastrous consequences to themselves, assuming that they achieved the capability to market and transport their oil. In view, however, of the improbability that such a cartel could endure as an effective instrument, the countries would appear unable to dispense with the functions now performed by the international companies without serious costs to themselves.

Within the producing countries, those who publicly discuss the relations between governments and companies do not all speak with one voice. Some express a strident nationalism and propose tough measures which, if adopted,

countries on the income they obtain from the sale of oil is of a different order of urgency. Many of them could absorb a reduction of their royalty and tax income for some period of time by drawing on accumulated funds or by reducing less essential expenditures." ("Issues in International Oil Policy," *Foreign Affairs*, April 1957, pp. 454–469.)

[14] In the event of a concerted shutdown of supplies for a long period of time, six months of stocks would probably permit consumption at normal levels for a period of perhaps three months or so. At some point, unless increased supplies from other regions appeared to be forthcoming in adequate amounts for the period ahead, action would need to be taken, based upon advance planning, to begin reducing consumption of petroleum products in those uses which are less essential.

would make the position of the major companies intolerable. There are, however, especially among officials now in positions of responsibility in OPEC and its member countries, men who are thoroughly aware of the problems facing their countries. It would be wrong to assign any consensus on detailed questions of national policy among the latter group; but, as we interpret their views, they entertain certain common attitudes in their approach to the broader problems of policy.

As would be expected, they are concerned to secure from the companies the best possible financial terms. Also, they are concerned, as fully as possible, with securing responsible country participation in the affairs of the companies—a concern perhaps dictated as much by national pride as by economic motives, but also envisaging effective national control in the course of time, at latest when the concession agreements expire. They exhibit some ambivalence between the principle of intercountry cooperation, especially with respect to collective production control programs to support prices, and the active promotion of single country interests—though the latter appears to predominate.

In spite of aims antipathetic to the attitudes of the companies, with their ingrained desire for freedom of action, presently responsible spokesmen in the producing countries express the intent of avoiding a rupture in the basic relations with the companies. The basis for this attitude is the prudent recognition that there exist no near term substitute methods for carrying out the international marketing functions and for assuring the stabilization of revenues that is afforded by taxation of the companies. It appears safe to say that while no one in the host countries likes the degree of economic dependence on the major concessionaires, no one knows how to do without them.

In the field of opinion concerning company-country relations, an oddly anomalous situation exists: some "outside" Western observers who have recently put themselves on record are inclined to think that a process of disruption is under way; on the other hand, the "inside" spokesmen appear in the main not to contemplate any such outcome, in the belief that the countries will find the will and the way to avoid cutting their own throats. The outside opinion is based upon two related presumptions: (1) that competitive forces will drive international prices down, and (2) that the countries, either by miscalculation and/or failure to understand the realities of the world oil market, or under popular political pressure, will make demands that the companies will find intolerable.

Two other factors in the situation have to be mentioned, though not much can be said about them. One is the attitudes entertained within the major international companies. Unlike the members of oil-country governments, they do not engage in public discussion of their problems. It is therefore impossible to report the range of views that exists concerning the future status of the oil concessions in one country or another, nor is it possible to know at what point concessionaire status would no longer be judged profitable.

The other factor is the policies that may be pursued by the governments of consuming countries. They have, on the whole, made no persistent attempt to modify the existing market structure which permits producing countries, and to a diminishing degree, companies, to enjoy the great economic rents which arise from oil production at going prices. Observers have pointed out that they probably could obtain their oil supplies at much lower cost if they cared to use the bargaining power they possess. If in practice they were to pursue such a policy, however, a new set of tensions would be added to the international oil situation—the parties to the controversies being the governments of consuming and producing countries, with the companies poised uncomfortably somewhere in between. There is evidence that the producing countries are giving thought to this danger, and hope in a measure to stabilize their markets by direct agreements with consumer countries. There is no evidence that such action on an important scale is imminent; but the possibility exists in embryo, and adds another element of uncertainty concerning the future status of the large international companies in the producing countries and their role in the international market structure.

The present facts of international oil can be made to support various inferences concerning the outcome of the evolutionary process. On a rather deterministic economic analysis, it may be inferred that unleashed competitive forces will eventually undermine the economic advantages currently possessed by the producing countries and, in combination with unruly political forces, disrupt the existing structure of relations of countries and companies. On the other hand, it is possible to imagine various sorts of negotiated solutions which, while bending somewhat to the economic and political winds, may retain much of the substance of present arrangements. What informed observers, both inside and outside the producing countries, appear to agree on is that an abrupt termination of existing relationships is unlikely if the countries are motivated by a concern for the economic consequences of their actions. Economic considerations may not, however, turn out to be the decisive factors.

POSTSCRIPT: A NOTE COMPARING SOME FINDINGS OF THIS STUDY WITH THOSE OF THE CABINET TASK FORCE REPORT ON OIL IMPORT CONTROL[1]

IN March 1969, President Nixon created a Cabinet Task Force to conduct a comprehensive review of oil-import restrictions in the United States. The report of this group was made public by the President on February 20, 1970. It appeared after the present study had gone to publication, our manuscript having been completed in draft in July 1969 and in final form in December 1969.

The two studies were, of course, directed to different questions, but they do cover the same ground on certain points. In this postscript we wish to do no more than to compare the findings of the two studies on a few significant matters covered in both. On recommendations regarding future U. S. oil-import policy, which will, of course, be a prime determinant of the course of future petroleum developments in the United States, the members of the Task Force were divided, although a majority favored the substitution of tariffs for import quotas as the mechanism of control. The President released the report, but deferred any decisions regarding basic changes in oil-import policy.

United States Dependence on Oil From the Middle East and North Africa in the Next Decade

A major reason for analyzing the oil position of the United States in the present study is to assess the possible dependence of the United States on oil imports from the Middle East and North Africa over the next decade. The degree of dependence in the recent past has been minimal; the conclusion we reach is that the degree of dependence can be kept at a very low level relative to consumption if the United States chooses to continue a policy designed to achieve such a result. The option open to the United States is, our study points out, in sharp contrast to the position of the countries of Western Europe and Japan, in which projected increases in oil consumption are almost wholly dependent on increased imports, mainly from the Middle East and North Africa. Our study directs attention to this important policy option enjoyed by the United States, but does not recommend the policy course which the United States should follow.

To make recommendations regarding United States import policy was the main purpose for which the Cabinet Task Force was established. In order to arrive at its policy recommendations the Task Force examined the question of whether, and at what cost, the United States could maintain its present degree of independence of oil supplies from the Eastern Hemisphere. On this subject, the Task Force concludes that at current price levels in the United States, none of this country's imports over the next decade *need* originate in the Eastern Hemisphere; that oil supplies from Canada and Latin America *could* easily satisfy—and more than satisfy—United States import requirements (*The Oil Import Question,* p. 48, Table D-1). It should be emphasized that the foregoing is a research finding, not a policy recommendation.

Although the two studies do not differ on the question of the degree to which the United States *need* rely on oil from the Middle East and North Africa over the next decade, the conclusion is reached via different routes which do reveal certain differences. In the present study it is argued that the United States can, over the next decade, hold the ratio of *total* oil imports to *total* liquid fuels consumption constant at the late 1960's level of about 20 percent at a cost penalty to the economy comparable to that being incurred, *if it chooses to do so.* (See "Summary and Conclusions," p. 2.) We conclude that production in the lower forty-eight states, plus the output from the Alaskan North Slope should make this possible, with fallback support provided by liquid fuels from unconventional sources, if deemed necessary and if accelerated research and development (R&D) were to be activated early enough. The conclusion of the Task Force report differs: at current domestic oil prices, it foresees imports rising to 27 percent of United States liquid fuels consumption. As the 1980 consumption projections are practically the same in the two reports, about 18.6 million barrels/day for comparable concepts of measurement (all liquid fuels uses—see "Summary and Conclusions," p. 2, and the appendix to Chapter 1 of this study[2]), the difference results from dissimilar views regarding the prospects for domestic supplies at going prices.

The absolute difference between the estimates of United States oil imports in the two studies amounts to 1.3 million barrels/day in 1980; 3.8 million barrels/day in the present study and 5.1 million barrels/day in the Task Force report. The latter figure is derived as the difference between estimated United States total liquid fuels consumption of 18.6 million barrels/day and total liquid fuels production of 13.5 million barrels/day. Estimated production in turn consists of (1) 11.9 million barrels/day of crude oil, an average of estimates submitted to the Task Force (mainly by oil companies, but including also the U. S. Department of the Interior) ranging from 11.1 to 12.7 million barrels/day at the present real price (*The Oil Import Question*, p. 39, para. 228 and pp. 228–229, Appendix D) and (2) an estimated output of natural gas liquids of another 1.6 million barrels/day. In the submissions to the Task Force the difference between the high and low estimates of crude oil production at the present real price is greater than the difference between the 1980 import estimate of this study and of the Task Force report. As the Task Force report comments:

[1] Cabinet Task Force on Oil Import Control, *The Oil Import Question: A Report on the Relationship of Oil Imports to National Security* (Washington: Government Printing Office, February 1970).

[2] In particular, table on p. 24; million metric tons per year of liquid fuels total consumption in energy and nonenergy uses divided by 50 for conversion to million barrels per day.

"Variations in individual estimates are due to inescapable uncertainties such as differences in opinion on the probable development of undiscovered production in Alaska" (p. 229, Appendix D). It is noteworthy, too, that the Task Force report characterizes its estimate of natural gas liquids output as a "minimum" (see p. 221, Appendix D).

Idle Oil Productive Capacity in the United States in the Future

As its starting point for estimating future idle oil productive capacity in the United States, this study cites the estimate appearing in the U.S. Department of the Interior's 1968 study, *United States Petroleum Through 1980*. That report's estimate of 3 million barrels/day for the United States in 1980 is judged by us to be unrealistically high ("Summary and Conclusions," p. 8) and consequently to be heavily discounted in assessing the extent to which United States excess productive capacity might serve as a security reserve to offset oil supply interruptions that may strike importing countries in the future. In arguing that substantially larger oil stockpiles are required in Europe and Japan for security purposes than are presently thought to be adequate by those countries, we observe "that stocks [in Western Europe and Japan] will increasingly need to act as a substitute for dwindling excess capacity in the Western Hemisphere relative to imports in Europe and Japan" ("Summary and Conclusions," p. 16).

The Task Force report expects that idle capacity that exists as the result of the oil industry's normal operations (the source of the idle capacity referred to in the foregoing paragraph) will decline to the neighborhood of 0.5 million barrels/day in 1980 (p. 51, para. 240). The report also estimates that with a maximum effort (requiring some form of subsidization) extra production in the United States in 1980 could be increased to about 1 million barrels/day at the end of one year (p. 52, para. 241a). (The Task Force estimates are predicated on a wellhead price of $2.50 per barrel, and assume that market demand prorationing would no longer be significant. This price is some 80 cents less than the report's base estimate of current wellhead price.)

Emergency Oil Stocks: The Cost of Storage

In estimating the capital cost of storage capacity in Western Europe in steel tanks such as are conventionally used, the present report presents a range of $1.25 to $2.00 per barrel (excluding land) and settled on a figure at the midpoint of this range (somewhat more than $1.60 per barrel). (See Chapter 6, pp. 80–81.) The oil company submissions to the Task Force estimated per-barrel capital costs of conventional tankage (including land) in the United States at somewhere between $1.80 and $2.75 (*The Oil Import Question*, p. 299, Appendix J). The midpoint of the latter range is roughly $2.30 per barrel, or about 70 cents (of which about 25 cents would represent the cost of the land) per barrel higher than the capital cost used in the estimate of the present study. (The Task Force apparently settled on $2.50 per barrel without indicating its reasons for selecting this particular capital cost.)

Substitution of the average of the estimates submitted to the Task Force for the capital costs of conventional storage facilities used in this report would affect our estimates of the cost of stockpiling oil roughly as follows. Stocks equal to, say, 180 days of estimated imports into Western Europe in 1975 from the Middle East and North Africa would require an investment, including oil and tanks for storage (in billion dollars):

Stocks of oil (this study's estimate)	$3.6	$3.6
Storage tanks (this study's estimate)	3.6	
Storage tanks (at average of cost estimates submitted to the Task Force by oil companies)		5.1
Total capital cost	$7.2	$8.7

The extra capital cost would be roughly $1.5 billion, as of 1975 (in prices of the late 1960's). Following the steps shown in Chapter 6, pp. 80–81, this higher investment would result in increasing the "surcharge" from 16 cents to 20 cents per barrel of oil consumption.

PART I

World Oil Production, Consumption, And The Problem Of Supply Interruptions

CHAPTER 1

World Oil Consumption and Major International Flows: Present and Projected

The Situation in 1968[1]

IN 1968, the year after the most recent major crisis affecting the international flow of oil, the world's consumption of oil came to approximately 1.9 billion metric tons or 39 million barrels/day.[2] (See Table 1.1.) Within this total, consumption by the United States was far and away the largest of any country, accounting for approximately one-third, about 13 million barrels/day. The countries of Western Europe, taken as a group, were the second largest consuming area, with a 25 percent share in the world's total, almost 10 million barrels/day. The remaining 40 percent or so of the world's total consisted of the 17 percent accounted for by the U.S.S.R., Eastern Europe, and Communist Asia; 9.5 percent for non-U.S. Western Hemisphere (including 3.5 percent in Canada); and another 15 percent for the rest of the world, with Japan alone (in the latter group) consuming 7 percent of the world's total. Japan's oil consumption (about 140 million tons per year, or 2.8 million barrels/day),[3] although well below that of the United States, placed it second only to the United States among the non-Communist countries of the world.

In the tabulation below, taken from Table 1.1, consumption in major areas is compared with the amounts of liquid fuels produced in those same areas in 1968.[4]

	Consumption		Production	
	Million tons/year	Million barrels/day	Million tons/year	Million barrels/day
United States	636	13.08	507	10.61
Western Europe	495	9.98	23	0.45
Japan	139	2.81	1	0.02
U.S.S.R., Eastern Europe, and Communist Asia	283	5.69	339	6.78

Whereas the U.S.S.R., Eastern Europe, and Communist Asia—as a group—produced more liquid fuels than they consumed, the other major consumer countries had to import substantial amounts of oil to meet their needs. The United States, although satisfying a very large share of its consumption from domestic supplies, received about 2½ million barrels/day in net imports. Western Europe and Japan, on the other hand, imported practically all they consumed.

Where did the imported supplies originate? In the case of the United States, the bulk of the imports came from other parts of the Western Hemisphere, mainly the Caribbean and, to a lesser extent, Canada. Some supplies were shipped from the Middle East and North Africa, but these came only to about 400,000 barrels/day, in the neighborhood of 15 percent of U.S. oil imports. (The Middle East and North Africa contributed, respectively, 9.5 and 5.5 percent to this figure.) For Western Europe, on the other hand, the Middle East was the most important supplier, accounting for almost one-half the total (more than 5 million barrels/day), followed by North Africa with about one-third (about 3⅓ million barrels/day). Taken together, the Middle east and the North African states (mainly Libya and Algeria) supplied over 80 percent of total imports into Western Europe, with the remainder scattered mainly among the U.S.S.R. and Eastern Europe (about 900,000 barrels/day), the Caribbean (760,000 barrels/day), and West Africa (220,000 barrels/day). Japan's supplies originated overwhelmingly in the Middle East, about 2.7 million barrels/day out of a total of 3.1 million.

Dependence is, however, a two-way street. Just as Western Europe and Japan are crucially dependent upon the Middle East and North Africa for their oil supplies, so the latter two areas are dependent upon Europe and Japan as customers. As Table 1.2 shows, out of total Middle Eastern exports of around 10½ million barrels/day in 1968, Western Europe took nearly one-half and Japan 25 percent. The remaining amounts were scattered worldwide, the United States itself accounting for less than 3 percent of the area's total exports. North Africa's exports of 3.6 million barrels/day were sold almost entirely to Western Europe.

The Outlook for the Future

There is no shortage of projections of oil production and consumption in the future. In this study, the analyses made by others have been used and supplemented by estimates of

[1] The consumption figures for 1968 refer to oil in all energy uses (including bunkers in maritime trade and international air traffic) as well as nonenergy uses, such as petrochemical feedstocks. Elsewhere in this study, data limitations compel us from time to time to refer to the following more restricted definitions of oil consumption: "inland" consumption of oil in energy uses (which excludes both bunkers and nonenergy consumption); and consumption of oil including bunkers, but excluding nonenergy uses. (See appendix to this chapter.)

[2] Hereafter, all tonnage figures are understood to refer to metric tons.

[3] British Petroleum Company, *Statistical Review of the World Oil Industry, 1968* (London, 1969) shows Japan's oil imports slightly above the country's consumption. The apparent anomaly is not explained in the publication; it may be partially due to some imports going into stocks rather than consumption.

[4] "Liquid fuels" and "oil" are used interchangeably throughout the study. The former is the more precise term since natural gas liquids (NGL), important primarily in North America, are also included.

Table 1.1

Oil Production and Consumption, by Major World Regions, 1968

	Production			Consumption		
	Million metric tons	Thousand barrels/day	Percent of total	Million metric tons	Thousand barrels/day	Percent of total
North America	564.2	11,775	29.3	700	14,410	37.1
United States	507.2	10,610	26.4	636	13,080	33.7
Canada	57.0	1,165	2.9	64	1,330	3.4
Latin America	263.5	5,120	12.7	118	2,370	6.1
Western Europe	22.9	450	1.1	495	9,980	25.7
Japan	0.8	16	–	139	2,810	7.2
Middle East	557.0	11,185	27.8	41	800	2.1
North Africa	183.8	3,815	9.5	14	275	0.7
Other Africa	12.1	245	0.6	27	545	1.4
Other Asia	40.6	810	2.0	67	1,380	3.6
Oceania	1.7	34	0.1	26	530	1.4
Communist countries	338.7	6,785	16.9	283	5,690	14.7
World total	1,985.3	40,235	100.0	1,910	38,790	100.0

Source: British Petroleum Company, *Statistical Review of the World Oil Industry, 1968.* The Japanese and Oceania production figures come from *Petroleum Press Service,* January 1969, p. 13, and consumption in North Africa and Other Africa is estimated on the basis of the areas' production and net trade (as shown in the British Petroleum publication).

Note: Consumption = production + net imports of liquid fuels. Consumption includes bunkers as well as fuels consumed in nonenergy uses. The conversion factors by which barrels per day are equated to metric tons per year may vary slightly among the areas shown, although on the basis of average specific gravity of crude oil around the world, 1 barrel/day equals approximately 50 metric tons/year. The percentage distribution is based on barrels per day.

our own. All of this has been incorporated within a framework in which the past, present, and future are handled in a consistent fashion, and in which oil is treated as one element within the total energy economy of the world and its major regions. Our estimates are presented and explained in detail in the Statistical Appendix (and summarily described in the appendix to this chapter). Here, and in subsequent chapters, we present only selected statistics which help to bring the main subjects of our inquiry into sharper focus. A summary comparison of some of the relevant statistics of 1965, 1975, and 1980 referred to subsequently in this chapter is presented in Table 1.3. The consumption figures in this table include oil for energy uses only.[5]

1. *Salient Features in 1975*

As shown in Table 1.3, the projected total world oil consumption in 1975 is 2.6 billion tons, equivalent to about 51 million barrels/day as compared with 28 million barrels/day in 1965. (These figures exclude oil in nonenergy uses. A rough allowance for the latter–based on past trends–would raise 1975 world oil consumption in *all* uses to perhaps 2.8 billion tons or 56 million barrels/day. See appendix to this chapter.) The United States, with an estimated consumption of almost 14 million barrels/day would continue as the world's largest consumer of liquid fuels, but its relative importance in the world total is projected to decline from approximately 35 percent to about 27 percent. Western Europe, with a projected total of close to 13½ million barrels/day, would have practically drawn even with the United States in its oil consumption. Together, the United States and Western Europe are projected to consume about 53 percent compared to the 60 percent of the world's total they accounted for in 1965.

Japan's consumption in 1975 is projected at about 4.4 million barrels/day. The Soviet Union, Eastern Europe, and Communist Asia are estimated at a level of well over 400 million tons, equivalent to more than 8½ million barrels/day.

As indicated, the foregoing consumption figures refer to energy uses only. Estimated 1975 net imports into the major consuming areas yield the picture summarized below:

Net imports into:	Million tons/year	Million barrels/day
Western Europe	709	14.2
Japan	237	4.8
United States[6]	166	3.3

Consequently, among the leading importing countries, import requirements will have risen from about 11⅓ million barrels/day in 1965 to about 22⅓ million barrels/day in 1975–an absolute increase which is about equal to their *total* import requirements in 1965. (The U.S.S.R., Eastern Europe, and Communist Asia–taken as a

[5] See Table 1.3, n.d, and appendix to this chapter.

[6] On the assumption that in the United States imports continue to hold at their present-day level of relative importance as a percentage of consumption–an assumption which is examined in subsequent chapters.

group—are assumed to be essentially self-sufficient; net trade with the rest of the world being projected at zero.)[7]

Where will these import requirements originate? As Table 1.4 shows, for the United States the imports are estimated to continue to come, in the main, from other parts of the Western Hemisphere, in particular the Caribbean and Canada, with the Middle East accounting for about 10 percent.[8] For Western Europe, primary reliance will still rest on the Middle East, although—in line with trends that have continued into 1969—North Africa's rising relative importance (23 percent of Europe's imports in 1965, 38 percent in 1975) will cut measurably into the preeminence enjoyed by the Middle East in the past. (All of this is estimated on the assumption that the U.S.S.R. and Eastern Europe remain essentially self-contained, and do not make either imports or exports of enough significance to affect the main movements of world flows.) As for Japan, the Middle East continues to be of overwhelming importance as a supply source. All told, the sharply rising quantitative importance of imports from the Middle East and North Africa to the major consuming countries of the Western world stands out boldly.

2. *Salient Features in 1980*

World oil consumption by 1980 is projected at a level of about 3.4 billion tons, equivalent to about 67 million barrels/day, an increase over the 1975 figure of somewhat more than 16 million barrels/day. (See Table 1.3.) (If the figure were to include an allowance for nonenergy uses of oil, worldwide consumption in 1980 might reach over 3.8 billion tons or 75 million barrels/day.) The United States, with an estimated consumption of about 16 million barrels/day, would still rank as the single largest consuming country, but the combined total of Western Europe's consumption is estimated at about 17 million barrels/day and would exceed that of the United States. In terms of percentage shares in the world's total, the United States will have fallen to somewhat less than one-quarter, compared to about 27 percent in 1975; and Western Europe's share will stand at just about one-quarter, slightly below its percentage in 1975. Together the United States and Western Europe are estimated to account for a bit under one-half the world's total, compared to around 53 percent in 1975 and 60 percent in 1965. In absolute terms, their consumption of more than 32 million barrels/day will exceed the *world's* total consumption in 1965, about equaling the world's total in 1967, the year of the Arab-Israeli war.

[7] Views vary regarding the future net oil trade position of the Communist countries. In the mid-1960's they were on a net export basis of about 1 million barrels/day, largely to Western Europe. Most recent analyses by Western observers suggest that the export balance will decline and that the Communist group, as a whole, may be net importers, although not in large enough amounts to affect the broad picture here drawn.

[8] Because of source and definitional differences, the figures on direction of trade in Table 1.4 differ somewhat from those in Table 1.3 just cited. Table 1.3 refers to total *net* trade balances while the data in Table 1.4 refer to an area's *gross* interregional imports or exports.

Japan's consumption in 1980 is projected at close to 7 million barrels/day, which is almost the same as the combined consumption of all of Western Europe in 1965. This vast absolute figure will, in 1980, account for just over 10 percent of the world oil total. Consumption in the Soviet Union, Eastern Europe, and Communist Asia is projected at over 12 million barrels/day, or about 18 percent of the world's total.

Taken together, the United States, Western Europe, Japan, and the Communist countries are thus projected to consume about 52 million barrels/day in 1980, out of a world total of 67 million barrels/day, down only slightly from the percentage they are estimated to account for in 1975. The combined consumption of these countries, which together will continue to dominate world con-

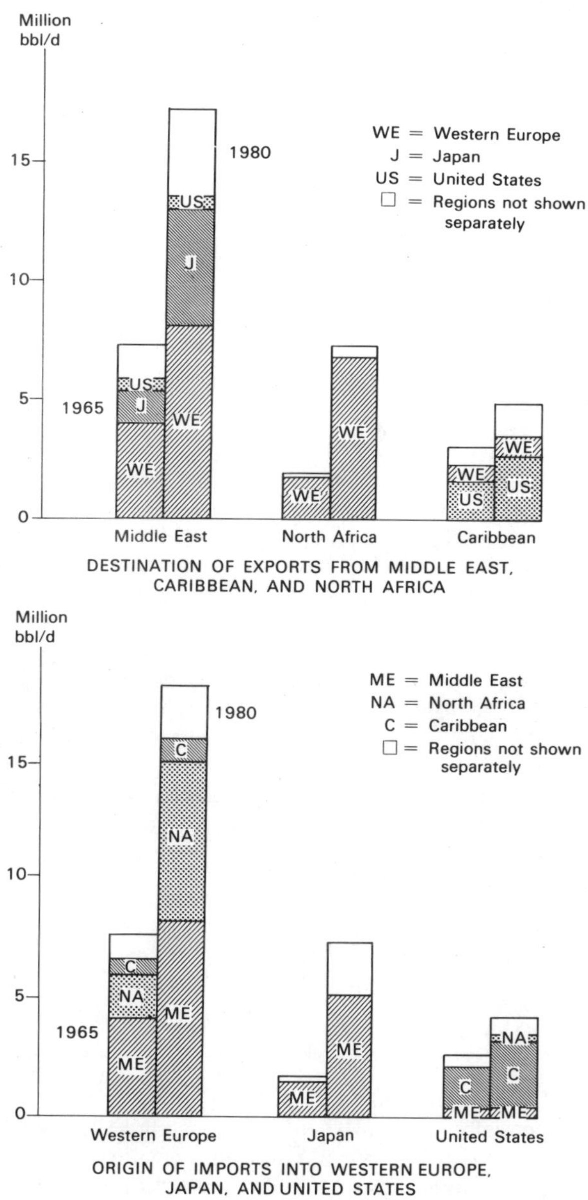

Fig. 1.1 World Oil Flows Among Principal Areas, 1965 and 1980.

Source: Table 1.4.

Table 1.2

Inter-Area Total Oil Movements, 1968

	Importing areas									
Exporting areas	U.S.A.	Canada	Other Western Hemisphere	Western Europe	Africa	South East Asia	Japan[a]	Australasia	Other Eastern Hemisphere	Total exports[b]
	Million metric tons									
U.S.A.	–	1¾	2¼	2	½	¼	1¾	¼	¼	9
Canada	23½	–	–	–	–	–	–	–	–	23½
Caribbean	86½	23¼	12¾	39½	3½	–	4	–	1	170½
Other Western Hemisphere	7	–	1¼	¼	–	–	–	–	–	8½
Western Europe	5¾	–	–	–	2¾	–	–	–	¾	11½
Middle East	13	6¼	9	253	22¾	31½	132¼	17¾	16¾	519¼
North Africa	7¼	–	1¼	159¾	½	–	¼	¼	1½	174¼
West Africa	½	½	¼	11	–	–	–	–	–	12¼
South East Asia	3½	–	–	¾	–	–	12	5¼	–	21½
U.S.S.R., Eastern Europe	–	–	6	45½	2¼	–	2¼	–	–	56
Other Eastern Hemisphere	–	–	–	¼	¼	8¼	¼	½	¼	11¼
Total imports	147	31¾	32¾	512	32½	40	152¾	24	20½	1,017½
	Thousand barrels/day									
U.S.A.	–	50	45	40	10	5	35	5	5	195
Canada	470	–	–	–	–	–	–	–	–	470
Caribbean	1,580	480	260	760	60	–	85	–	20	3,245
Other Western Hemisphere	125	–	25	5	–	–	–	–	–	155
Western Europe	120	–	–	–	60	5	–	–	10	240
Middle East	270	125	175	5,065	465	645	2,650	365	345	10,445
North Africa	155	–	30	3,315	10	–	5	5	35	3,625
West Africa	15	10	5	220	–	–	–	–	–	250
South East Asia	75	–	–	15	–	–	240	105	–	435
U.S.S.R., Eastern Europe	–	–	115	900	40	–	45	–	–	1,100
Other Eastern Hemisphere	–	–	–	10	5	175	5	10	5	240
Total imports	2,810	665	655	10,330	650	830	3,065	490	420	20,400

[a] See n. 3, p. 23.
[b] Including minor flows not otherwise shown and destinations not known.
Source: British Petroleum Company, *Statistical Review of the World Oil Industry, 1968*, p. 11.

sumption, is estimated to just about equal *world* consumption in 1975.

The foregoing consumption figures refer to energy uses only. Estimated 1980 net import requirements, including oil consumed in nonenergy uses, yield for the major non-Communist oil consuming countries the following approximate picture (shown in detail in Table 1.3):

Net imports into:	Million tons/year	Million barrels/day
Western Europe	893	17.9
Japan	376	7.5
United States[9]	188	3.8

(As in 1975, the Communist country group is assumed to be essentially self-sufficient.)[10] Among the leading importing countries, net import requirements of oil will have risen from around 11⅓ million barrels/day in 1965 to about 22⅓ million barrels/day in 1975, to over 29 million barrels/day in 1980. If these projections are near the mark, the magnitude of estimated import flows into these countries in 1980 *would exceed total world oil consumption in 1965.*

The projected 1980 imports of the major Western world consuming countries are very large. Though West Africa's share of European imports is assumed to rise somewhat, Western Europe will, if existing trends persist, continue its overwhelming import reliance upon the Middle East and North Africa, taking more than 15 million barrels/day from these two areas out of a total import requirement of 18 million barrels/day, as indicated in Table 1.4.[11] (Because of the rapid increase of Libyan exports since 1965, the relative rise of North Africa shows up most conspicuously in a comparison of 1975 with 1965; after 1975, it is assumed that Middle Eastern and North African exports will grow at about the same rate.) Japan is projected to continue its heavy dependence on the Middle East, taking approximately 5 million barrels/day out of a total import

[9] See n. 6, p. 24 above.
[10] See n. 7, p. 25 above.

[11] See n. 8, p. 25.

requirement of 7.3 million barrels/day; the remainder is accounted for by other Asian sources–mainly Indonesia, whose relative importance can be assumed to rise in the years ahead. As for the United States, its projected imports of about 4.3 million barrels/day are estimated to originate mainly in Western Hemisphere countries, and only to a small extent in the Middle East and North Africa.

Turning the picture around, Table 1.4 also reveals the shifting relative importance of different markets in the oil shipments of the principal exporting areas. Here the most conspicuous change foreseen during 1965–80 is some shift in the relative importance of Western Europe and Japan in the exports of the Middle East: Western Europe's share is projected as declining from 55 percent to about 47 percent,

Table 1.3

Liquid Fuels: Consumption and Net Trade,[a] the World and Major Consuming Regions, 1965, 1975, and 1980

	Consumption			Net imports	
	Million metric tons	Million barrels/day	Percent of world total	Million metric tons	Million barrels/day
			1965		
United States	494.4	9.89	35.7	117.8	2.36
Western Europe	343.0	6.86	24.8	364.3	7.29
Japan	76.7	1.53	5.5	83.5	1.67
Communist countries[b]	193.7	3.87	14.0	−47.2	−0.94
Total, above	1,107.8	22.16	80.0	565.6[c]	11.31[c]
World total	1,384.7[d]	27.69	100.0	[e]	[e]
			1975		
United States[g]	692.3	13.85	27.1	165.6	3.31
Western Europe	670.2	13.40	26.2	709.4	14.19
Japan	217.9	4.36	8.5	237.4	4.75
Communist countries[b]	434.6	8.69	17.0	[f]	[f]
Total, above	2,015.0	40.30	78.8	1,112.4[c]	22.25[c]
World total	2,557.2	51.14	100.0	[e]	[e]
			1980		
United States	799.5	15.99	23.8	188.4	3.77
Western Europe	837.3	16.75	24.9	893.1	17.86
Japan	346.2	6.92	10.3	375.6	7.51
Communist countries[b]	612.6	12.25	18.2	[f]	[f]
Total, above	2,595.6	51.91	77.3	1,457.1[c]	29.14[c]
World total	3,358.9	67.18	100.0	[e]	[e]

[a] Consumption includes liquid fuels for energy uses only, while net imports refer to all liquid fuel imports, including those consumed as petrochemical feedstocks, and in other nonenergy uses; net imports may, therefore, exceed consumption, as here measured.

[b] U.S.S.R., Communist East Europe, and Communist Asia.

[c] Sum of net imports only; i.e., net exports of the Communist countries in 1965 have not been subtracted.

[d] The 1968 world total presented earlier in this chapter in Table 1.1 (1.9 billion tons) includes nonenergy uses. The 1965 figure here shown refers to energy uses only; the 1965 total comparable to the 1968 figure of 1.9 tons is 1.5 billion tons.

[e] Not applicable.

[f] Net trade with rest of world projected at zero, although there will undoubtedly be some trade (see p. 25, n. 7).

[g] U.S. oil consumption for energy uses only is projected at 13.85 million barrels/day in 1975, whereas in Table 1.1 a level of 13.08 million barrels/day is shown as having been reached in 1968. The latter figure comprises nonenergy uses as well. Even so, our projected estimate of U.S. oil consumption, including nonenergy uses, in 1975 (see Statistical Appendix, Panel H of the table on p. 169) yields a 1965-75 interpolated trend-line figure for 1968 of only 12.33 million barrels/day compared to the above realized figure of 13.08 million barrels/day. The reason lies in the unusually rapid growth of U.S. oil consumption since 1965; a rate of 5 percent yearly, as against a trend rate of 3.5 percent yearly between 1965-75 projected in this study and in other analyses of long-term trends in U.S. oil demand.

Source: Statistical Appendix Table V.
Note: Figures in barrels per day represent metric tons per year divided by 50; see Statistical Appendix, p. 167.

while Japan's rises from under 20 percent to nearly 30 percent. By 1980, Japan—as here projected—will be taking a far greater quantity of Middle Eastern oil than Western Europe did in 1965.

As for the other principal exporting regions shown in Table 1.4, the proportionate positions of different import markets are not assumed to undergo major shifts to 1980—Europe continuing to take virtually all of North

Table 1.4

World Oil Flows Among Principal Areas, 1965, 1975, and 1980

	Exporting areas				
Importing areas	Middle East	North Africa	Caribbean	Other[b]	Total imports[a]
			A. Million metric tons		
1965					
United States	19.8	3.8	83.2	20.5	127.3
Western Europe	202.2	85.2	38.0	47.0	372.4
Japan	67.8	0.1	1.9	11.5	81.3
Total, above three	289.8	89.1	123.1	79.0	581.0
Other world imports	75.4	1.8	36.1	31.1	144.4
Total world[a]	365.2	90.9	159.2	110.1	725.4
1975					
United States	18.3	6.4	123.0	39.2	186.9
Western Europe	339.0	275.6	35.8	68.1	718.5
Japan	170.7	–	6.0	55.2	231.9
Total, above three	528.0	282.0	164.8	162.5	1,137.3
Other world imports	140.7	8.1	53.0	32.1	233.9
Total world[a]	668.7	290.0	217.8	194.7	1,371.2
1980					
United States	17.9	8.8	141.0	47.6	215.3
Western Europe	407.9	347.1	40.3	107.6	902.9
Japan	253.4	–	7.3	105.8	366.5
Total, above three	679.2	355.9	188.6	261.0	1,484.7
Other world imports	182.6	14.2	63.4	40.9	301.1
Total world[a]	861.8	370.0	252.9	302.0	1,785.8
			B. Million barrels/day		
1965					
United States	0.40	0.08	1.66	0.41	2.55
Western Europe	4.04	1.70	0.76	0.94	7.45
Japan	1.36	c	0.04	0.23	1.63
Total, above three	5.80	1.78	2.46	1.58	11.62
Other world imports	1.51	0.04	0.72	0.62	2.89
Total world[a]	7.30	1.82	3.18	2.20	14.51
1975		–			
United States	0.37	0.13	2.46	0.78	3.74
Western Europe	6.78	5.51	0.72	1.36	14.37
Japan	3.41	–	0.12	1.10	4.64
Total, above three	10.56	5.64	3.30	3.25	22.75
Other world imports	2.81	0.16	1.06	0.64	4.68
Total world[a]	13.37	5.80	4.36	3.89	27.42
1980					
United States	0.36	0.18	2.82	0.95	4.31
Western Europe	8.16	6.94	0.81	2.15	18.06
Japan	5.07	–	0.15	2.12	7.33
Total, above three	13.58	7.12	3.77	5.22	29.69
Other world imports	3.65	0.28	1.27	0.82	6.02
Total world[a]	17.24	7.40	5.04	6.04	35.72

[a] Total interregional shipments (i.e., excluding flows *within* the areas shown in Statistical Appendix Table VI).

[b] Includes net exports from the Communist area in 1965 as pointed out on p. 25, net Communist countries oil trade with rest of the world is projected at zero.

[c] Less than 5,000 barrels/day.

Source: Statistical Appendix Table VI.

Africa's exports, as in 1965; and the United States's share of Caribbean exports persisting at well over 50 percent. However, Western Europe's share of Caribbean exports is projected as falling rather sharply.

More detailed interregional flows are shown in Statistical Appendix Table VI. Among the more marked shifts which are projected, there is Oceania's disappearance as a Middle Eastern market (it accounted for 15 million tons or 4 percent of Middle Eastern exports in 1965), a change arising from presumed Australian self-sufficiency in oil. The same development would eliminate Oceania as a market for other Asian (principally Indonesian) oil, as a result of which those exports can be expected to be increasingly dominated by the Japanese market.

APPENDIX: A NOTE ABOUT THE STATISTICAL FRAMEWORK OF THIS STUDY

It was necessary in the course of this study to rely upon oil production, consumption, and trade statistics from a variety of sources, even though there were definitional and other differences among them. This brief note is intended to clarify some of these differences. A fuller discussion of many of the points appears in the Statistical Appendix, particularly on pp. 168–170.

The most comprehensive measure of consumption, and the one that we would have preferred to use throughout, includes all oil used, whether in energy or nonenergy uses (e.g., as petrochemical feedstocks). (This is the measure employed in Table 1.1, in which 1968 world statistics are summarized.) The least comprehensive measure is that of *inland consumption for energy uses:* it excludes (1) bunker fuels for maritime use (mainly international) and international air traffic, and (2) oil consumed in nonenergy uses. The broad background data on energy consumption by region and by energy sources which are basic to the projections in this study (see Statistical Appendix Table IV) employ this measure, because it permitted the projection of oil consumption starting with an historical base that includes all other primary commercial energy sources–particularly coal, natural gas, and hydro. The historical base is provided by United Nations statistics (*World Energy Supplies,* Series J), in which this narrower definition of consumption is employed.

The foregoing differences will present a possible source of confusion to anyone comparing statistics appearing at different points in this study, although it is a problem we have tried to minimize through the use of explanatory footnotes. Unfortunately, the confusion is compounded to a degree by other aspects of the statistics presented. In order to compensate for the less comprehensive coverage of our underlying projections of oil consumption, we have, at various places (e.g., Table 1.3 of this chapter and Table V, line 2, of the Statistical Appendix), included bunkers, although the nonenergy uses of oil are still excluded. The availability of regional historical data on bunkers permitted us to make rough projections of consumption of bunker oil; although desirable from an analytic standpoint, this practice carries the unfortunate consequence of introducing still another measure of oil consumption–one inclusive of bunkers, yet still excluding nonenergy uses of oil. (Again we have attempted to minimize the possible confusion through the use of explanatory footnotes.)

Yet another aspect which yields an analytical advantage, at the possible cost of some additional confusion, is that the statistics on world trade in oil (consisting largely of shipments of crude oil prior to its refinement into end-products destined for energy and nonenergy uses) conform to the most exhaustive definition of consumption. Since international oil movements between major supplying and importing areas are central to the analysis, the comprehensive nature of these statistics is a decided advantage. However, it does carry the complication that in comparing imports to consumption, as we do, there may be a difference between the range of oil uses covered in the two, a point referred to in some of our footnote references.

The quantitative importance of the differences resulting from the three alternative measures of consumption referred to in the preceding paragraphs is indicated by the following statistics for the United States and Western Europe:

Liquid Fuels Consumption[12]
(million metric tons)

	United States		Western Europe	
	1965	1980 projections	1965	1980 projections
Inland consumption in energy uses	482	778	308	767
Total consumption in energy uses (i.e., including bunkers)	494	800	343	837
Total consumption in energy and nonenergy uses	556	928	370	914

The methods used in making the projections for 1975 and 1980 also call for a word of explanation at this point, although they, too, are explained in detail in the Statistical Appendix. The projections are to a considerable extent based upon the extrapolation of trends, but not without recourse to judgment concerning possible departures from trends in particular cases–for example, nuclear power development, and the prospects for gas from the North Sea. It almost goes without saying that the projected figures, although developed in detail, are meant to provide no more than a broad basis for treating the main questions with which this study is concerned. Differences with other

[12] From Statistical Appendix, pp. 168–170, which may be consulted for a fuller explanation and derivation of these alternative measures.

projections, which might deviate in various ways from those here presented, would be a cause of concern only if they signified markedly different conclusions regarding a matter such as the future mutual dependence between Western Europe, Japan, and the United States, on the one hand, and the countries of the Middle East and North Africa, on the other.

One factor now emerging which could bring about a fundamental change—oil from Northern Alaska and the Canadian Arctic—was purposely omitted from the projections. It seemed far better at this stage to give separate examination of this possibility (as we do, principally in Chapter 7) in order to show explicitly to what extent a major new development could affect the trends that might otherwise be expected to unfold.

A word about the year chosen as the basic statistical bench mark for the projections: the use of 1965 permitted us to make the oil projections within the framework of overall international energy statistics available for that year in the previously mentioned United Nations source. In addition, the 1965 base facilitated relating our projections to several major analyses (e.g., those of the Organisation for Economic Co-operation and Development and the European Coal and Steel Community) which also use the mid-1960's as their point of departure.

Of course, developments in world oil often move at a rapid pace and it would have impaired the analysis to neglect some of the principal changes that have occurred since mid-decade—for example, the extraordinary increases in Libyan output or the sharply rising expectations regarding natural gas in Western Europe. Thus, even though 1965 serves as the statistical base for the projections, more recent events have, where judged appropriate, been taken into account both in the text discussion and in the Statistical Appendix.

CHAPTER 2

The Present and Projected Oil Position of the United States

THE preceding chapter provided an overall view of current international oil flows, and a projection of future volumes and directions of such flows. As indicated, the dominant consumer areas now are the United States, Western Europe, and Japan; their preponderance is expected to continue into the future, with some changes in their comparative positions in the world total. In this chapter and the next, the importance of oil imports in the energy supply of these major consumer areas will be examined more closely, as a stepping-stone to the subsequent analysis of the impact of possible interruptions in the flow of imported supplies. We shall begin with the United States, which is by far the largest single oil-consuming country.

Importance of Liquid Fuels Imports

With liquid fuels providing somewhat over two-fifths of total U. S. energy consumption, oil imports—which account for roughly 20 percent of liquid fuels consumption—constituted nearly 10 percent of total energy consumption in the mid-1960's.[1] This level of imports does not, however, place the United States in the highly dependent position of Western Europe and Japan (both current and as projected).

Nevertheless, the 20 percent or so of oil consumption accounted for by imports is of importance to those uses in which oil is employed. Were these imports to be suddenly stopped for political, military, or other reasons, the United States might find itself faced with difficulties. This problem has not loomed as an important one in the past for two major reasons: (1) because U. S. oil imports originate in the main in countries of the Western Hemisphere (Venezuela and Canada) which have proved to be dependable sources in times of crisis; and (2) because for a long time production of oil in the United States has been below capacity, and the excess capacity could be regarded as a security reserve, not just for the United States but to a certain extent for other consuming countries—particularly in Western Europe—as well.

The key question which must be examined for the United States is the extent to which the conditions that have characterized the United States oil supply position to date may be projected to continue in the future. Two subjects must be considered: (1) Is it realistic to assume that imports, particularly from the Middle East and North Africa, can continue to be held at current relative levels? and (2) Will the United States continue to have a cushion of excess capacity that will serve as a security reserve?

The Outlook for Domestic Oil Production and Reserves: As Viewed Prior to the Discoveries on the Alaskan North Slope

As the point of departure for our analysis of the future sources of supply to support domestic consumption, we shall utilize *United States Petroleum Through 1980,*[2] the 1968 report issued by the Office of Oil and Gas of the U. S. Department of the Interior (prepared prior to the apparently vast oil discoveries in Northern Alaska). There are two reasons for this procedure: (1) The report, commissioned by the Secretary of the Interior, was the latest official comprehensive government statement available on the subject at the time this study was completed. (See "Postscript: A Note Comparing Some Findings of this Study With Those of the Cabinet Task Force Report on Oil Import Control," p. 18 of this study.) It was two years in preparation and represented a combined effort of the various branches of the Interior Department concerned with energy matters, assisted by information provided by other agencies of the Federal government, state regulatory agencies, industry and banking sources, and trade associations and publications. (2) The report serves as a convenient point of reference because it is based on specific explicit assumptions, and includes a great amount of background information, thereby permitting its conclusions to be evaluated.

Our presentation of the conclusions of the report, as will be seen, does not reflect full agreement with its findings or with the process of analysis by which they were arrived at. Indeed, the report itself calls attention to weak points in the underlying analysis. Nevertheless, it provides the most useful comprehensive framework presently available (1968/69) within which to depict the salient features of the outlook for U. S. domestic oil, as viewed prior to the

[1] The 20 percent import share indicated in the Interior Department report (see n. 2) on which this chapter is largely based corresponds to a somewhat higher percentage (around 23 percent) used in the broad statistical framework of this study (e.g., Table 1.3 and Statistical Appendix Table V). The difference arises essentially from the use in the latter case of a narrower concept for the measurement of national consumption—i.e., one which excludes nonenergy uses. As for crude oil imports, which are of primary importance within the context of this study, they are measured throughout as total liquid fuels irrespective of their disposition in energy or nonenergy uses.

[2] U. S. Department of the Interior, *United States Petroleum Through 1980* (Washington, July 1968). The projections of U. S. oil consumption and imports used in the present study (Statistical Appendix Tables IV–VI) do not conform precisely with those in the Interior Department Study, as noted in the preceding footnote. Discussion of Interior's projections seemed more appropriate in the present context, since their estimates are part of a much wider quantitative framework for the United States, including projections of reserves and recovery rates.

Northern Alaskan discoveries. Following the summarization of the report, the implications of the recent Alaskan discoveries will be considered to the extent that they appear to affect those conclusions of the report that are of particular pertinence to this study.

Let us look first at the major quantitative findings, which may be summarized as follows:

(1) Between 1965 and 1980, net liquid fuels consumption in the United States is projected to increase from 4.1 billion to 6.5 billion barrels/year (equivalent to 11.3 million and 17.9 million barrels/day, respectively).
(2) Liquids derived from domestic sources are projected to increase from 3.3 billion to 5.2 billion barrels/year over the same period of time (equivalent to 9.0 million and 14.2 million barrels/day, respectively).
(3) Net imports, the difference between domestic consumption and production, are accordingly estimated to rise from 0.8 billion barrels in 1965 to 1.3 billion barrels in 1980 (equivalent to 2.3 million and 3.7 million barrels/day, respectively). In both the beginning and terminal years, net imports account for approximately 20 percent of total consumption.

Thus, the assumption employed in Chapter 1 of this study in deriving an estimate of import flows into the United States to 1980 (i.e., a constant relationship to domestic consumption) is in line with the quantitative framework of the Interior Department's study. However, one cannot, on this basis alone, conclude that the assumption has been justified. The factors underlying the Interior Department estimates must be examined more closely.

Domestic liquid fuels production of 5.2 billion barrels/year in 1980 (14.2 million barrels/day) is estimated by Interior to consist of approximately 4.2 billion barrels (11.4 million barrels/day) of crude oil and about 1 billion barrels (2.8 million barrels/day) of natural gas liquids (NGL) and condensate. The liquids other than condensate are largely a by-product of natural gas output; and production heavily depends, therefore, upon the production of natural gas. Accepting for convenience the figure for NGL and the underlying estimate for natural gas production (which itself calls for separate appraisal not undertaken in this study), we will concentrate our attention upon the estimate of crude oil output in 1980.

On what basis is the conclusion reached that the United States in 1980 might be able to produce about 4.2 billion barrels of crude oil from domestic reserves? The chain of calculations runs as follows:

(1) The assumed increase in domestic crude oil production is 54 percent between 1965 and 1980; this implies a cumulative production of 52 billion barrels between 1965 and 1980.
(2) The crude oil recovery rate (i.e., the amount brought to the surface in the past as a percentage of the total amount of oil in place underground in known deposits) is estimated at 30 percent as of the end of 1965, and is believed to be increasing at an annual rate of 0.5 percent. On the assumption that this rate of increase can be sustained, the rate of past recovery as of 1980 will have reached 37.5 percent.
(3) *Application of the 37.5 percent rate (instead of 30 percent) to oil in place underground in deposits known as of 1965 would yield an additional 29 billion barrels of economically recoverable reserves even in the absence of new discoveries of oil in place between 1965 and 1980.*
(4) In addition, there will be new discoveries of oil in place underground. On the basis of a calculated trend, *it is estimated that such new discoveries of oil in place between 1965 and 1980 will come to 72 billion barrels. Application of the 37.5 percent recovery factor to these discoveries yields an estimated 27 billion barrels of economically recoverable reserves from new discoveries.*
(5) *Hence, the total of additions to reserves between 1965 and 1980 is estimated at 56 billion barrels: 29 billion from improvements in recovery in already known deposits and 27 billion from the discovery of new deposits.*
(6) Additions to reserves of 56 billion barrels would be sufficient to offset cumulative production requirements of 52 billion barrels. They would also yield a net addition to reserves of 4 billion barrels between 1965 and 1980 (i.e., proved reserves as of 1980 would exceed those in 1965 by 4 billion barrels).

How dependable are the links in this chain? The Interior Department's report itself calls attention to one of the links which was obviously weak on the basis of evidence already at hand when the report was written. The new discoveries of oil in place of 72 billion barrels between 1965 and 1980 imply annual average discoveries of 4.8 billion barrels. However, discoveries of oil in place reported since 1957 (adjusted to compensate for deposits only partially developed owing to the recency of their discovery) are estimated at an average of only 3.3 billion barrels annually. Consequently, the discovery rate experienced since 1957, if projected into the future, would not be sufficient to achieve the estimates of the Department's report for the period to 1980.

The calculated trend that was employed in making the projection of 72 billion barrels of discovery of oil in place between 1965 and 1980 (which, in turn, would yield 27 billion barrels of additions to reserves if a 37.5 percent recovery rate were to prevail) was based upon a curve fitted to data for cumulative discoveries over the years 1920 through 1966. In commenting on the actual record for the more recent years, the report states that the last great wave of discoveries crested in 1950 and that the trend has been fairly steadily downward since. (See Figure 2.1.) According

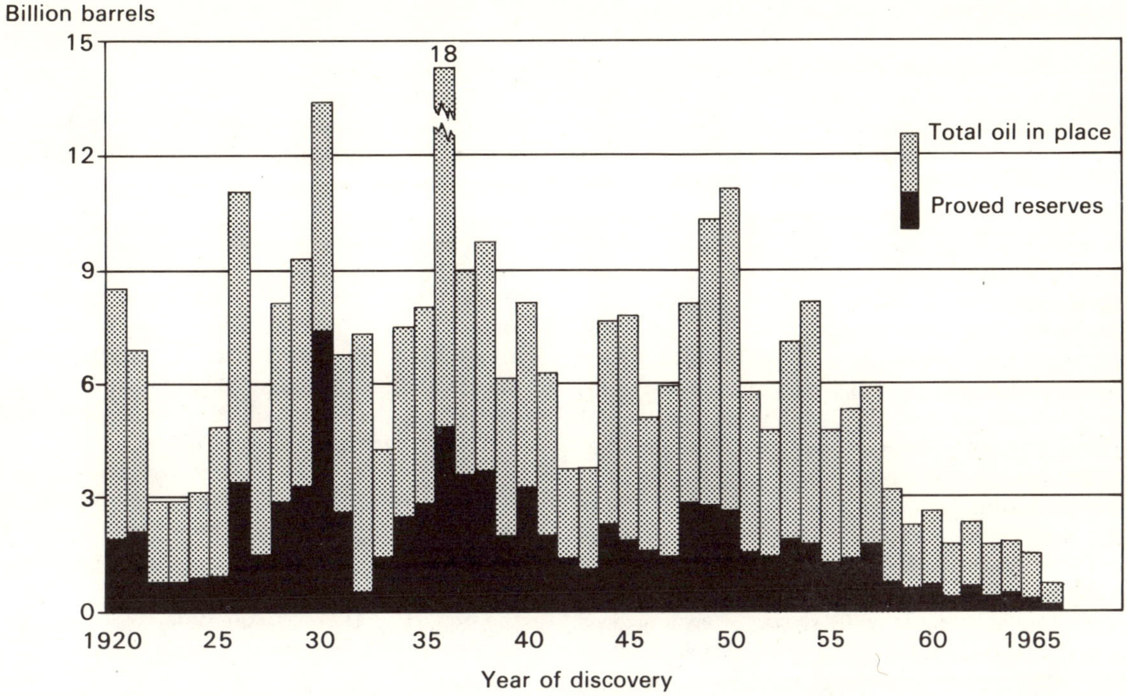

Fig. 2.1. Crude Oil Discoveries in the United States Show Long-Range Decline[a]

[a]The quantities shown have been attributed to the years in which the original discoveries were made. This is done by dating back additions to reserves (which stretch out over a period of years) to the year in which the field was first found. For the most recent years shown in the chart not enough time has elapsed for the "dating back of additions" to reveal the full extent of reserves (or oil in place) to be attributed to discoveries in a particular year. Consequently, an estimating procedure, based on past experience, is employed and the data are less reliable. Total oil in place is dated back in accord with the approach followed for proved reserves. (See *Reserves of Crude Oil, Natural Gas Liquids, and Natural Gas in the United States and Canada as of December 31, 1967,* [New York: published jointly by the American Gas Association, American Petroleum Institute, and Canadian Petroleum Association, May 1968], XXII.)

Source: American Petroleum Institute. From U.S. Department of the Interior, *United States Petroleum Through 1980* (Washington: July 1968), p. 16, Figure 7.

to the report: "The picture suggested is that of a mature extractive industry well into the decline phase of the discovery cycle. The questions to be answered are whether the rate and extent of the decline are sufficient to create a problem of supply during the period under survey, and whether the decline is inherent in our resource base, or merely reflects a slowing down in exploration while recovery from known reservoirs is being increased."[3]

The report is of two minds on interpreting the recent record. Noting that a decline in exploratory drilling has carried through for a decade since 1957 and that discoveries have followed drilling in its downward course, the report states: "Indeed, it may be that the record of the domestic producing industry since 1957 is more truly indicative of what to expect in the future than all the experience prior to that date."[4] A major reason for believing that this may be the case, according to the report, is that the decline in drilling is the largest in the domestic industry's history and that the decline had actually steepened during the most recent three-year period examined, 1964–67. The latter fact is the more remarkable in that during this period domestic production had risen by over one million barrels/day. As prices remained firm, annual revenues to producers increased by over a billion dollars between 1964 and 1967, a result which would ordinarily be expected to lead to an increase in drilling.

However, another interpretation of the recent record is also offered in the Interior Department report—one which is far more hopeful regarding future prospects for oil production in the United States. Essentially, this boils down to the view that efforts to discover new reserves, or to increase reserves by developing new discoveries, are being deferred.

Events in the South Louisiana region, including the offshore portion, are cited in support of the interpretation that field development is being deferred. In this region, the reported total of oil in place discovered from 1957 through 1966 was 70 percent below the total reported for the decade 1947–56. Despite this lackluster showing, exploratory drilling in South Louisiana reached an all-time high in 1966 (while the total for the nation reached a nineteen-year low), and bidders were placing a strongly rising value on

[3] *Ibid.*, p. 16.
[4] *Ibid.*, p. 19.

offshore leases in the region.[5] The report concludes from the anomalous behavior of the most recent indicators that the new fields that have been found in the South Louisiana region since 1957 are actually much richer than their reported evaluations would suggest. If this is the case, their subsequent development will reveal that the total amount of oil in place reported as having been discovered during the past ten years will turn out to be substantially greater than now indicated.[6]

The only evidence offered is for South Louisiana, and this is capable of different interpretations. In any event, even if the more optimistic interpretation is accepted as valid, and the inference drawn that a substantial amount of drilling has been deferred over the past several years, this shortfall will have to be made up if the projected total of new discoveries is to be achieved. It has been estimated that to make good the gap between current completion rates and a rate commensurate with the estimated future discoveries could amount to $10 billion in additional drilling and equipping costs alone over a fifteen-year period.[7] On the other hand, the report suggests, the oil industry may find it less costly to strive for an improved recovery rate (beyond the projected 37.5 percent), as opposed to spending the additional funds that increased completion rates would require. As the arithmetic of future additions to proved reserves is the product of improved recovery rates and new discoveries of oil in place, a superior performance in the former would offset the estimated requirements in the latter category.[8]

Thus, apart from the North Alaskan discoveries, the report's quantitative projections point to the future capability of domestic supply to meet the same percentage share of U. S. liquid fuels consumption as at present. However, no effort was made to conceal the fragile foundation upon which the broad quantitative conclusions rest. It is probably correct to say that the quantitative projections themselves are not to be regarded as a forecast; but rather a statement of one set of possible circumstances under which, with imports maintaining their current relative importance, domestic production would continue to be adequate to fill the remaining demand. Although the report does include a wide-ranging examination of factors which may influence the discovery and production of oil in the United States, it does not assign to them any quantitative values which would give results alternative to those presented.

In his statement accompanying the report's release (about a week prior to the first public announcement of the apparent size of the first major Alaskan North Slope discovery), Secretary Stewart L. Udall called attention to the fact that no conclusive judgment could be made concerning the future and that "the report ends on a note of challenge to the industry, either to mount the additional effort that appears to be necessary to provide the required additions to proved reserves of crude oil, or to demonstrate that the things it is now doing will prove sufficient to supply all the oil needed to meet the rising production requirements of the years ahead."[9] In the closing words of the report: "There is still time to wait for a clear signal either of the efficacy of present industry efforts, or of a turnaround from the historic decline in activity which began in 1956. But it ought not be too long in coming."[10]

The Outlook for Domestic Oil Production and Reserves: As Modified by the Outlook for the Alaskan North Slope

The 1968/69 discoveries on the North Slope of Alaska and their implications for the future are dealt with in Chapter 7, which is concerned more broadly with the North American Arctic, including Canada, and with possible effects not just on the oil supply for the United States, but also on the supply prospects for Western Europe and Japan. Certain of the estimates presented in that chapter will be introduced at this point to indicate how sharply the outlook for domestic oil may have been affected by developments in Northern Alaska.

The 1968 report of the Interior Department draws a picture in which the United States might just manage to squeeze through if, over the next decade, it were to attempt to maintain the same relative level of liquid fuels self-sufficiency as in the late 1960's. A grand total of 56 billion barrels of additions to oil reserves between 1965 and 1980 was postulated as obtainable through the combination of: (1) 29 billion barrels yielded by continuous improvements in recovery of oil already known to be in place; and (2) 27 billion barrels yielded by new discoveries of oil in place, an estimate based upon a calculated trend acknowledged to be highly optimistic in the light of experience over the previous decade.

Although much remains to be learned about the richness of the oil deposits on the North Slope, as of 1969 it

[5] The rising value in offshore leases in South Louisiana is also consistent with the hypothesis that the lack of new good places to drill prior to the Alaskan North Slope discovery gave this region a scarcity value compared to other prospects then available.

[6] U. S. Department of the Interior, *United States Petroleum Through 1980, op. cit.* (above, n. 2) p. 19.

[7] *Ibid.*

[8] Excess productive capacity within the United States is the subject of the next section of this chapter, but it is necessary to anticipate the subsequent discussion because the presence of idle productive capacity may be an important factor in explaining the industry's incentives to search (or not to search) for new oil reserves. As will be seen below, a large amount of idle productive capacity presently exists, and has persisted for many years. It may be theorized that with idle capacity overhanging the domestic market, the incentive to invest additional funds in developing reserves has been weakened. Indeed, a striking phenomenon of the past decade or so has been a redirection of effort into secondary recovery projects (which would account for the improvement in recovery rates), made possible by an improvement in secondary recovery techniques such as water flooding. A significant feature of such projects is that the investments they require can be recovered rapidly, owing to their exemption from production restrictions such as those applied to flush primary producing wells under state proration systems. (See pp. 35 to 38 below.)

[9] U. S. Department of the Interior Press Release, July 9, 1968.

[10] U. S. Department of the Interior, *United States Petroleum Through 1980, op. cit.* (above, n. 2), p. 92.

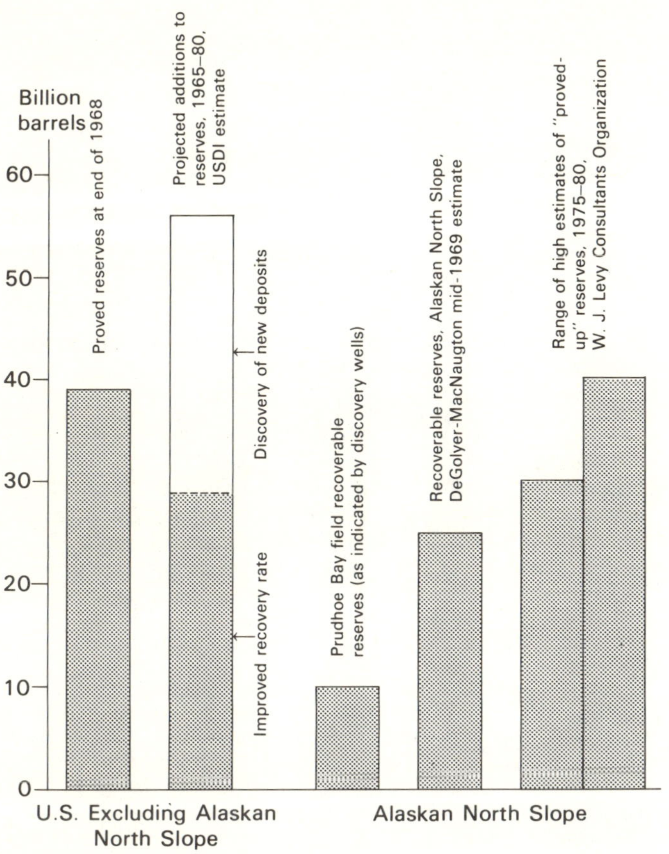

Fig. 2.2. Estimated U.S. Oil Reserves[a] and Projected Discoveries, 1965-80

[a]For a description of the different reserve concepts underlying the alternative estimates presented above, see the text pages cited in *Source.*

Source: Text, pp. 32, 33 to 35; and *Oil and Gas Journal,* December 30, 1968.

appeared likely that one field (Prudhoe Bay) contained at least 10 billion barrels of recoverable reserves. Beyond this, highly reputable analysts were speaking of "proved up" reserves of up to 30 or 40 billion barrels within the decade of the 1970's. Thus, a massive new element appears to have come upon the scene which could convert the cliffhanger conclusion of the Interior Department report to a virtual certainty. (See Figure 2.2.)

This is not to say that the 40 billion barrels of recoverable reserves, should this level be achieved, would be in addition to the quantities of additions to reserves estimated in the Interior Department report. A sizeable portion of the investment that would otherwise have gone into the search for new reserves in the "lower forty-eight" states and into improvements in recovery from known oil reservoirs, will now be diverted to exploration and development of reserves in Northern Alaska. The important point is that the prospects in Alaska add a new, highly significant factor to the picture, a factor that appears greatly to enhance the assurance with which one may contemplate the outlook for oil supply in the United States over the next decade. The billions of dollars which have already been committed by the oil companies to the acquisition of leases in Northern Alaska; to the achievement of a pipeline transport capability expected to begin operations in late 1972; to exploration and development of reserves; to the testing of an icebreaking tanker for year-round transportation in Arctic waters, etc., point to a high degree of confidence on their part in the future commercial potential of this region.

It should also be kept in mind that developments in Northern Alaska are taking place within an economic decision-making framework strongly influenced by a particular set of policies now operative in the United States, especially the quantitative restriction on imports, which serves to support an oil price level in the United States well above that of the world market. The prospective profitability of ventures in Alaska could be significantly affected by policy changes resulting in price reductions in the United States. However, if Alaskan oil turns out to be substantially lower in cost (both at the wellhead and delivered to important markets in the United States) than a sizeable portion of the output now being produced in the contiguous forty-eight states, the United States would be in a position to maintain the same, or even a higher degree of oil self-sufficiency, at a lower cost penalty than in the past.

The Outlook for Excess Productive Capacity in the United States

In 1948, as in the preceding war years, oil was produced in the United States to the full extent of the country's capacity. With the exception of those reserves which had been set aside as an emergency reserve—the Elk Hills Naval Petroleum Reserve in California—there were no crude oil production facilities that were idle or "shut-in" in 1948, barring conservation measures to prevent damage to reservoirs. In that year a single quantity, 5½ million barrels/day, was, therefore, the measure both of production and productive capacity.

Since that time, however, productive capacity has increased more rapidly than production, and indeed more rapidly than the number of producing wells and the amount of proved reserves. For example, as of December 31, 1967, the idle crude oil productive capacity in the United States was estimated by the American Petroleum Institute at about 4 million barrels/day, though, as we shall see, this figure greatly exaggerated the effective deliverable capacity available for temporary emergencies.

The existence of idle crude oil productive capacity in the United States has been an important factor in meeting supply crises that have occurred in the past, most recently in the June 1967 war and its aftermath. However, it is necessary to realize that, except for the Naval Reserve, the presence of idle capacity is not the result of planning; it is an accidental by-product of certain technological and institutional factors that have been at work. Consequently, the excess capacity has been a security reserve only in the sense that it existed and could be drawn upon in case of emergency, but not in the sense that it had been brought into being to cope with emergency situations.

The fact that the excess capacity is not the result of emergency planning gives rise to several important questions: (1) What factors account for its existence? (2) What portion of the idle productive capacity could be converted into needed supply in an emergency situation? And (3), looking to the future, is it possible to make a reasonable prediction concerning the amount of excess capacity that is likely to exist in the years ahead?

1. *Factors Accounting for the Existence of Excess Capacity*

Although a satisfactory explanation of this phenomenon would require a more thorough treatment of the economic, technological, and institutional aspects of the domestic oil industry than can be undertaken within this study, certain significant factors can be identified.

On the institutional side, the most important factor has been the past operations of the state regulatory agencies. These regulatory practices had their origins in the fact that, as a migratory liquid, oil will issue at the surface according to the number, location, and rates of production of the wells which are drilled to tap the reservoir. With divided surface ownership of petroleum lands and a legal principle that oil is held to belong to the owner of the particular surface area at which it issues from the ground, the stage was set for chaotic drilling and production practices in the industry's early history. Aside from their price consequences, which were undesirable from many standpoints, these practices also had unfortunate effects on the conservation of oil resources in that natural pressures of the reservoir were being prematurely dissipated through excessive drilling and production. As a consequence, much of the oil would never be recovered, or recovered only at much higher costs.

To counteract these effects, a system evolved which requires the regulatory authorities in most of the important oil-producing states to decide such matters as the total amount of crude oil which may be produced within the state; the allocation of this total among reservoirs and individual wells; and specific production and development practices such as the spacing pattern of wells within a reservoir.

In practice, the interaction of the rules governing the amounts of oil that may be produced (the proration system) and the rules governing the spacing of wells led to a situation in which there were incentives to the overdevelopment of productive capacity. Faced with the existence of more current capacity than the market is able to absorb at the going price, the regulatory authorities assign production quotas designed to yield the amount of production that will satisfy expected demands. The remainder of the potential production is "shut-in" and constitutes the nation's idle productive capacity. Although available as a security reserve, the excess capacity, and the factors underlying its creation, have been subject to strong economic attack on the grounds that oil production in the United States is far more costly than need be.[11]

2. *Portion of Excess Capacity That Can Be Converted to Needed Supply*

The measure of excess productive capacity at year-end 1967 of about 4 million barrels/day, cited earlier, was derived by subtracting actual production from the estimate of productive capacity made by the American Petroleum Institute.[12] The API estimate of productive capacity is one of several prepared by various official or semiofficial groups. Although no independent estimate has been developed by an agency of the federal government, the Department of the Interior has, in its report, chosen to employ the API figure in deriving its estimate of excess productive capacity.

The API estimate was prepared for the first time in 1968, and will continue to be derived for subsequent annual publications of the Institute.[13] In concept, it represents an

[11] For a good account of the operation of the regulatory system and its consequences, see Wallace F. Lovejoy and Paul T. Homan, *Economic Aspects of Oil Conservation Regulation* (Baltimore: The Johns Hopkins Press, 1967).

While recognizing that more drilling was done in the past than was necessary for developing the discoveries made, the 1968 Interior Department report argues that improvements in the industry's efficiency also contributed to the existence of excess capacity. It states that "large gains in well productivity magnified the effects of the excessive investment in developmental drilling" (*op. cit.* [above, n. 2], p. 30). In support of this view, it is shown that between 1944 and 1948, years when crude oil capacity and production were practically identical, the ratio of proved oil reserves to both production and capacity fluctuated between 11.6 and 12.1, with a five-year average of 11.8. If there had been no increase in the rate at which oil could be withdrawn, the proved reserves available in 1967 would have been able to yield no more than 7.3 million barrels/day in that year, when, in fact, actual production exceeded this amount by about 1 million barrels/day. The ratio between proved reserves and production has, in fact, declined between 1959 and 1967, from 12.7 to 10.4, and the Interior Department report projects a further decline into the future. The report argues that the smaller the volume of proved reserves needed to sustain a given level of output, utilizing what are considered to be sound production practices from the standpoint of reservoir engineering, the greater the industry's efficiency in that less drilling to prove reserves is required.

The argument of the Interior Department report that improving efficiency contributed to the growth of excess capacity can hardly be accepted without serious reservation, if at all. Excess capacity is the result of drilling by operators who find it to their advantage to engage in development drilling under the proration rules set by the regulatory authorities. No obvious causal relation can be drawn between the incentives to such drilling and the declining ratio of proved reserves to production.

[12] *Reserves of Crude Oil, Natural Gas Liquids, and Natural Gas in the United States and Canada as of December 31, 1967* (New York: published jointly by the American Gas Association, American Petroleum Institute, and Canadian Petroleum Association, May 1968), XXII, p. 82.

[13] The 1969 report shows a total productive capacity of 12.1 million barrels/day as of the end of 1968, compared to 12.3 million barrels/day, year-end 1967.

attempt to approximate the volume of oil that could be produced on relatively short notice. The definition employed is as follows:

> The crude oil productive capacity is the maximum daily crude production rate at the point of custody transfer that could be achieved in ninety days with existing wells, well equipment and surface facilities—plus work and changes that can be reasonably accomplished within the time period using present service capabilities and personnel—and with productivity declining as it would under capacity operation. It is assumed that there would be no change in crude oil prices or costs of materials, equipment and/or labor, no statutory restrictions on production rates (but no relief from surface regulations on gas or water production), no restrictions on storage or transportation beyond the point of custody transfer, and no marketing constraints.[14]

In summary, to quote the Interior Department report, the API definition "contemplates a practical limit of oil availability at the point where it enters the transportation system, given ninety days' effort to increase production under current economic conditions."[15]

From the definition it is immediately apparent that the estimate of excess capacity refers to *producibility* and not to *deliverability*, since the latter depends upon the ability of the storage and transport system to handle additional amounts of production. Equally important is the fact that even as a measure of producibility the quantity is an estimate and one that cannot be evaluated realistically in the absence of any need to produce to the limit of capacity. As a practical matter it has been about twenty years since the industry in the United States has produced at capacity, and much has happened to change its size and composition over this time period. Consequently, it is very difficult to determine, in advance, what portion of the capacity that is estimated to be idle can be counted upon if additional supplies are needed in an emergency situation.

A partial test of the ability of the U. S. industry to activate idle capacity was provided by the June 1967 war and its immediate aftermath, when there was an abrupt but short-lived interruption in oil supplies from the Arab world. Between May 1967, the month prior to the outbreak of hostilities, and August 1967, when peak U. S. output was achieved, production rose by somewhat more than 1 million barrels/day. In its report, the Interior Department concludes that if further increases beyond the August rate had been required, transportation and storage might soon have posed a limit, and that even the achievement of the August rate of oil production taxed the ability of the facilities available to handle the water and gas associated with it.[16] Nevertheless, the report comments that "... the experience of 1967, occurring in the context of peacetime routine and lasting barely three months, cannot be compared with the response that can be expected in a national emergency where the safety of the nation is clearly at stake."[17]

If the supply interruption problem, as in 1967, is not one "where the safety of the nation is clearly at stake," but rather one in which necessary supplies are cut off to major oil importing countries, at what point between 1 million barrels/day and 4 million barrels/day (estimated idle capacity in 1967) would the real limit be encountered to U. S. capabilities to expand deliverable production to meet the needs of others? One thing is plain from the actual 1967 experience, which lasted for roughly ninety days: *the actual increase in deliverable production to be expected during a 90-day period is far below the excess capacity implied in the API 90-day productive capacity estimate.* To be sure, the latter refers to producibility rather than deliverability, but, as indicated above, quite apart from transport and storage problems, the production levels achieved pressed against the capacity of facilities to handle the water and gas associated with it.

As for the idle capacity that might have been activated in 1967 over a longer period of time, assuming the investment of additional funds in field and transport facilities, a rough effort at estimation is made in the appendix to this chapter. Beginning with something like the 90-day level actually achieved in 1967 (about 1 million barrels/day), these estimates show increases to 1.8 million barrels/day in six months; 2.4 million barrels/day in nine months; and 3.5 million barrels/day at the end of a year.

3. *Excess Capacity in the United States in the Future*

As a concomitant to its estimates of domestic reserves and production between 1965 and 1980, which were presented earlier in this chapter, the Interior Department report includes an estimate of the growth in productive capacity in the United States over the same period. However, the latter estimate appears to be little more than a rough guess, expressed as follows: "... the projected increase for the 15 years between 1965 and 1980 [is] about equal to the growth experienced in the five years ended in 1965. *If this is a reliable forecast,* then the steady rise in production will, over the years, reduce the amount of idle capacity from 39 percent in 1965 to 15 percent in 1980. At that time, reserve productive capacity will be just under 3 million barrels a day."[18] (Italics added.)

[14] U. S. Department of the Interior, *United States Petroleum Through 1980, op. cit.* (above, n. 2), p. 29.

[15] *Ibid.*

[16] *Ibid.*, pp. 22–28. "On the average, U. S. oil wells produce about three barrels of water for every barrel of oil, and since most of the water is saline, it must be disposed of in a manner that will not damage the soil or water resources of the surrounding countryside." In addition, it is estimated that about three times the normal volume of gas had to be flared (rather than put to productive use) during the period of increased production.

[17] *Ibid.*, p. 28.

[18] *Ibid.*, p. 32.

No acceptable rationale is provided for this estimate. Two points should be made, however, which have a bearing on its value: (1) There is, as already noted, reason to believe that the Department's estimate of new reserves—which does not encompass prospective developments on the Alaskan North Slope—is too high; as the productive capacity estimate is meant to be consistent with the reserves estimate, it should, therefore, also be judged to be on the high side. (2) There is reason to believe that with the ongoing modification in state regulatory practices in the direction of greater efficiency in drilling and production practices, excess productive capacity as an unintended by-product of state regulation is likely to decline.[19] On balance, the Interior Department's estimate of about 3 million barrels/day of excess capacity in 1980, excluding the North Slope of Alaska, appears to be substantially higher than can be accepted.

Some Provisional Conclusions Regarding Imports and Excess Productive Capacity

Earlier in this chapter it was pointed out that the key questions so far as the United States oil position is concerned are (1) whether it is realistic to assume, as has been done in Chapter 1, that imports can be held down to current relative levels; and (2) whether the United States will continue to have a cushion of excess capacity that could serve as a security reserve in case of possible supply interruptions that might occur in the future affecting imports into the United States or other countries.

On the basis of the ground covered to this point, what provisional conclusions may be reached on these questions? The qualifying adjective "provisional" must be kept in mind because certain aspects that may bear importantly on the conclusions will be dealt with in later chapters. One such subject is the outlook for Northern Alaska that has been briefly summarized at an earlier point in this chapter and will be treated more fully in Chapter 7. Another concerns the possible development of technological substitutes for crude oil to be derived from oil shale, tar sands, or coal, if the crude oil reserves situation does not shape up satisfactorily.

1. *The Relative Level of Imports*

Taken at face value the quantitative framework of the Interior Department report would confirm the belief that imports during the next decade or so can be held at their current relative level, that is, at about 20 percent of domestic consumption. However, as is evident from a closer examination of the report, the reserves and production projections depend upon the attainment of levels of domestic oil discovery or improvements in crude oil recovery which are markedly better than those that have been realized over the past decade (except for recent developments in Northern Alaska). To be sure, the report offers arguments for the belief that the record of the recent past may not be a valid guide to the future. Nevertheless, the case has not been established that the ratio of imports to consumption can be held at its present moderate fraction. The opposite case could be argued with at least equal force: that imports levels will rise substantially (when and by how much it is difficult to say) unless essentially new developments not allowed for in the Interior Department's study come into play with considerable force.[20] Oil discoveries in Northern Alaska appear to be such a development. As indicated earlier, the proving of new reserves during the next decade in that region in amounts being spoken of in 1968/69 by highly regarded analysts would, if they materialize, virtually assure the ability of the United States to hold the import/consumption relationship to present levels, or even to reduce the relative importance of imports, should the government choose to do so as a matter of policy.

2. *Excess Productive Capacity*

Compared to Western European imports from the Middle East and North Africa of nearly 7 million barrels/day in

[19] For example, proration formulas for allocating production determine how many wells it is profitable to drill in an established field. In the past, four wells on 40 acres would have received much larger combined allowable than one well on 40 acres. This bias has been partially removed in most current formulas by making the allowable proportion to *acreage* at any depth, at least up to 40 acres. Thus there is no incentive to drill two or more wells to secure the same allowable as from one well. (See *U. S. Energy Policies: An Agenda for Research* [Washington: Resources for the Future, Inc., 1968], p. 36).

[20] In a report issued in October 1968, several months after the release of the Interior Department's study, the Energy Division of the Chase Manhattan Bank concludes that in the United States in the period to 1980: ". . . the future discovery of oil and natural gas reserves—excluding those found in Alaska—is expected to fall short of the nation's rising level of consumption. And, consequently, the ratio of reserves to production for both oil and gas will continue to fall. To some extent, lower ratios can be tolerated, but eventually the declines will have to be arrested by limiting production. And at that time the need for proportionally higher imports will develop." This study, it should be noted, is oriented far more heavily to the financial aspects of the oil industry's operations than is the Interior Department report. (See John C. Winger, John D. Emerson, and Gerald D. Gunning, *Outlook for Energy in the United States* [New York: Chase Manhattan Bank, October 1968], p. 57.)

In communications from industry sources we have received some opinions that recent evidence indicates that the 1980 projections of production and productive capacity of the Interior Department study are too high. In an article in the *Oil and Gas Journal* (April 28, 1969, pp. 33-38) industry views are reported which are consistent with certain opinions expressed privately to us. This article contains some projections which are seriously at odds with those in the Interior Department study. But while the Interior Department projections are contained within a systematic study which provides a highly visible basis for critical appraisal, the basis for the projections attributed to industry sources are not equally visible, particularly for those estimates which are in most serious disagreement with the Interior Department's projections. The April 28 article appeared in the wake of the oil rush that began in Northern Alaska in 1968 and in the shadow of newly-launched governmental inquiries in the Congress and in the Cabinet into oil-import policy and other aspects of the system of incentives applicable to the oil industry.

1967, and Japanese Middle Eastern imports of 2 million barrels/day, U. S. estimated idle productive capacity of about 4 million barrels/day (almost 50 percent of the combined imports of Japan and Western Europe from the Middle East and North Africa) could be considered a security reserve of some significance. (Add the estimated idle productive capacity in Canada and Venezuela [see Chapter 6] and the total Western Hemisphere security reserve looms even larger.)

Only a fraction of the idle productive capacity in the United States—about 1 million barrels/day—was, as has been seen, converted into additional supplies during the three months of the 1967 Middle East crisis. It does not seem likely that significantly larger amounts could have been made available without the expenditure of funds on additional field and transport facilities. With more time and extra expenditures—if justified by the seriousness and duration of the crisis—the additional amounts that could have been made available over a period of up to a year are estimated by us at about 3½ million barrels/day as of 1967. (See appendix to chapter.) Although less of a security reserve than the API estimate of productive capacity would indicate, the ability of the United States to expand production in short order was nevertheless a factor of some consequence in coping with the 1967 supply crisis.

However, the 1967 situation will not long serve as a prototype of what to expect from the United States in the event of a major supply interruption elsewhere, for the comparative (and absolute) importance of idle capacity in the United States will diminish greatly as time passes. Compared to projected Western European imports from the Middle East and North Africa of about 12½ million barrels/day in 1975 and more than 15 million barrels/day in 1980, and Japanese imports of 3½ million barrels/day in 1975 and 5½ million barrels/day in 1980, even excess productive capacity of the United States, as estimated in the Interior Department report, of about 3 million barrels/day would not provide much of a security cushion. And, as indicated, a realistic estimate of future U. S. excess capacity in the contiguous forty-eight states is likely to be much less than 3 million barrels/day.[21]

APPENDIX: ESTIMATION OF EXCESS OIL PRODUCTION DELIVERABLE TO WESTERN EUROPE TO MEET EMERGENCY NEEDS, AS OF 1967

As shown in the chapter, it is apparent that "true" 90-day deliverability in 1967 was much below the API estimate of producibility at the point of custody transfer. Table 2A.1 represents an attempt to estimate 1967 deliverability after ninety days and up to one year.

Column (6) in Table 2A.1 represents an effort to develop a relatively conservative estimate of such deliverable excess capacity at the end of ninety days. For Texas and Louisiana (and also New Mexico) no more is estimated than the actual increase which did take place during the three months, May–August 1967. Estimates of 30,000 barrels and 20,000 barrels/day for Montana and Wyoming, respectively, are based upon statements in the *Oil and Gas Journal,* as noted in the footnotes to the table. No extra capacity was attributed to Oklahoma from this source because of lack of transport facilities in the short run. Of the 23,000 barrels/day of spare capacity in Mississippi, it was estimated that some 20,000 barrels/day could be made available in ninety days, on the basis of the ample pipeline facilities in the area and the availability of barge capacity in the Mississippi River and Intracoastal Canal areas. The total 90-day estimate for deliverable spare capacity is only 1.091 million barrels/day, or only about 5 percent more than the increase in production actually experienced. In a crisis like that of 1967, domestic producers were not motivated to take any and all steps possible to increase the production rate, due both to the costliness of such steps and to the presumably temporary nature of the added demand. As the *Oil and Gas Journal* stated editorially in the August 14, 1967, issue:

> Now the failure of many fields, particularly in Texas, to make their new high allowables makes it clear that the published estimates were far from "instant capacity." Perhaps the full . . . extra capacity could be attained in a few months with additional field work and equipment. But operators are showing no great haste to do this beyond projects previously scheduled and some temporary improvisations.
>
> Even if the added crude demand continues unabated for several months (which it probably won't), few producers will start crash programs to expand capacity. First they want to be sure what their present facilities can do. And they don't want to go through another costly cycle of carrying much excess capacity.[22]

Another factor which retarded increases in the production rates in the summer of 1967 was the reluctance of, and statutory restraints upon, state regulatory commissions to employ methods of increasing production as rapidly as possible. To accomplish this latter goal efficiently, the most productive wells should be granted the largest increases in allowables. Instead, the application of the traditional formula of allowing each well to produce the same percentage of its "legal allowable" kept total production under physically attainable production since no individual well was permitted to produce more than its allowable, even though the less productive wells might not be able to produce their share of an increased allowable. For example, total production in Texas might be estimated

---

[21] As for Northern Alaska, the legal framework to be imposed upon oil operators by the state is not yet established. (In the "lower forty-eight," idle capacity, as indicated, has been an inadvertent by-product of state regulatory practices.) It is too early to say whether other elements, such as the production practices pursued in Alaska, or the competitive framework within which operations will take place, will result in the creation of a substantial amount of idle capacity that could be quickly activated, if required.

[22] *Oil and Gas Journal,* August 14, 1967, p. 67.

Table 2A.1

U.S. Crude Oil Production and Estimated Productive Capacity and Deliverable Spare Capacity for Meeting Western European Shortages: Major Oil-Producing States, 1967

(Thousand Bbl./Day)

State	1967 Production	Productive capacity 12/31/67 (API)	Spare capacity (90-day basis API) Amount	Percent of national total	Percent of state capacity	Deliverable spare capacity (90 days)*	Deliverable spare capacity (assuming production at 1967 rate) At end of 6 months	At end of 9 months	At end of one year**
	(1)	(2)	(3)	(4)	(5)	(6)	(7)	(8)	(9)
Texas	2,878	5,251	2,373	59.7	45.2	612[a]	973[b]	1,150[d]	1,500[c]
Louisiana	1,863	2,765	902	22.6	32.3	402[a]	646[c]	1,000[d]	1,695[c]
California	985	1,283	298	7.8	30.3	(R)	(R)	(R)	(R)
New Mexico	327	441	114	2.8	25.9	7[a]	100[e]	110[e]	114[e]
Alaska	79	222	143	3.6	64.4	(R)	(R)	(R)	(R)
Montana	96	200	104	2.6	52.0	30[f]	50[f]	68[d]	104
Wyoming	380	401	21	0.5	5.1	20[f]	21	21	21
Oklahoma	610	619	9	0.2	1.5	0[f]	30[f]	45[d]	75[f]
Kansas	270	267	−3	–	−0.9	–	–	–	–
Mississippi	157	180	23	0.6	12.7	20	23	23	23
Other states	677	660	−17	−0.4	−2.6	–	–	–	–
Total	8,322	12,289	3,967	100.0	32.3	1,091	1,843	2,417	3,532

(R) Not available for aiding Western Europe to meet emergency shortages.
* Assumes motivation no greater than was present during crises of 1956-57 and 1967.
** Maximum estimate; assumes all-out efforts such as might be obtained only during a more serious and prolonged crisis.
[a] Actual production increase, May-August 1967.
[b] Louisiana estimate for six months multiplied by ratio of 90-day capacity for Texas relative to Louisiana (column 3).
[c] Statements of Texas and Louisiana Regulatory Commissioners, as reported in *Oil and Gas Journal*, June 12, 1967, pp. 46-47.
[d] Six-month estimate plus one-third of estimated total expansion in last six months.
[e] Estimates based on study of available transport facilities.
[f] *Oil and Gas Journal* estimate, June 12, 1967, p. 46.

Source: Reserves of Crude Oil, Natural Gas Liquids, and Natural Gas in the United States and Canada as of December 31, 1967 (text n. 12), p. 82, cols. (1) and (2).

at 3 million barrels/day as the desired level. If all wells in the state had a joint 100 percent legal proration capacity of 6 million barrels, regulatory concepts of "equity" would dictate that each well produce at 50 percent of its allowable. It might be known that the inability of lower-productivity wells to produce as much as 50 percent of their nominal "allowable" would result in underproduction of the quota by perhaps 10 percent. Hence the regulatory commission would increase the proration percentage to perhaps 55 percent, although the logical approach would be to allow the more productive wells the higher outputs. (Toward the end of the summer, the Texas Commission did allow the more productive wells on a lease to produce more in order to make up for deficiencies in the ability of the poorer wells on the lease to produce, but this was only a temporary expedient and was further limited to the individual lease as the unit of reconciliation.)

Estimates of productive capacity for levels that might have prevailed at the end of six, nine, and twelve months after the outbreak of an emergency supply situation in 1967 are given in columns (7), (8), and (9) of Table 2A.1. These columns assume great motivation on the part of the industry to leave no stone unturned in order to increase production as rapidly as possible.

Let us look first at the data for Louisiana, since petroleum industry economists appear to believe that the conclusions of a Louisiana state commission study made in 1966 regarding productive capacity were worthy of considerable respect. This study concluded that maximum spare capacity in the state, in terms of sheer ability to produce, was 1.695 million barrels/day, while limitations on deliverability reduced effective spare capacity to 646,000 barrels/day. It is assumed that this 646,000 barrels/day estimate would have been valid for a six-month projection (possibly even for three months if maximum motivation to produce were assumed), and that the full 1.695 million barrels could have been produced at the end of one year. This would have required considerable additional investment in well, lease, and field facilities, and in pipelines; but since the pipeline distances involved to reach the coast would be relatively minor, it is assumed that within a year this entire additional output could have been placed on tankers in the Gulf ports.

The Louisiana estimate is perhaps on the high side, but it is accompanied by what is perhaps a slightly conservative estimate for Texas. For Texas for a six-month horizon, deliverable spare capacity is assumed to have had the same relation to Louisiana six-month capacity as the ratio of actual production for the two states during three months of the 1967 emergency; hence the estimate of 973,000 barrels/day for Texas six months after the materialization of a serious emergency. The estimate of 1.5 million barrels at the end of one year is based upon statements by the Texas Railroad Commission in June 1967 to the effect that over a longer period Texas might have produced up to 1.5 million barrels/day more crude. This would have required much in the way of additional field and transport facilities, but if one assumes that these expenditures were to be made, 1.5 million barrels/day of spare capacity in 1967 may be a somewhat conservative estimate. It seems all the more conservative when compared to a 1.7 million barrels/day figure for Louisiana, where the API-based estimate of 90-day excess capacity in 1967 was only about 40 percent that for Texas. Nevertheless, the larger and newer Louisiana fields may more readily be able to increase their output rates than the bulk of the older and often individually smaller Texas fields. (It is to be noted that during 1967, Texas fields appear to have been more troubled than Louisiana fields by inability to fulfill their allowables.)

New Mexico is estimated to have been able to produce and transport 100,000 of its 114,000 barrels/day of API-estimated spare capacity at the end of six months, primarily because of the relative adequacy of existing pipeline facilities. During the 1967 crisis, New Mexico output did not increase greatly, in part because the excess capacity of the closer West Texas fields was not fully used during the crisis. But the West Texas fields were among those which had trouble meeting higher percentage allowables, and it is likely that in a more serious and prolonged crisis the full extent of New Mexico's spare capacity would have been used. The prediction for the end of twelve months is the full 114,000 barrels/day.

It is estimated that some 50,000 barrels/day capacity in Montana and Wyoming could have been moved to Kansas and other Midwest markets after ninety days, freeing Gulf Coast crudes for movement overseas. Of these 50,000 barrels, 20,000 barrels represents essentially the entire API-estimated 90-day spare capacity of Wyoming in 1967, and 30,000 barrels represents a much smaller part of Montana's 104,000 barrels/day of API-estimated 90-day spare capacity. By the end of a year of all-out effort it is estimated that the full 104,000 barrels/day of Montana spare capacity could have been delivered to Midwest markets, allowing diversion of an equal amount of Gulf Coast crudes. Mississippi spare capacity of 23,000 barrels/day, it is assumed, could have been made available in full at the end of six months, and 20,000 barrels in three months, as estimated earlier. Although in 1967 it was estimated that Oklahoma would not have been capable of producing and shipping any spare capacity oil within three months, it was also estimated that, given time, up to 75,000 barrels/day more oil could have been produced. Reversing the direction of flow in existing pipelines might have allowed production of up to 30,000 barrels/day of Oklahoma crude oil by the end of six months; all-out efforts to increase production, plus the construction of new pipelines, would have made 75,000 more barrels available at the end of a year—although this may be regarded as a somewhat optimistic estimate in view of the declining long-run trend in Oklahoma reserves and production.

The resulting six-month total, as of 1967, was 1.843 million barrels/day; 2.417 million barrels for a nine-month period; and 3.532 million barrels after twelve months. The last figure may be somewhat on the high side, even apart from the dubious nature of the concept of an "emergency" supply available only after twelve months.

CHAPTER 3

The Present and Projected Oil-Import Positions of Western Europe and Japan

Western Europe

WESTERN EUROPE, as a region, is by far the leading market in the international oil trade. In 1968, the region's imports accounted for approximately half of the world's total inter-area oil movements (about 510 out of 1,020 million tons, or 10.3 out of 20.4 million barrels/day—see Table 1.2), and of the amounts flowing into the area, over 80 percent originated in the Middle East and North Africa. (Roughly similar relationships had prevailed in 1965, as shown in Statistical Appendix Table VII.) It is estimated, as indicated in Chapter 1, that the region will continue to be the world's preeminent consumer of imported oils over the next decade or so.

This chapter will examine the importance of oil supplies, practically all of which originate in the Middle East and North Africa, to the internal energy economy of the region now and as projected into the future. To do so, it is necessary to begin by placing oil within the context of the region's total energy consumption.

The Relative Importance of Oil in Total Energy Consumption: Present and Projected

The rapidly changing importance of oil within Western Europe's energy economy is clearly indicated in Table 3.1. In 1965, oil accounted for less than half, and in 1980, it is projected to account for about three-fifths of total energy consumption. Looking back to the mid-1950's, oil's importance within the total was less than 20 percent.

The rising importance of oil consumption in Europe (and in Japan, see next section) is perhaps the most striking development on the international energy scene in the years since the end of World War II. Prior to that war, Western Europe was essentially a coal-based economy, and, therefore, one in which practically all energy supplies were indigenous to the region. Coal did, of course, cross international borders and the desire to possess coal deposits was at the root of many boundary problems leading to international frictions within Europe. However, in the postwar years, the growing unity of Western Europe (beginning with the creation of the European Coal and Steel Community) promised to eliminate coal as a source of national conflicts. At the same time, European coal exploitation, intensive in its use of labor, became increasingly more difficult and more expensive. Under the impact of changing price relationships among fuels and structural shifts in use patterns, coal declined steadily. It was not in a position to withstand the growing competition from oil which moved quickly to the forefront. Coal's postwar decline can be measured against its prewar importance in total energy consumption: 95 percent in 1929 and 90 percent in 1937.[1]

Oil was a latecomer as a major energy source in Europe compared to the United States. The level of almost 20 percent which was attained in Western Europe in the mid-1950's had been reached in the United States in the mid-1920's; and the 30 percent level which characterized Europe in 1960 had been reached in the United States about 20 years earlier. Despite its later start, oil's share in total consumption in Western Europe is greater than its corresponding level in the United States—45 percent as compared to 40 percent in 1965. And in the projections for 1975 and 1980, oil's relative importance within Europe is estimated substantially to exceed that in the United States: oil's share in the total is estimated to rise to over 60 percent in Western Europe by 1980, compared to an actual slight decline to a little under 40 percent in the United States. A major factor accounting for this difference in the role of oil in the two areas is the estimated share of natural gas, which is set at about 35 percent for the United States (essentially its present-day level of comparative importance) and at about 14 percent in Western Europe.[2] (See Figure 3.1.)

The fact of overwhelming importance in the future energy situation of Europe, as currently projected, is this heavy reliance on oil, practically all of it to be supplied from imports. (In the United States, as projected, not only is oil's share lower, but also not much over one-fifth of oil consumption is estimated to be supplied by imports.) As an estimate, the 60 percent oil share for Western Europe can claim no more than to reflect today's best judgment about the comparative importance of oil in Western Europe's energy position over the next decade or so. Indeed, other sources—coal, gas, and nuclear power—may play a larger role than is presently thought likely, thereby lessening the estimated future importance of oil imports. The remainder of this chapter is devoted to an examination of this possibility.

Potential Energy Supplies from Coal, Natural Gas, and Nuclear Power

The consideration of each of the major alternative sources—coal, natural gas, and nuclear power—will be made in the light of what might be possible in the absence of major breakthroughs in technology or reserve discoveries

[1] Organisation for Economic Co-operation and Development, *Energy Policy: Problems and Objectives* (Paris, 1966), p. 160.

[2] On these Europe-United States comparisons, see Statistical Appendix Table IV.

Table 3.1

Western Europe's Inland Energy Consumption[a]

Year	Total (thousand metric tons of oil equivalents)	Percent of total consumption				
		Solid fuels	Liquid fuels	Natural gas	Hydro[b]	Nuclear[b]
1955	484,204	78.2	18.1	1.0	2.7	
1960	555,568	64.8	29.8	1.9	3.5	
1965	696,880	49.3	44.3	2.7	3.5	0.2
1975	1,031,380	23.9	58.2	13.1	3.1	1.7
1980	1,254,730	18.1	61.2	13.5	3.0	4.2

[a] For definitional adjustments which permit comparability of the above data with other, somewhat different concepts of energy consumption, see Statistical Appendix, pp. 168-170.

[b] These percentages result from the use of conversion factors in which original units are translated into oil (or coal) equivalents on the basis of heat value of the hydro or nuclear electricity produced. When hydro and nuclear are converted on the basis of equivalent fuel-input requirements at fossil-fueled thermal power stations, their percentage shares rise markedly, as seen in the following figures for 1965, 1975, and 1980 (taken from percentage distribution of panels (C) or (D) in Statistical Appendix, p. 168):

	Solid fuels	*Liquid fuels*	*Natural gas*	*Hydro*	*Nuclear*
1965	45.8%	41.1%	2.5%	9.9%	0.7%
1975	22.1	53.8	12.1	7.8	4.2
1980	16.1	54.6	12.1	7.1	10.0

Source: Statistical Appendix Table IV.A.

and without major policy changes on the part of governments. Even so, a range will be introduced, to allow for uncertainties, and it will be of particular interest to examine, in the final section, the effect on estimated oil imports produced by employing what are regarded as the highest feasible levels for these other energy sources.[3]

1. *Coal*

The production of coal,[4] on which Europe's earlier industrial supremacy had been built, has in recent years been declining steadily in both absolute and relative terms. Output, which stood at about 540 million tons in 1957, had declined to 480 million tons in 1965, and falling further, went down to approximately 430 million tons in 1967.[5] It is highly significant that, in the words of the OECD: "The decline of coal production would probably have been even greater but for the various measures of protection that all the coal producing countries introduced in one form or another—aids, subsidies or duties and restrictions on competing imported fuels."[6]

Indigenous coal has been losing ground not only because of the comparative growth of sectors of consumption—mainly transportation—which are specifically dependent upon petroleum products, but more so because of coal's loss of market to oil products in general heating uses in industry, power plants, commercial establishments and residences. It is estimated that those uses in which petroleum products are specifically required accounted for less than 20 percent of total consumption of primary energy sources in Europe in the mid-1960's,[7] while petroleum, according to Table 3.1, accounted for about 45 percent of the total. To be sure, the domestic refining of crude oil to produce gasoline and other specifically required fuels inevitably yields some additional oil products that are placed upon the market, but the relative amounts of different refinery products can usually be managed, within rather wide limits, according to the demands of the market. This is evident in Figure 3.2 in which the United States, Western Europe, and Japan are compared.

Neither the OECD nor the ECSC is optimistic about the prospects of coal improving its competitive position in the general heating market in the future. Coal mining is a labor-intensive industry and although worker productivity has been growing in response to heavy investments in mechanization, hourly earnings have been rising faster.[8] After taking all of the relevant factors into account, including the political consideration that there is a limit to the rate at which indigenous coal production will be allowed to run down, the OECD, in its 1966 publication, estimated that solid fuel production by 1970 would be in the range of 410–440 million tons, and in 1980, 340–440 million tons (measured in hard-coal equivalents);[9] while the ECSC, in what it plainly presents as an unrealistically high estimate, places the peak possible production at 435 million tons (HCE) in 1980.[10] As it turns out, 1967 output already came close to the OECD's *low* 1970 projection, while 1968 production actually fell below the low end of the 1970

[3] The remainder of this chapter will draw heavily on the energy projections contained in two reports prepared by international agencies that are deeply involved in Western European energy matters. The reports are: *Energy Policy: Problems and Objectives*, by the Organisation for Economic Co-operation and Development (Paris: OECD, July 1966); and *Review of the Long-Term Energy Outlook for the European Community*, published as a Bulletin of the European Coal and Steel Community (Bulletin No. 61 [Luxembourg: ECSC, April 1966]). The former will be referred to as the OECD study, the latter as the ECSC study. It should be noted, however, that the latter report represents a combined effort of the High Authority of the ECSC, the Commission of the EEC (European Economic Community), and the Commission of the EAEC (European Atomic Energy Community).

[4] For convenience, "coal" covers both hard coal (essentially equivalent to bituminous coal in the United States) and brown coal, the latter always being figured at its hard-coal equivalent (HCE).

[5] United Nations, *World Energy Supplies*, Series J, various numbers.

[6] OECD, *Energy Policy*, *op. cit.* (n. 3), p. 82.

[7] *Ibid.*, p. 32.

[8] *Ibid.*, pp. 82–83; and ECSC, *Long-Term Energy Outlook*, *op cit.* (n. 3), pp. 38–39.

[9] OECD, *Energy Policy*, *op. cit.* (n. 3), p. 177.

[10] ECSC, *Long-Term Energy Outlook*, *op. cit.* (n. 3), p. 46. An indication of the acknowledged lack of realism in this high coal estimate may be seen in the fact that the countries belonging to the ECSC account for 225 million tons of this total, whereas the ECSC report states that 140 million tons would be in excess of the strictly competitive coal production of these countries in 1980 (pp. 39–40).

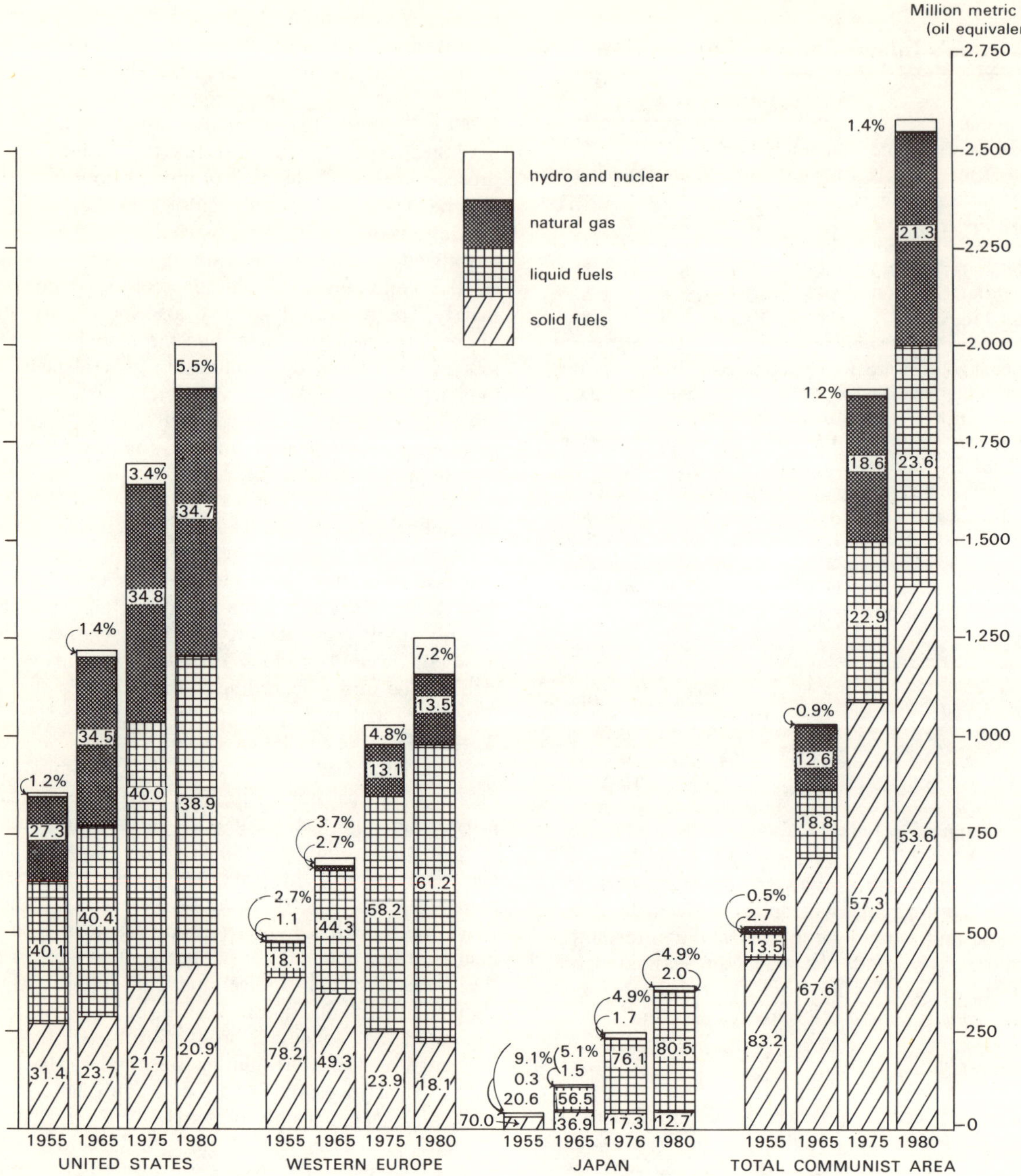

Fig. 3.1. Energy Consumption Patterns, Selected World Regions, Selected Years, 1955-80

Source: Statistical Appendix Table IV.A.
Note: Numbers represent percent share of the indicated energy source in the total energy consumption of each area.

OECD range. The present study, taking these recent developments into account (as well as projecting higher levels of natural gas and nuclear power than does the OECD volume; see the discussion below) foresees 1980 solid fuel output of around 280 million tons (Statistical Appendix Table XII).

In addition to indigenous coal, there might also be considered the prospects for coal imports. Within the context of this discussion, in which the range of possibilities for reducing the reliance on oil imports is the question being examined, the possible future role of coal imports from the United States requires attention. Western Europe in recent years imported somewhat less than 10 percent of its coal consumption, about equally divided between the United States and Eastern Europe (Statistical Appendix Table X); and as the OECD points out, if comparative internal and external coal prices were the sole criterion, the volume of imports would be greater.[11] Again,

[11] OECD, *Energy Policy*, *op. cit.* (n. 3), p. 86.

making what it acknowledges to be a plainly unrealistic estimate (unrealistic, that is, by virtue of European political factors involving the rate at which their domestic coal industries can be phased out) the ECSC conjectures that 115 million tons of coal might be bought from North America in 1980.[12] However, calculations in the report indicate that this hypothetical level of imports, in the extremely unlikely event that it were to occur, would be inconsistent with the very high (435 million tons) level of European output, referred to above.[13]

2. *Natural Gas*

As noted above, the larger estimated place of oil in the future European total compared to that in the United States reflects mainly the comparatively greater role of natural gas in the United States, where the production and use of natural gas are well established. Without any change in its share between now and 1980, the relative importance of natural gas in the U. S. energy total in 1980, would be about 35 percent, compared to approximately 14 percent in the estimated Western European energy total by that time.

However, even a 14 percent natural gas share in total energy consumption represents, for Europe, a major prospective change. The discovery of large natural gas reserves in Europe is of relatively recent origin—a fact reflected in the small share of natural gas in the energy total in recent years (less than 3 percent in 1965). Are there prospects for larger discoveries and still more rapid adoption of this fuel than the roughly 14 percent allowed for in Table 3.1?

To explore this question briefly, it is useful to look at the way in which developments in the most recent two-three years have given rise to revised thinking about future prospects. In its 1966 report, the OECD pointed out that the estimates for indigenous production must be regarded as quite uncertain because of the lack of knowledge concerning the possible amounts of reserves that might be discovered, and the timing of these discoveries. And even following discovery, much time would be required for drilling to prove up reserves and bring the deposits into production; for the development of pipelines for gathering and transporting gas; and for the putting into place of facilities in which to use it.[14] Taking these factors into account, the OECD stated: "We have taken as the best estimate we can make that 1970 might see some 40–60 billion cubic metres (say 35–55 Mtoe) from indigenous resources, rising by 1980 to about 80–120 billion cubic metres (say 70–110 Mtoe). To achieve the lower figures would involve some additions to reserves with normal rate of development, while the higher would involve substantial additions to reserves and very rapid development of them."[15] The ECSC, in its most optimistic 1980 estimate exceeded the OECD somewhat, envisaging the possibility of 200 million tons coal equivalent, which converts to about 113 Mtoe.[16]

In a revised analysis, the OECD in 1969 reviewed more recent trends in European natural gas development and examined their implications for the future.[17] First, the latest reserve figures suggest a more promising outlook for future production, the 1966 estimate of 2 billion tons oil equivalent (or close to 15 billion barrels) in proved reserves having more recently been increased by 35 percent to 2.7 billion tons (or nearly 20 billion barrels). It should be noted that large-scale exploration efforts on the North Sea continental shelf were only being started at the time of the earlier OECD analysis. Second, and relatedly, on the basis of the latest trends, European natural gas production seems quite certain to approach, and perhaps even exceed, the high end of the 1970 output range (35–55 million tons oil equivalent) earlier envisaged. Output in 1968 already surpassed the low end of this range. Third, there appears now to be a fairly solid basis for assuming reasonably rapid

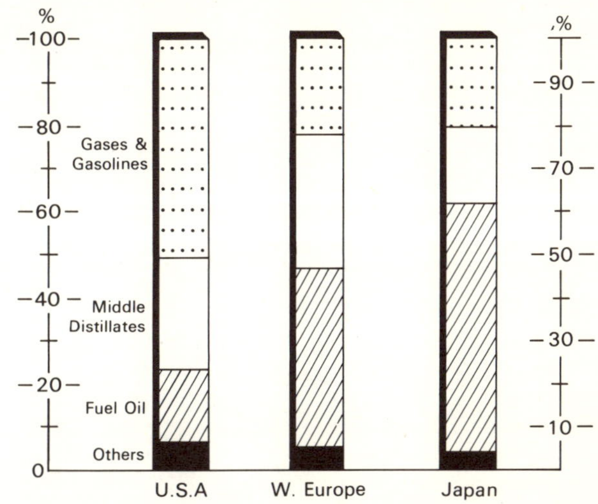

Fig. 3.2. Pattern of Inland Demand for Major Petroleum Products, 1967, United States, Western Europe and Japan

Source: British Petroleum Co., Ltd., *Statistical Review of the World Oil Industry, 1967* (London, 1968), p. 8.

[12] ECSC, *Long-Term Energy Outlook, op. cit.* (n. 3), p. 42. The 115 million tons would consist of 35 million tons of coking coal (over one-third of Western European estimated consumption of such coal) and 80 million tons of coal for power stations (20 percent of the estimated coal consumption in this use).

[13] *Ibid.*, pp. 43 and 46.

[14] OECD, *Energy Policy, op. cit.* (n. 3), pp. 87-88.

The "proving-up" of the Groningen field in the Netherlands in 1959-60 may be taken to mark the new period of major gas discoveries in Western Europe. Prior to that time relatively small deposits were exploited in France, Austria, and Italy. Groningen, on the other hand, is said to be the largest gas field in the world, with reserves officially estimated at a maximum of 1,800 billion cubic meters. This discovery has led to intensive efforts, with some success to date, to discover gas reserves in the North Sea, and on its continental boundaries. (See M. E. Orlean, "Impact of Natural Gas on West European Energy," *World Petroleum,* April 1968, pp. 44-46.)

[15] OECD, *Energy Policy, op. cit.* (n. 3), p. 88. (Mtoe = million tons oil equivalent.)

[16] ECSC, *Long-Term Energy Outlook, op. cit.* (n. 3), p. 36.

[17] OECD, *Impact of Natural Gas on the Consumption of Energy in the OECD European Member Countries* (Paris, 1969).

Table 3.2

Western Europe: Projected Oil Imports in 1980, on Various Assumptions

	This study[d]		OECD 1966 study[e] Most likely		OECD 1966 study[e] Low-import model	
	Quantity (Mtoe)	Percent of total	Quantity (Mtoe)	Percent of total	Quantity (Mtoe)	Percent of total
Total primary energy consumption[a]	1,498	100.0	1,440	100.0	1,440	100.0
Solid fuels[b]	227	15.1	275	19.1	310	21.5
Natural gas[b]	170	11.3	90	6.2	110	7.6
Hydro	48	3.2	50	3.5	50	3.5
Nuclear	140	9.3	140	9.7	150[g]	10.4
Indigenous oil	f	f	30	2.1	40	2.8
Subtotal	584	38.9	585	40.6	660	45.8
Net oil imports[c]	914	61.0	855	59.4	780	54.2

[a] Including bunkers and primary fuels entering into nonenergy uses.

[b] The figures relating to this study refer to total consumption (whether domestically produced or imported), while those taken from the OECD report relate to consumption from indigenous sources. See also n.c, below.

[c] The 914 million ton figure relating to this study covers total liquid fuel consumption from imported as well as indigenous sources—the latter not having been explicitly calculated due to its relative unimportance. The 855 and 780 million ton OECD figures *are* limited to net imports, but they implicitly include net imports of natural gas and coal in addition to oil. The figures are presented in the OECD report with comment that "Most of these imports will undoubtedly be in the form of oil." (OECD, *Energy Policy, op. cit.* [text, n. 3], p. 90).

[d] Taken from panel (H) in the Statistical Appendix, p. 169. Those figures, in turn, represented modifications of Statistical Appendix Table IV so as to permit greater comparability with OECD definitions and energy-conversion practices. This adjustment explains the differences between figures in the present table and those in Table 3.1.

[e] Taken from OECD, *Energy Policy, op. cit.* (text n. 3), p. 90. What is termed "Most Likely" represents midpoints of a range in the OECD report. What is called "Low-Import Model" is presented in the OECD report as high end of range of indigenous resources. See discussion in the text of this chapter regarding these figures.

[f] Not explicitly estimated due to relative unimportance.

[g] Shown in OECD report, but derivation not explained. However, this figure is roughly consistent with "high consensus" figure referred to in the text of this chapter.

growth in natural gas imports, much of it presumably in the form of liquefied gas from North Africa, although shipments from the Middle East and (pipeline) imports from the U.S.S.R. are also held to be a possibility.[18] (The OECD estimate of 12 million tons oil equivalent in natural gas imports by the mid-1970's compares with barely a million tons in 1967.)

In view of these and other points, and depending on a number of alternative hypotheses, overall "natural gas availabilities may be expected to cover between 10 and 14% of the area's total primary energy requirements around 1975–1980; this corresponds to a minimum of 115 Mtoe and a maximum of 160 Mtoe in 1975, and to a minimum of 140 Mtoe and a maximum of 195 Mtoe in 1980."[19] Based upon this latest OECD thinking, we have, in the present study, adopted the approximate midpoint in these ranges, and assumed 130 million oil equivalent tons in natural gas consumption in 1975 and 170 million tons in 1980. (See Statistical Appendix Table IV.A.)

In view of the fact that oil and gas frequently are found together, the additional question arises whether the formations that will be tapped for natural gas will not also yield quantities of oil. Neither the OECD nor the ECSC sees much promise that Western Europe's essentially barren oil position (about 20 million tons produced in each of the years between 1964 and 1967) will undergo noticeable change. The ECSC report remarks that "... present geological knowledge does not suggest any major increase in [oil] reserves. The exploration of certain off-shore areas might produce discoveries of greater value, but there is no means of knowing in advance whether they will find oil associated as usual with natural gas or natural gas alone ...."[20,21]

[18] See n. 31, p. 48.

[19] OECD, *Impact of Natural Gas, op. cit.* (n. 17), p. 102.

[20] ECSC, *Long-Term Energy Outlook, op. cit.* (n. 3), p. 36.

[21] However, a major North Sea oil discovery was announced while this book was in press. A field, said in preliminary estimates to contain 7 billion barrels of reserves, was tapped in Norwegian territorial waters. The Ekofisk field appears to be roughly comparable in size with the original Prudhoe Bay strike in Northern Alaska (see Chapter 7) although the estimates for that field were stated in terms of recoverable reserves, while the early Ekofisk estimates are for total resources, with the recovery factor unstated. Even so, a rate of output of about 1 million barrels/day from this one field was suggested as a possibility in these early reports.

Obviously, many such fields would be required if Europe's dependence on the Middle East and North Africa is to be greatly reduced. And much time would be required, at best, to find and develop such fields and construct the necessary production and transportation facilities.

As of mid-summer 1970, a letter to us from the OECD stated

3. *Nuclear Power*

The level of nuclear power production over the next decade or so depends upon the growth of electricity within the total energy economy, and the choices that are made as between conventional and nuclear plants in the provision of new capacity. The latter choice will depend, in turn, upon estimates of comparative costs (and reliability in performance) and on this subject there continues to be considerable uncertainty even in the United States, where, even though nuclear power appears to have crossed the competitive threshold, the rate of new plant orders has fallen off rather markedly between 1966-67 and 1969, partly in response to a reassessment of comparative costs in a light more favorable to the conventionally-fueled plants. The continuing uncertainty with respect to its commercial outlook must be kept in mind in connection with all attempts to forecast nuclear power's future prospects.

In the OECD report, it is estimated that nuclear capacity in Western Europe will reach 40 million kw in 1975 and 90 million kw in 1980 (out of a total electrical generating capacity in those years of 400 million and 540 million kw, respectively).[22] The nuclear electricity that would be generated, assuming an average utilization of 7,000 hours per year (80 percent plant factor), comes to 280 billion kwh in 1975 and 630 billion kwh in 1980, which correspond to required heat input of 64 and 140 million tons of oil equivalents, respectively.[23]

The OECD, in presenting these nuclear estimates, pointed out that it took into account the various assessments made in the official reports emanating from countries and international agencies within Western Europe and indicated that some then current estimates ran higher than those it adopted.[24] A convenient summary of many of these European-made estimates is contained in a subsequent report issued by the RAND Corporation.[25] These indicate a range of 1980 estimates running from about 70 million kw to about 100 million kw.[26] The high end of the range is taken from a report prepared by an American consulting firm (A. D. Little, Inc.) for the U. S. Atomic Energy Commission, and is described as the "high consensus" derived from interviews with "key decision-makers" in the governments and industries of the countries concerned and in various international agencies. This "high consensus" estimate runs about 10 percent above the OECD estimate of 90 million kw for 1980.

Finally, a still more recent set of nuclear projections appeared in a 1969 study of the European Nuclear Energy Agency.[27] In a range of estimates which conform reasonably to the OECD study of three years earlier, the Agency foresees European nuclear output capacity by 1975 ranging between 29 million and 43 million kw; and by 1980, between 88 million and 118 million kw. On the basis of this most recent study, we have assumed in our projections a nuclear generating capacity of 30 million kw in 1975 and 90 million kw in 1980, deliberately adopting figures near the lower end of the indicated ranges so as to allow for the kinds of uncertainty that have characterized nuclear power expectations in the past. (Using an 80 percent plant factor, these projected figures correspond to 210 billion kwh or 47 million tons oil equivalent, on a heat input basis, in 1975; and to 630 billion kwh or 140 million tons oil equivalent in 1980.)

Relative Importance of Oil Imports in 1980: Using Various Assumptions Regarding Other Energy Sources

In the preceding section the range of potential energy supplies from coal, natural gas, and nuclear power has been examined, with an indication in each case of what has been, in various reports, regarded as their highest feasible future output levels. Drawing upon materials already referred to, Table 3.2 compares three sets of projections for 1980: those used in this study, based on what we consider to be the most prudent assumptions; and two sets from the 1966 OECD report, one representing what may be called its estimate of the "most likely" situation and the other representing what might be regarded as a "low-import" model.[28] The ECSC figures are not summarized in the table because of difficulties in conforming the categories employed in its report to those used in this study. However, the general implications of the ECSC "low-import" model will be referred to in this discussion, consistent with the materials from the ECSC report presented earlier.

In brief summary the results are as follows. This study estimates imports of oil in 1980 at a bit over 60 percent of total energy consumption. The 60 percent level compares closely with the results of the OECD's set of median figures (which we have labeled "most likely") for the same year, in spite of the fact that consideration of developments since appearance of the OECD report has prompted us to depart from the OECD in some of the other fuel shares.[29] When the OECD's estimates of maximum indigenous supplies from coal, natural gas, and nuclear power are used, this

that the Organisation did not have "any estimates as yet on the possible importance of the recent oil discoveries in the North Sea for OECD Europe over the next decade."

[22] OECD, *Energy Policy, op. cit.* (n. 3), p. 61.

[23] *Ibid.*

[24] *Ibid.*

[25] W. E. Hoehn, *The Economics of Nuclear Reactors for Power and Desalting* (RM-5227 [Santa Monica: The RAND Corporation, November 1967]). (Updated to allow for major changes in an August 1968 prefatory note by the author.)

[26] *Ibid.*, p. 105.

[27] European Nuclear Energy Agency, *Uranium–Production and Short Term Demand* (Paris, January 1969), p. 21.

[28] It should be emphasized that our own projections, which consider, where possible, trends discernible through 1968, are being compared with those of the partly dated OECD study. It is not unlikely that, as in the case of the cited natural gas report, a more current analysis would prompt the OECD to adopt somewhat different views.

[29] The comparison is in addition slightly impaired by the fact that our estimate of oil import requirements has not been reduced by the estimated amount of indigenous oil, while the OECD figure includes net imports of gas and coal along with oil. (See, in particular, n. c, Table 3.2.)

results in import requirements (essentially oil) amounting to about 55 percent of Western Europe's total energy consumption. Utilization of the ECSC "low-import" model results in a reduction in the oil import level to about 50 percent of European consumption.[30]

These computations indicate that there is some room for maneuver within Western Europe in affecting import requirements, and that government policy directed towards the preservation of a higher degree of self-sufficiency could meet with some success. To the extent that the improved indigenous position depends upon higher coal production, the higher level of self-sufficiency would be bought at a cost that the European countries give no current sign of being willing to incur. Even the OECD's "most likely" level, as indicated earlier, involves coal production well beyond the amounts considered to be competitive with imported energy supplies.

If the enhanced self-sufficiency could be achieved through substantial increases in natural gas output, this would not appear to involve an economic sacrifice and, as indicated in the earlier discussion, natural gas prospects have been taking a steadily more optimistic turn.[31]

Under presently foreseeable circumstances, with all due allowance for indigenous production, Europe's level of dependence on the Middle East and North Africa for oil supplies in 1980 could drop to as low as 50 percent of total energy needs. So low a level is, however, highly unlikely. But, whether it be 50 percent or (roughly) 60 percent as estimated in this study, the projected level of dependence is very heavy and indicates the very great importance of providing for protective measures against the possibility of supply interruptions.

Japan

Japan is today the world's largest oil importing country. In 1968, its imports amounted to about 15 percent of the world's inter-area oil trade—153 million out of a world total of 1,017 million tons (or 3 million barrels/day out of 20 million barrels/day). Of its total imports, over 85 percent originated in the Middle East; around 8 percent came from Asiatic sources—primarily Indonesia; and the remainder originated among a scattering of countries including the Soviet Union.[32] (See Table 1.2.) As indicated in Chapter 1, Japan is estimated to continue to rank first among oil-importing countries for the period covered in the projections, with a continued heavy reliance on the Middle East.

The purpose of this section is to examine the importance of oil supplies to the internal energy economy of Japan now and in the future. As in the preceding section, this will be done by placing oil within the context of the country's total consumption of energy.

The Relative Importance of Oil in Total Energy Consumption: Present and Projected

As in Western Europe, oil is a rapidly growing component of Japanese total energy supply. Table 3.3 shows that in 1955 solid fuels accounted for 70 percent of the country's energy total, and liquid fuels for 20 percent, with hydroelectric power providing practically all of the remaining 10 percent. By 1965, the shares had changed sharply: in that year oil constituted 56 percent compared to coal's 37 percent of the total and hydro's share had shrunk to 5 percent. This pattern of transformation in the country's energy consumption is estimated to continue and by 1980 the projected relative shares are around 80 percent for oil, 13 percent for solid fuels, and less than 3 percent for hydro. In the latter year, natural gas, which provided a scant 1½ percent in 1965, is estimated to account for only a slightly higher (2 percent) share, and nuclear energy, a newcomer, is estimated to provide about 2½ percent of the total.[33]

In Japan, as in Western Europe, oil is a recent arrival compared to its emergence as a major source in the United States. However, oil's share of total Japanese energy consumption in 1965 had already surpassed its relative importance in the United States and in Western Europe. Oil's portion of the total in 1965 was about 40 percent in the United States, about 45 percent in Western Europe, and about 55 percent in Japan. Looking to the future, oil's relative importance in Japan's energy consumption in 1980 is estimated at 80 percent, compared to 60 percent in Western Europe, and somewhat less than 40 percent in the United States.

In Western Europe, as indicated, oil's projected larger share compared to that in the United States reflects the relatively smaller importance envisaged for natural gas in Europe's 1980 energy position. In the case of Japan, oil's even heavier share of the projected total reflects, in the main, a smaller contribution from coal than in either the United States or Europe, and a relatively minor contribution from natural gas. In addition, hydropower, which had been a factor of some importance in Japan in the past, is projected to diminish rather rapidly in relative importance in the years ahead.

Even more so than for Western Europe, imported oil is projected as the mainstay of Japan's future energy supply. But it should be noted that the overwhelmingly important role projected for oil reflects the comparatively small

[30] ECSC, *Long-Term Energy Outlook, op. cit.* (n. 3), p. 46. On the other extreme, the ECSC "high-import" model, which it regards as economically more rational and probably more realistic, results in an oil import level amounting to almost 65 percent of Western European consumption (p. 47).

[31] To the extent, however, that natural gas consumption depended partly upon imports from North Africa or the Middle East (see remarks on p. 46, above), supply dependence on that region would persist, and therefore need not be considered as a separate element in our analysis.

[32] These 1968 figures are taken from British Petroleum Co., *Statistical Review of the World Oil Industry, 1968, op. cit.* p. 11. Data for 1965, 1975, and 1980 appear in Table 1.4 and Statistical Appendix Tables VI and VII.

[33] As shown in a footnote to Table 3.3, the hydro and nuclear shares are higher where a different procedure for converting original units into fuel equivalents is used.

Table 3.3

Japan's Inland Energy Consumption[a]

Year	Total (thousand metric tons of oil equivalents)	Percent of total consumption: Solid fuels	Liquid fuels	Natural gas	Hydro[b]	Nuclear[b]
1955	43,916	70.0	20.6	0.3	9.1	
1960	72,430	57.3	35.0	0.9	6.8	
1965	116,630	36.9	56.5	1.5	5.1	
1975	249,350	17.3	76.1	1.7	3.5	1.4
1980	364,680	12.7	80.5	2.0	2.5	2.4

[a] For definitional adjustments which permit comparability of the above data with other, somewhat different, concepts of energy consumption, see Statistical Appendix, pp. 168-170.

[b] These percentages result from the use of conversion factors in which original units are translated into oil (or coal) equivalents on the basis of the heat value of the hydro or nuclear electricity produced. When hydro and nuclear are converted on the basis of equivalent fuel-input requirements at fossil-fueled thermal power stations, their percentage shares rise markedly, as seen in the following figures for 1965, 1975, and 1980 (taken from percentage distribution of panels (C) or (D) in Statistical Appendix, p. 168):

	Solid fuels	*Liquid fuels*	*Natural gas*	*Hydro*	*Nuclear*
1965	33.4%	51.2%	1.4%	14.0%	–
1975	15.9	70.2	1.6	8.7	3.5%
1980	11.8	74.5	1.8	6.0	5.9

Source: Statistical Appendix Table IV.A.

shares estimated to accrue to the other energy sources—coal, natural gas, hydro, and nuclear. These projections are based upon current indications, and it is, therefore, important to inquire whether the other energy sources might play a larger part than has been estimated: the remainder of the chapter examines the possibilities.

Potential Energy Supplies from Coal, Natural Gas, Hydro, and Nuclear

As in the discussion on Western Europe, one proviso must be introduced at the outset: the potential contribution of each of the other sources will be made in the light of what might be possible in the absence of major breakthroughs in technology or in the discovery of indigenous reserves, and without major policy changes on the part of government beyond those already announced.[34]

1. *Coal*

Japan's major indigenous fuel resources are its coal deposits upon which its earlier, coal-based economy was built. Coal resources are, however, small when compared to those of many other countries. It is estimated that measured reserves equal 6 billion tons, about one-half of which is judged recoverable. These would last for something like fifty years at current rates of production. Total resources of coal, not usually subject to dependable assessment, are estimated at about 20 billion tons.[35]

For a period following World War II, Japanese coal production rose, peaking at about 55 million tons in 1961. Since then production has declined (reaching around 46 million tons in 1968), and markets have been lost in the face of severe competition from imported oil. Special efforts have been made to strengthen the position of coal through long-term agreements between coal mines and large consumers in the iron and steel industry and electric power generation. In addition, coal production has been rationalized through government financial assistance and as a result labor productivity has increased at a rapid rate.[36]

Despite these special efforts to improve its competitive position, it is estimated that coal will continue to lose markets. As in Western Europe, this is not merely a matter of losing ground in those uses in which petroleum products are specifically required, as in transportation. Actually, in 1964, specifically required petroleum products accounted for about 25 million tons, whereas oil consumption came to some 75 million tons; and in 1980, as estimated in the 1966 OECD report, the same 3:1 relationship between those oil products which are specifically required and total indigenous oil consumption is assumed to obtain.[37] As indicated in the earlier discussion of Western Europe, the domestic refining of crude oil to produce such specific products as gasoline inevitably yields additional oil products that compete in more general energy uses. Nevertheless, the relative amounts of different refinery products can be controlled within rather wide limits according to the demands of the market, as shown in Figure 3.2, comparing the United States, Western European, and Japanese patterns for different refined products.

The special measures on coal's behalf, on the one hand, when joined with the realities of price competition in the market place, on the other, might be expected to result in the stabilization of coal's output at recent levels. In its 1967 report, the Japanese Advisory Committee on Energy had stated: "We advise as a reasonable scale . . . to maintain a level of about 50 million tons a year . . . ."[38] Even so, imports of some coal, especially for coking purposes, would be required.[39] And, according to the OECD, even if it were possible to maintain stability in output, this would involve the continuation of all the present forms of assistance to coal.[40]

Recent policy developments, responding further to comparative fuel costs, point to continued contractions rather than stable levels of coal output. A newly formulated

[34] The remainder of the chapter draws heavily on the OECD report, *Energy Policy, op. cit.* (n. 3), and *Report on Energy Policy* by the Advisory Committee for Energy, a consultative body to Japan's Ministry of International Trade and Industry (Tokyo, February 20, 1967; translated and published by the Institute of Energy Economics).

[35] OECD, *Energy Policy,op. cit.* (n. 3), p. 44.

[36] *Ibid.*, p. 91.

[37] *Ibid.*, pp.38 and 93.

[38] *Report on Energy Policy, op. cit.* (n. 34), p. 13.

[39] OECD, *Energy Policy, op. cit.* (n. 3), p. 44.

[40] *Ibid.*, p. 93.

coal mining reconstruction program, designed to take effect early in 1969, provides for a slash in yearly production from recent levels of about 45 million tons to 35 million tons in the early 1970's.[41] Our own projections of Japanese coal consumption and imports conform to the steady declines in indigenous output envisaged in this latest policy statement. (See Statistical Appendix Table XII.)

2. *Natural Gas*

Natural gas accounts for a miniscule fraction of Japan's energy total today, and only minor changes are anticipated in the future. Reserves of natural gas are exceedingly small, and although exploration is being carried out there appear to be no expectations of significant discoveries. The same situation prevails with respect to oil, whose natural occurrence is often in association with gas.[42] It is likely, however, that overseas importation of natural gas in liquefied form to be reconverted to gas on arrival in the country will become a factor of some importance. To the extent that these gas supplies originate in the Middle East, the supply dependence on that part of the world would remain the same and need not be considered as a separate element in our analysis. If, however, much larger liquefied gas supplies should originate in Alaska than has been assumed in our projections, the geographic pattern of supply dependence would be somewhat affected.[43]

3. *Hydro*

Unlike natural gas, hydro has been a major factor in the Japanese energy economy, having as recently as 1955 accounted for about 10 percent of total consumption or 25 percent if hydro is figured on the basis of the quantity of fossil fuels that would have been necessary to produce the same amount of electricity (see footnote b to Table 3.3). However, because of the shortage of potential development sites, the outlook for growth in hydropower production beyond present levels is not favorable. The OECD estimates that most of the suitable sites will be harnessed by 1970.[44]

In its report, the Japanese Advisory Committee on Energy also observes that the conditions for developing hydropower have deteriorated continuously, and advises that the development of electricity generating plants should place the heaviest emphasis on thermal power with future hydro construction limited to a supplemental role, for example, to help meet peak loads in particular systems. Nevertheless, it recommends that about one million kw of hydro capacity be developed annually for "some years to come."[45] In the concluding section of this chapter, an effort will be made to translate this recommendation into an element in the estimated quantitative supply pattern for Japanese energy in 1980.

4. *Nuclear Power*

Owing to the extremely poor prospects for indigenous conventional energy sources, there is a great interest in Japan in the possibilities for nuclear power development. Although it is recognized that at first nuclear power equipment and fuels will need to be imported, there is confidence in the ability of Japanese industry to master the technology and thus lead to a future situation in which nuclear power will be an essentially indigenous energy resource—both in terms of equipment and fuel elements.

How large a factor could nuclear power become over the next decade or so? As indicated earlier, there have been numerous ups and downs in the optimism with which the future of nuclear power has been regarded in the United States and elsewhere. Consequently, estimates must be viewed with caution. In Japan, there is the additional problem of susceptibility to earthquakes. This, coupled with extremely high population densities, could lead to a go-slow approach in order not to run the risk of disastrous accidents inherent in an early and unproved technology employing potentially lethal materials.

In the OECD report, it is estimated that nuclear power capacity in Japan will amount to 5 million kw in 1975 and 10 million kw in 1980 (out of a total electric generating capacity of 90 million and 130 million kw, respectively, in those years).[46] Assuming an average utilization of 7,000 hours per year (80 percent plant factor), the kwh generated would come to 35 billion in 1975 and 70 billion in 1980, which corresponds to a required heat input of 8 million and 16 million tons of oil equivalents, respectively.[47] These

[41] Institute of Energy Economics (Tokyo), *Energy in Japan* (Quarterly Report No. 7, April 1969), p. 6.

[42] OECD, *Energy Policy, op. cit.* (n. 3), pp. 45 and 47. However, there has been some speculation recently about the prospects for oil discovery in Japanese offshore areas. As put in a trade journal report:

> Without drilling one wildcat, the U.S. Navy's Oceanographic Office has announced that "potentially one of the most prolific oil reserves in the world" has been discovered in the East China and Yellow seas near Japan, Taiwan and Korea.
>
> The prospects were discovered by a United Nations-sponsored survey team working on a research ship contracted by the U. S. Navy .... According to Navy reports, the most promising area extends along three ridges on the Continental Shelf between China and the Ryukyu Islands in the East China Sea. The Yellow Sea prospects are in three basins between China and Korea.
>
> Determinations were made on the basis of airborne geomagnetic surveys, followed by seismic work and gathering of sediment samples. The sediments, containing up to 1.5% organic matter, were carried into the seas from China's mainland by the Yangtze and Yellow rivers, it was reported. The East China Sea sediments cover an area the size of Texas, New Mexico and Oklahoma. (*World Oil*, July 1969, p. 86.)

Reports like this, referring to various regions of the world, are not unusual. Much more exploratory work would be needed to provide a reasonable degree of information. Beyond this, even if circumstances were propitious, a lengthy time period and heavy investments would be required to prove up reserves and develop productive capacity. It is, therefore, unlikely that developments in this region, if they do occur, would significantly affect the Japanese oil situation within the 1970's.

[43] There are also plans to import liquefied natural gas from Brunei and elsewhere.

[44] OECD, *Energy Policy, op. cit.* (n. 3), p. 93.

[45] *Report on Energy Policy, op. cit.* (n. 34), p. 14.

[46] OECD, *Energy Policy, op. cit.* (n. 3), p. 61.

[47] *Ibid.*, p. 61.

Table 3.4

Japan: Projected Oil Imports in 1980, on Various Assumptions

	This study[b]		OECD 1966 study[c]		Low-import model[d]	
	Quantity (Mtoe)	Percent of total	Quantity (Mtoe)	Percent of total	Quantity (Mtoe)	Percent of total
Total primary energy consumption[a]	473	100.0	464	100.0	464	100.0
Solid fuels	46	9.7	64	14.0	64	14.0
Indigenous	23	4.9	46	10.0	46	10.0
Imports	23	4.9	18	4.0	18	4.0
Natural gas	7	1.5	6	1.5	6	1.5
Hydo	11	2.3	9	2.0	13	2.8
Nuclear	23	4.9	16	3.5	32	6.9
Indigenous oil	e	e	1	0.2	1	0.2
Subtotal	87	18.4	96	20.0	116	25.0
Oil imports	385[b]	81.4	368	80.0	348	75.0

[a] Including bunkers and primary fuels entering into nonenergy uses.

[b] Taken from panel (H) in Statistical Appendix p. 168, and from Statistical Appendix Table XII.A. The panel (H) figures, in turn, were adapted from Statistical Appendix Table IV so as to permit greater comparability with OECD definitions and energy-conversion practices. This adjustment explains the differences between figures in the present table and those in Table 3.3. Note that the 385 million tons implicitly include indigenous oil supplies, which were not separately estimated.

[d] Same as OECD except that (1) nuclear represents high end of latest Japanese power plan (see text, below); (2) hydro represents interpolated 1980 figure based in 1975 and 1985 projections appearing in the Japanese *Report on Energy Policy, op. cit.* (text n. 34), Tables 4 and 6 (these hydro figures are consistent with the aforementioned "high" nuclear estimate); and (3) oil imports are computed as residual.

[e] Not estimated, due to relative unimportance.

estimates correspond very closely to those contained in the report for the U. S. Atomic Energy Commission referred to earlier (prepared by the American consulting firm of A. D. Little, Inc.).[48]

More recently, a New Long Range Program has been adopted by the Japanese Atomic Energy Commission and submitted to the Government.[49] In this new program, the sights for future development are set higher, and a capacity of 15–20 million kw is envisaged for 1980, with "at least" 30–40 million kw by 1985.[50] This range of capacity estimates for 1980 corresponds to a required heat input of 24 to 32 million tons of oil equivalent (using the same factors as for the OECD estimates, above). In the following section, the upper estimate of the New Long Range Program estimates will be taken to represent the highest feasible nuclear power levels in 1980, while, for purposes of this study, a somewhat more modest assumption projects a nuclear capacity of 6 million kw in 1975 and 15 million kw in 1980, which translates, respectively, into 9 million and 23 million tons oil equivalent of heat inputs.

[48] See p. 47 above. The "high consensus" for Japan in 1975 is put at 5.3 million kw and the "low consensus" at 4.3 million kw. For 1980, the "high consensus" is put at 11 million kw and the "low consensus" at 9.2 million kw. (See Hoehn, *op. cit.* [n. 25], p. 105.) In a prefatory note written in August 1968 after the completion of the study, the author acknowledges that the nuclear outlook in Japan was exceeding his earlier expectations (p. iv).

[49] See Victor Gilinsky and Paul Langer, *The Japanese Civilian Nuclear Program* (RM-5336 [Santa Monica: the RAND Corporation, August 1967]). The Japanese *Report on Energy Policy, op. cit.* (n. 34), uses these long-range nuclear estimates.

[50] *Report on Energy Policy, ibid.*, p. 15.

Relative Importance of Oil Imports in 1980: Using Various Assumptions Regarding Other Energy Sources

In the discussion just concluded, potential Japanese energy supplies from coal, natural gas, hydro, and nuclear power have been assessed on the assumption that they are produced at what studies in the past few years have regarded as their highest feasible levels. In Table 3.4, the results are summarized and compared with the estimates of this study, as they affect the 1980 projections.

Three sets of figures are compared: those used in this study; those of the 1966 OECD report; and those of the latter report, adjusted to allow for higher estimates of nuclear and hydropower as presented in subsequent Japanese reports.

The main results may be summarized as follows: the 1980 reliance on oil imports shown in this study's projections differs almost not at all from the OECD—about 81 percent in the former case and 80 percent in the OECD study. When the OECD figures are adjusted to allow for the substantially higher level of nuclear power estimated in more recent Japanese studies (double the OECD estimate), and for the hydro projections, which are about half again as large as those of the OECD, the level of oil imports is reduced to 75 percent.

The major absolute difference is in the nuclear estimate. There are grounds for skepticism on this score because, as indicated earlier, estimates of nuclear power costs and potentials have behaved in an erratic fashion over the years.

If, however, the Japanese should fulfill their latest nuclear power plans, and at competitive costs, this would be a favorable sign for the long-run future of indigenous energy supplies.

In any case, the projected level of dependence on imported oil supplies is extraordinarily high and points to the great importance of taking measures to guard against the possibility of supply interruption.

CHAPTER 4

Prospects for Liquid Fuels from Unconventional Sources in North America

THE United States and its continental neighbor, Canada, are in an unusually strong position with respect to unconventional sources from which liquid hydrocarbons could be derived. Within North America lie vast deposits of oil shale, tar sands, and coal, all of which could provide an immense supply of liquid fuels for generations to come.

Technologies for the exploitation of each of these sources are known. What is far from clear, however, is how costly liquid fuels from these unconventional sources would be; how far in the future the commercial development of liquids from shale and coal lies (tar sands are already being produced in one plant on a commercial scale, although apparently not yet competitively); and how rapidly any of them could become an important supplement to conventional oil supplies after commercial technologies are developed.

The prospects for the several sources in North America, with which this chapter is concerned, are of necessity based on evaluations made prior to a reliable economic assessment of the apparently rich oil strike in Northern Alaska. It is almost needless to point out that the motivation, both public and private, for pushing forward on technology for synthetics is particularly susceptible to the degree of success that will be experienced in the development of commercial oil supplies in Alaska and other parts of the North American Arctic. The pace of unconventional liquid fuels research and development has already slackened as a result of the Alaskan oil discovery. Additional successes in Alaska and in the Canadian Arctic would almost certainly further depress the outlook for unconventional sources. A situation in which the determinants are characterized by so much uncertainty is virtually impossible to pin down with any precision. The quantities here assigned to productive capacity from the various sources that might be achieved during the 1970's must, therefore, be viewed as highly conjectural.[1]

Shale Oil[2]

The known resources of oil shale, regardless of the precise basis of measurement selected, are enormous and some 60 percent of the estimated world total is located within the borders of the United States.[3] The advent of a shale oil industry in the United States has seemed tantalizingly close on past occasions, mainly after World War I, but, to date, commercial production is still a promise, not a reality.

Resource Base

While other areas of the United States also contain oilbearing shale deposits, the sedimentary rocks of the Green River formation in Colorado, Utah, and Wyoming are believed to have the greatest commercial potential (see Figure 4.1). They total some 16 million acres, of which about 11 million appear to have potential commercial value. Slightly less than one-half of the latter area has been explored.[4]

Deposits vary widely in quality and physical characteristics. High-quality deposits (yielding over 25 gallons/ton and running at least 10 feet thick) contain an estimated 590 billion barrels, or roughly fifteen times U. S. proved reserves of crude oil and natural gas liquids. In addition, there are 1,150 billion barrels contained in lower-grade deposits yielding 15-25 gallons/ton. The unappraised resources may be roughly similar in magnitude.[4] Of the three states, Colorado has the smallest acreage of oil-shale lands, but these are believed to contain the largest deposits (some 60 percent of the total and an even higher percentage of the richest deposits); Utah ranks second; and Wyoming last. An estimated 72 percent of the land containing about 80 percent of the oil shale is owned by the federal government, including over 80 percent of the best lands with resources of an estimated 480 billion barrels. Even so, deposits on high-quality nonfederal acreage are huge compared with total proved U. S. reserves of crude oil.[5]

Several complexities which would affect any development of the oil shale lands should be noted. (1) Their location is in a fairly isolated part of the Rocky Mountains where population is sparse and facilities to support large-scale industrial development are very limited. (However, crude oil, too, is located in some of the least hospitable parts of the world; despite this, development does take place.) Water supply in this section of the country might

[1] Additional uncertainty is of course an inevitable by-product of the effects on domestic crude oil prices of possible changes in U. S. oil import policy.

[2] For the factual information in this section heavy reliance has been placed on the U. S. Department of the Interior report, *Prospects for Oil Shale Development–Colorado, Utah, and Wyoming* (Washington, May 1968).

[3] *Petroleum Press Service*, September 1967, pp. 328–330.

[4] "Explored" implies that at least enough geologic evidence has been assessed to *assure* that the oil shale is physically present. "Unappraised resources" are those whose extent and physical characteristics are based upon geological inference.

[5] See U. S. Department of the Interior, *Prospects for Oil Shale Development, op. cit.* (n. 2), Table A.2, for some detailed estimates. However, much of the privately-held land is in parcels too small to provide the amount of oil shale required if plants are to be large enough to achieve full economies of scale. See Harry Perry, "Oil Shale Development Non-Technologic Factors" (paper presented at the 1969 meetings of the American Institute of Mining, Metallurgical, and Petroleum Engineers; Preprint No. 69-AIME-77), p. 17.

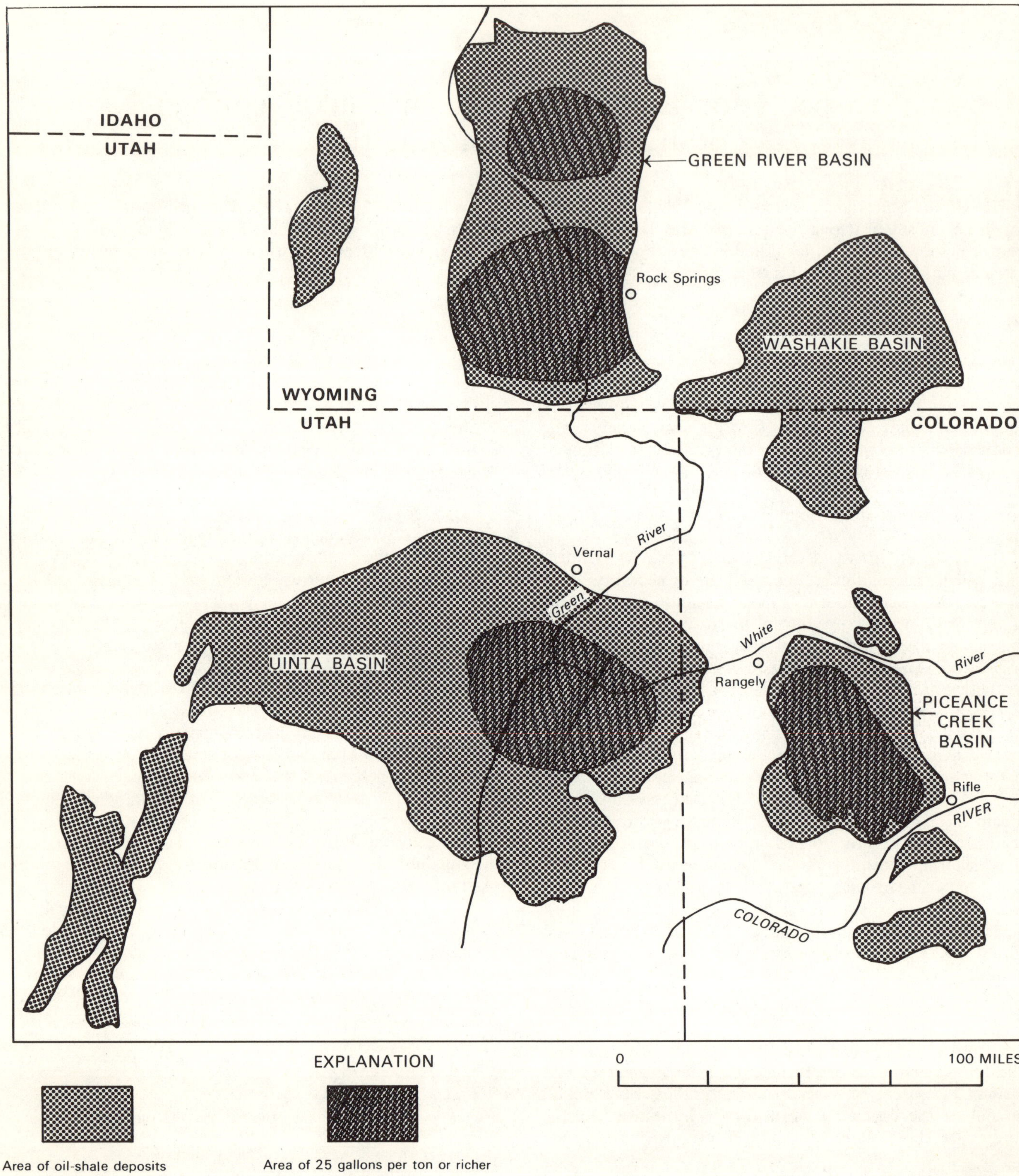

Fig. 4.1. Oil-Shale Deposits in Colorado, Wyoming, and Utah

Source: U. S. Department of the Interior, *Prospects for Oil Shale Development–Colorado, Utah, and Wyoming, op. cit.* (Washington, May 1968), p. 12.

become a problem if the scale of operations were to become very large. Transportation facilities, including pipelines to link with existing major crude oil lines, would have to be constructed. (2) Federal lands were withdrawn from leasing in 1930, and new policies to make them available to private operators have not yet been formulated; a further complication is the dispute over preexisting and disputed mining claims covering much of the federal acreage (see below). (3) Finally, the oil-shale formations frequently also contain other associated minerals, including dawsonite (an aluminum carbonate) and various sodium carbonates from which soda ash and sodium chloride may be derived. The commercial value, if any, of these minerals is not known, but their separation from the oil-bearing kerogen may become a problem depending on the type of recovery process used.

Technology

"Shale oil" does not occur as liquid oil but is embedded in sedimentary clay and silt deposits as a waxy substance, known as kerogen, which is solid at normal temperatures. Conversion into a liquid of crude oil quality requires one of two series of processes: (1) mining, crushing, surface processing, and waste disposal; or (2) underground (*in situ*) processing, recovery of the liquid, and semirefining. At the present time, only the first approach has been fully developed to the point where operations on a commercial scale are technologically feasible. Underground recovery work has been limited to various kinds of experimental research.

If mined, two stages of processing are required. The shale must be retorted and the shale oil refined. In addition, the spent shale must be disposed of. Retorting involves heating the crushed shale to 850°-1,000°F., which transforms the solid kerogen into an oily vapor ("pyrolyzing"); the latter product is then removed and condensed as shale oil.[6] Several retorting processes have been tested, and others are at various stages of laboratory development.

After emerging from the retort, the shale oil requires further refining to upgrade it to normal crude oil standards (or, if desired, to produce finished petroleum products). This involves the addition of hydrogen to improve the quality of the shale oil and to make possible the removal of sulfur and nitrogen. Normal techniques widely used in refining crude oils are applicable. The precise process used depends on the type of retorting employed and the proportions and qualities of the finished products desired.

An enormous amount of rock must be handled and disposed of in the operation of even a minimum-sized commercial shale oil plant. A 50,000 barrels/day retorting operation would require 66,000 tons, or more, of 30 gallons/ton shale. On an annual basis, this would total at least 24 million tons of rock. A million-barrel-a-day industry would have to cope with some 480 million tons or more of materials, an amount almost as large as the current production of bituminous coal in the United States.

[6] Charles H. Prien, "The Future of U. S. Oil Shale Resources," reprinted from *Geoscience News*, I, No. 4 (March-April 1968), p. 6.

Nearly all of the spent shale would constitute a waste material requiring disposal to avoid adverse environmental and aesthetic effects, but the costs of disposal for a functioning shale oil industry are unknown at this time. In addition, the retorting process and movement of spent shale could lead to dust and other air pollution problems. The salts contained in exposed spent shale would be subject to leaching with subsequent potential contamination of water supplies. Moreover, the amounts of water required for surface handling and retorting are formidable. In view of these problems, the alternative *in situ* approach would be very attractive, if an economical technique could be developed.[7] But at the present time most phases of *in situ* processing must be considered highly experimental, or perhaps no more than conceptual.

Outlook for Commercial Development

The outlook for commercial development will be considered at this point on the assumption that no extraordinary factors leading to an acceleration in the pace of development will come into play. Governmental support for the development of a shale oil industry is assumed to continue to be limited to a modest effort on methods for improving technology. (In the appendix to this chapter, development possibilities will be evaluated on the assumption that special efforts are put forth.)

Major financing decisions will be required before the first commercial plant can be constructed. Even if technical feasibility has been thoroughly tested in pilot plants, the "scaling-up" and "running-in" of a full-size unit may encounter unforeseen difficulties which could delay its profitable operation. Recent experience with the first Athabasca tar sands plant (discussed below) bears ample witness to what may occur. In addition, there are at this time major uncertainties regarding the long-run competitive strength of shale oil, since the prospects for U. S. crude oil for the next decade are still in doubt, as is oil import policy upon which the level of domestic oil prices depends (see Chapters 2 and 7); and the comparative costs of other sources of synthetic liquids, such as those from coal (see below), have not been established.

Since the bulk of total oil shale acreage is in the public domain, federal government policies for opening up public lands are of critical significance for a large shale oil industry. Federal oil-shale lands were withdrawn from leasing in 1930, and the question of making them available for private exploitation, and the terms on which they should be leased, has engendered heated controversy. Further complicating the situation is the fact that title to only 92,000 acres (out of 662,000 acres with an overburden of 1,000 feet or less) of federal oil-shale lands in Colorado (plus a similar fraction in Utah and Wyoming) is clear; and most of these clear title lands are located in "scattered, isolated, small tracts that would not support commercial produc-

[7] A major drawback of any *in situ* technique might be the inability to recover by-products.

tion."[8] Part of the title dispute arises over the validity of claims filed under the old 1872 mining law which was in effect before additional legislation halted further claiming in 1920. The latter action followed the staking of many claims in a "shale-rush" between 1915 and 1920. A series of test cases is now in litigation but a final decision may not come for several years. Another facet involves mining claims for associated minerals intermingled with oil shale, filed under the Multiple Mineral Development Act of 1954. Although the government believes that these claims are invalid, it is engaged in adjudication; and these claims therefore introduce another element of uncertainty into the picture.

In 1968, the federal government ended a 38-year freeze on the leasing of oil-shale lands, by offering three test leases in Colorado. One of the tracts offered holds an estimated 500 million barrels of recoverable liquids, the other two are each estimated to contain 900 million barrels. The lease sale was conducted by competitive bonus bidding with sealed offers, but the bids were so disappointing that none of them was accepted by the Interior Department.[9] (These bids, it should be noted, were made when optimism was already running high concerning oil prospects in the North American Arctic.) The leases had been intended as a test of the market, meant, if successful, to lead to eventual large-scale leasing of U. S. oil shale lands five to ten years hence. The test leases were intended to determine whether any company or group of companies was prepared to go ahead with a shale oil plant immediately, given access to good deposits. The results were negative.

As to the privately-held shale lands, the opinion has been expressed that little or no production will occur there either, until policies for public lands are clarified. It is possible, however, that an industry of moderate size could be established quite independently of production from federal lands. Titles to private lands are clear, reserves are adequate and of a type which lend themselves to exploitation by established mining techniques, and sufficient water rights are being acquired by active developers.[10] Construction of the first plant would thus appear to depend on the technical feasibility of recovery processes and the economic attractiveness of the venture.

In the study which laid the groundwork for the Interior Department's test lease policy, a schedule was presented for illustrative purposes to indicate the rate at which commercial-scale development might proceed, consistent with cost and technological estimates made in the report. A first-generation plant, producing 35,000 barrels/day of liquids, was thought to be marginally competitive with domestic crude oil in the United States in 1972 (say, about $3 per barrel); and an improved first-generation plant, using a larger retort and benefiting from operating efficiencies resulting from the experience with the first plant was envisaged for 1976 (producing, say, within 10 percent of $2.15 per barrel).[11,12]

A second-generation facility was thought possible by 1980. For the second-generation system, the Interior Department postulated major technical improvements in the retorting process. In addition, a new "cut-and-fill" mining method to replace the traditional "room-and-pillar" method was envisaged which should allow a mining recovery of over 90 percent of the total oil shale in place as well as permitting effective waste disposal. Large, high-capacity machines would be utilized in drilling, blasting, and loading, and would permit a high degree of automation.[13] It was estimated that the combined savings would result in reducing the cost of semirefined shale oil to $1.58 a barrel (plus or minus 10 percent). At this level, shale oil costs would compare favorably with those of domestic crude oil within the present structure of costs and prices. In addition, of course, exploratory risks would be virtually nonexistent for shale oil as compared with crude oil.

All of the figures cited are, of course, estimates representing a scaling-up of pilot plants, or, at most, "semi-works." Since no commercial-sized plant has yet been built, or even fully designed, all cost estimates must be regarded with strong reservations which can only be removed by actual operating experience over a period of time.

To proceed from the cost and timing estimates in the Interior Department study to the sketching out of a pos-

---

[8] U. S. Department of the Interior, *Prospects for Oil Shale Development, op. cit.* (n. 2), p. 17.

[9] *Oil and Gas Journal*, December 30, 1968, p. 97.

[10] Prien, *op. cit.* (n. 6), p. 10.

[11] For the assumptions underlying these cost estimates, see the U. S. Department of the Interior report, *Prospects for Oil Shale Development, op. cit.* (n. 2), Appendix B, Section A, "Method of Calculation," pp. B-1 through B-15.

[12] It should be noted that lower near-term costs than those assumed for early plants in the Interior study are claimed for an alternative process being developed by The Oil Shale Corporation (TOSCO). TOSCO management states that the presently estimated operating cost to produce, in a commercial complex, a barrel of hydrogenated crude shale oil ready for shipment at the plant gate is $1.55. This cost is said to include depreciation taken on a straight-line basis. With allowance for credits for by-products of 30 cents per barrel, TOSCO estimates a net operating cost of $1.25 per barrel of pipeline oil. (*Summary of the Annual Meeting of Stockholders* [The Oil Shale Corporation, July 10, 1969], pp. 4–5.) Capital investment for the plant envisaged by TOSCO (processing 66,000 tons of oil shale a day and yielding about 55,000-58,000 barrels of shale oil, assuming 35 gallons per ton shale) are estimated at about $130-$140 million. The TOSCO plant would thus have a much lower capital requirement per unit than the first-generation plant envisaged by the Interior Department (about $2,550 versus nearly $4,000 per barrel/day). We cannot reconcile this wide difference, but it appears to be due, in some measure, to the substantially larger yield assumed by TOSCO.

[13] For details, see U. S. Department of the Interior, *Prospects for Oil Shale Development, op. cit.* (n. 2), pp. 54–55.

Room-and-pillar mining removes the shale from the rooms separated by pillars. Customarily in this method of mining, after the rooms have been opened, the pillars are removed with subsequent subsidence. This method is not very adaptable to mining the thicker beds of shale and would involve mining losses of up to 50 percent of the shale. The cut-and-fill stoping method is expected to be more successful in the thick beds. The shale would be removed in slices working upward from the bottom of the bed. As the shale is removed, the opening is filled with waste material providing a platform from which to work on the next slice.

sible schedule of the introduction of shale oil production facilities over the next decade or so requires the imposition of another layer of independently-derived estimates on those made in the Interior Department's report. Such an attempt yields the following tentative picture, and one which (as of 1969) might be judged optimistic on the basis of the replies to the Interior Department's test lease proposal in late 1968.

During the early 1970's, one would expect much effort to be devoted to clearing up title controversies, working out policy essentials, and research and experimentation with a variety of mining and retorting techniques. Assuming that considerable progress in all of these areas is achieved during this period, one could look forward to measurable commercial progress beginning in the mid- or late 1970's, on the assumption that the prospects for shale oil versus alternative sources were still to look favorable at that time. This would mean the construction of perhaps two or three "improved first-generation" plants after 1975, the first of which could not be completed before about 1977.

In view of the magnitude of the physical problems involved–size of mining operations, requirements for ancillary and supporting facilities, etc.–annual capacity additions are unlikely to exceed a 100,000 barrel/day plant at best; at least until such time as major advances in mining and processing techniques have reduced costs to a level comparable to the Interior Department's "second-generation" unit. The outer limit of total shale oil capacity, combining facilities on both private and public lands, might thus be of the order of 350,000 barrels/day by 1980.

Whether this potential will actually be achieved is, of course, clouded by uncertainties. In addition to those already mentioned, there are others inherent in the interests of the various business entities from whom the initiative must come. Among companies which have been interested in shale oil and which have developed a position in land or research, some have a strong domestic crude oil base while others do not. The former may limit themselves largely to a defensive position in synthetic liquids until the comparative attraction of their domestic crude oil supplies versus shale oil can be more definitely assessed; those with little or no crude reserves would be guided by the commercial prospects for synthetic fuels versus the cost and quality of U.S. crudes that would be available to them through exploration or purchase. Further, some companies have a stake in two or even three of the unconventional crude oil sources. These are unlikely to proceed with a large-scale investment until the relative merits of shale oil versus liquids from coal (and possibly tar sands crude) are more fully known.

Coal Hydrogenation

The possibility of using coal as a liquid fuel source has been known for a long time. Even before the drilling of the first oil well, small plants were producing liquids from eastern bituminous coal, from which oils could be extracted rather simply by heating. The "coal oil" was then sold as an illuminant to supplement whale oil and other traditional lamp oils. The term "coal oil" later became attached to kerosene when it entered the market as a product of petroleum refining.

Throughout World War II, Germany without significant local crude oil sources, relied to a considerable extent on synthetic liquid fuels produced from its brown coals. However, the plants were a wartime necessity and could not survive commercially in the postwar period. At present, one commercial-sized coal-to-oil plant is in operation in South Africa, using the Fischer-Tropsch process.

Following the war, synthetic oil research was undertaken in the United States by both private industry and the U. S. Bureau of Mines. Conflicting reports regarding the costs associated with the processes under development in the federal program emanated from the government, on the one hand, and from the petroleum industry, on the other. These activities diminished rapidly after 1952. Although the change in national administrations played a part, a major factor was the growth in world supplies of crude oil. Except for a modest government effort, research on synthetic oil from coal remained dormant through the late 1950's. A revival set in in the early 1960's with the creation of the Office of Coal Research in the U. S. Department of the Interior, for the purpose of sponsoring research designed to create tonnage markets for the slumping coal industry.

Resource Base

Estimates of the U. S. Geological Survey indicate that there are approximately 1.6 trillion tons of bituminous and sub-bituminous coal in the United States in beds 14 inches thick in the case of bituminous coal or 2½ feet thick for sub-bituminous coal. Only beds less than 3,000 feet below the surface are included.[14] The deposits are widespread, with thirty-two states having at least some coal. Customarily, it is assumed that only half of the amounts contained underground will be recovered using present mining methods.

At the present time, the U. S. industry usually mines beds that are more than 2½ feet thick and located within the first 1,000 feet beneath the surface. The Department of the Interior has estimated that coal deposits within the first 1,000 feet and of minable thickness constitute approximately 220 billion tons, but there is said to be a wide margin of error in these figures.[15] Although much of this coal would be available at or near present-day costs, it is estimated that, on the average, cost would probably be somewhat higher, assuming the use of present-day mining methods.

[14]Paul Averitt, *Coal Resources of the United States, January 1, 1967* (U. S. Geological Survey Bulletin 1275 [Washington: Government Printing Office, 1969]), pp. 10–11.

[15]Known recoverable reserves were estimated by the Interior Department for the U. S. Atomic Energy Commission; see U. S. Congress, Joint Committee on Atomic Energy, *Nuclear Power Economics–1962 Through 1967* (Washington: Government Printing Office, 1968), Appendix I, pp. 177, 189.

Conversion techniques under investigation would yield an estimated 1½ to 4 barrels for each ton of coal. The liquid equivalent of 220 billion tons of coal, therefore, would be from 330 to 880 billion barrels. Even using the lower estimate and recognizing that not all of these coal reserves can be dedicated to synthetic liquid fuels, the quantity of minable coal that could be made available for synthetic liquid fuels still represents a very substantial supply. (High-grade shales, as indicated above, are estimated to contain 590 billion barrels.)

One of the attractive features of coal as a source of syncrude, when compared to tar sands or shale oil sources, is its availability in many locations close to current markets and refinery centers. However, there is very little knowledge as to how much of the available reserves are not already committed to other uses. Information on the ownership and plans for "proved" reserves of coal is very sparse. It is almost a certainty that in the high-quality, Eastern bituminous coal fields it would be difficult to purchase a block of 300 or 400 million tons of reserves needed to support a single coal plant of economical size to produce synthetic liquid fuels. The most attractive coal lands from Illinois east are already owned by coal and oil companies, railroads, or coal users such as steel and utility concerns.

It does seem reasonable to expect that some reserves could be acquired from the current owners or that some of the owners themselves would elect to use their coal for synthetic liquids production if it were an attractive venture. Therefore, those close to the situation do not seem concerned about total national reserve availability limiting the growth of synthetic liquid fuels at least to the extent of providing for perhaps the first fifteen plants, with a combined capacity of 1½ million barrels/day. However, there is a strong possibility that more than half of these plants would have to locate in the Western coal lands, where they will face the same problems as oil shale and tar sands—the cost of transportation of the liquid produced to distant markets, limited water supplies, and possible stringencies in support facilities.

Technology

The immediate post-World War II investigations of synthetic liquid fuel from coal encountered a variety of technical problems, most of which were connected with the difficulties and costs of reactions at high pressures, the heat requirements of the process, and the large amount of hydrogen needed to convert the low hydrogen to carbon constituents of the coal into hydrocarbons richer in hydrogen. Although the current approach to producing liquids from coal is basically similar to these earlier methods, there has been a shift away from attempting to convert a large portion of the coal to hydrocarbons, to other methods that permit a carbon residue to remain unconverted, thereby achieving a lower cost in the process.[16] One expert likens the current approach to early oil refining, when part of the crude was converted to gasoline with much being left over as a residual oil to be sold at a price below the cost of the crude. Similarly, the current coal liquefaction processes are geared to the production of a high-value synthetic liquid plus a residual char that can be sold as a low-cost solid fuel to customers such as nearby utility plants.[17]

Hydrogen continues to pose a difficult problem even though research has now cut in half the amount of hydrogen needed compared to the German plants. All syncrude processes for coal, tar sands, and oil shale require hydrogen. However, coal may require 2,000 to 5,000 cubic feet/barrel of product compared to the 1,000 to 1,500 required for shale oil or syncrude from tar sands.[18] With hydrogen at 20 to 40 cents per 1,000 cubic feet, its cost has been a critical element in the economics of coal liquefaction. Although natural gas is today the cheapest source of hydrogen, a continuing upward trend in its price may render natural gas uneconomic as a source for large-scale coal hydrogenation in the future. Consequently the coal projects supported by the Office of Coal Research take on an additional importance, for not only would they produce synthetic pipeline gas, but also they are being designed to provide hydrogen for syncrude plants. It appears essential to the long-run success of coal hydrogenation that a means be found for providing a low-cost, reliable hydrogen source from coal itself.

Although a number of coal syncrude processes have appeared attractive in bench-scale operations, only one has arrived at the pilot plant stage—a 50 to 60 barrel/day plant which has been constructed at Cresap, West Virginia. The Cresap plant uses a technique in which the coal is dissolved in a liquid produced in the process itself; then the solution is filtered and hydrogenated to produce a synthetic crude. This crude is then refined in the usual manner and yields up to 3 barrels of gasoline per ton of coal. (Other processes yield 1½ to 4 barrels per ton of coal.) Early, but inconclusive, results of the pilot plant operation were achieved in 1969.

Another approach, employing a simple thermal cracking process is also scheduled to go into the pilot plant phase. The plant is now being designed, with construction to be completed by 1970 and results not likely before 1971 or 1972.

A technical advantage of coal compared to other syncrude sources is that less material needs to be handled and smaller waste disposal problems are involved. Since coal can yield up to 3 or 4 barrels of product per ton of feed, compared to about three-fourths of a barrel per ton from shale, a 100,000 barrel/day coal liquids plant would only require 10 to 15 million tons of coal to be mined per year compared to the 45 to 60 million tons of shale that would be needed per year for the same plant size. *In situ* processes, a possibility for shale, are however not highly re-

[16] U. S. Department of the Interior, *United States Petroleum Through 1980,* (Washington, July 1968), p. 72.

[17] E. H. Reichl, "Liquefaction and Gasification of Coal" (paper presented at the U. S. Department of the Interior Symposium, Washington, D. C., March 10, 1967).

[18] R. J. Cameron, "A Comparative Study of Oil Shale, Coal, Tar Sands As Sources of Oil" (paper presented to the AIME Symposium, Dallas, Texas, March 4-5, 1960).

garded for coal because of poor results in past research on the underground gasification of coal.

Outlook for Commercial Development

As in the section on shale oil, the outlook for commercial development will be evaluated on the assumption that the factors affecting research and development will continue as in the past. In the appendix to this chapter, commercial development will be evaluated on the assumption that a level of effort beyond that made to date is put forth in order to expedite technical and commercial progress.

It was only during late 1969 that the first pilot plant using current processes for coal hydrogenation was operated on a full scale. Thus, what little has been known regarding costs was derived totally from scaling-up paper calculations based upon bench results into estimates of what investment and other costs might look like on a commercial scale.

The consensus appears to be that a coal syncrude plant of 100,000 barrels/day capacity may be the minimum size worth considering.[19] A plant of this size is estimated to cost $200 million if it is to produce syncrude ready for shipment to a regular refinery for final processing. If it is to be a fully integrated plant to carry the coal all the way to final products, then the cost would be $350 million. To this must be added anywhere from $50 million to $150 million to account for the coal mine investment. The variation in the mining investment is to allow for differences in the location, types of coal, and methods of mining that would be involved.[20]

Some estimates have been made of the cost of syncrude produced from coal, and these range from $2.50 to $3.00 a barrel depending upon the process being considered. This cost range is in the neighborhood of conventional crude prices, but it is unclear whether these rough estimates are meant to be approximations of cost or of selling price including an adequate rate of return.[21] In addition, the cost calculations allow for the use of the char residues as a power plant fuel. It is also expected that the income from surplus power sales, and the returns from pipeline quality gas, sulfur, inorganics, and other by-products will all improve the economics of production. Markets probably could absorb these by-products from a limited number of plants favorably located. But there remains the critical question of possible limitations on the rate of growth and eventual size of a synthetic oil industry based upon coal because of its sensitivity to the availability of markets for its by-products.

A crucial element in the development of synthetic liquid fuels from coal concerns the level and the length of time over which the government will expend R&D funds in its support. The government-supported research program of about $12 million per year (see p. 65) is large compared to that financed by industry. The initiative now rests with government, and if federal funds were to be cut back this would undoubtedly result in delay in the appearance of commercial-scale coal syncrude plants. Or if the government funds were not available for the research sequence to be completed, there is the question of whether industry would carry on alone.

The current state of demonstrated technology of using coal as a source of synthetic crude is not as far advanced as that for shale oil (discussed in the preceding section). Consequently, estimating a timetable for the building of commercial production facilities is even more difficult than for shale. To develop a timetable of commercial development, one must begin with the assumption that the results of the pilot plant operations will be favorable enough to point the way towards a continuation of R&D activities. Naturally if technological problems emerge that cannot be solved or if the process economics turn out to be unpromising the whole sequence might well halt at that point.

On favorable assumptions a reasonable sequence of events would seem to indicate first dependable pilot plant test results in 1970. If an immediate decision were then to be made to design a prototype commercial plant, that facility could probably be completed around 1974. The latter should begin to yield evidence as to commercial feasibility during 1975. Again assuming success, it does not appear unreasonable to estimate that one or two plants a year would come on-stream in the period 1977 through 1980. Assuming this favorable chain of events, there would be a total of at least three plants but not more than five by 1980. The total capacity, assuming plants of 100,000 barrels/day, would range from 300,000 to 500,000 barrels/day by 1980. It must be stressed, however, that everything would need to go just right, with no loss of time in making decisions to press ahead, in order to achieve commercial production within the range indicated.

Tar Sands and Other Bitumens

In many sedimentary rocks petroleum occurs as a heavy asphaltic material that will not flow through the rock in the fashion of conventional crude oil. These are actually petroleum deposits, but they cannot be produced in the normal way.

Tar sands and similar oil-impregnated rocks occur throughout the world and attempts have been made to extract their oil from time to time; but the extraction has been in small quantities and the activity has been sporadic in nature. The Athabasca tar sands of Alberta, Canada, have always attracted the most interest because of the vast quantity of petroleum contained in these beds. The oil sand in the beds is a quartz mixed with silt and clay, and extends over 21,000 square miles. The beds are as much as 200 feet thick in some places and vary in oil saturation. The sands are near the surface in the vicinity of the Athabasca River

---

[19] Howard W. Blauvelt, Continental Oil Company, speech before the Houston Chapter of the AIME, November 7, 1967; Carl G. Harrington, Humble Oil Company, speech at Wyoming Governor's Coal Conference, Gillette, Wyoming, May 25, 1968.

[20] E. H. Reichl, "Liquefaction and Gasification of Coal," *op. cit.* (n. 17).

[21] George Fumich, "Coal Conversion to Liquids and Gas at Competitive Costs Expected Soon," *Oil Daily*, January 29, 1968, p. 28.

but plunge to 2,000 feet along the southern and western edge of the oil-bearing portions.[22]

Construction of the first commercial plant to produce oil from the tar sands was begun in September 1964, the plant was completed in September 1967, and the first delivery of oil for processing in a refinery located in Toledo, Ohio, was made in March 1968.[23] The construction phase was preceded by eight years of financial studies, engineering work, development activities, and negotiations with the provincial government.[24] If the plant is judged economically successful, it will stand as the first competitive development of a major substitute for crude oil in North America. Also, it may lead to a better understanding of the possibility of using bitumens found in Canada, the United States, and elsewhere.

Resource Base

The Athabasca occurrence is one of the major oil deposits of the world. Although the total content of the entire formation is probably somewhat larger, recoverable pipeline-grade synthetic crude from the tar sands has been estimated at 285 billion barrels, with 85 billion barrels considered currently minable.[25] (This latter figure is more than twice as large as the U. S. proved liquid hydrocarbon reserves, as of the beginning of 1969.) The most important limiting factor is that perhaps 93 percent of the tar sands are too deep for open pit mining.[26]

The United States resource picture, as now known, is far less impressive. The latest U. S. Bureau of Mines report on oil-impregnated rocks indicates that known deposits in this country contain a minimum of 2½ billion barrels of recoverable oil and a maximum of 5½ billion barrels. The term "recoverable" as used in the report pertains to physical availability rather than economic recoverability.[27]

Although interest in U. S. tar sands has not reached the point of actual operation, oil companies have leased over a million acres in Missouri, Kansas, and Oklahoma, where 40 percent of the known reserves occur.[28] California has some large deposits in the 100 million barrel class and Utah has a few occurrences that may contain close to one billion barrels.[29] It is believed that other deposits may be discovered in the future, as there has never been a concentrated exploration effort to find tar sand deposits.

Despite their small size compared to Canada's, the United States tar sand beds do have the attractiveness of being comparable in many cases to the Athabasca sands in character and quality. The amount of reserves in particular locations in the United States might be large enough to support a single plant and would have the advantage, as compared to Athabasca, of being closer to markets, closer to established facilities, and having more favorable climatic conditions. As a result, a large number of projects, mainly steam flood, are operating or being planned within the United States.[30]

Technology

The tar sands present the least problem in processing of all of the replacements for conventional crude. In the Canadian plant, after separation from its host rock by washing, the oil is thinned and then processed to reduce its viscosity, to remove the diluent, and to extract any contaminants such as sulfur. The end product is similar to normal crude oil and may be charged directly into a standard refinery.[31]

The capacity of the Athabasca plant is 45,000 barrels/day. This is probably the minimum size that can be built with a reasonable expectation of profitable operation.[32] The size was dictated by Alberta's policy to restrict the amount of tar sands development owing to the fact that half or more of her conventional oil well capacity was shut-in. It is expected that the cost characteristics of the plant would improve if it were to be enlarged. Its design will permit expansion to 60,000 barrels/day without exceeding the capability of the pipeline and other facilities. It seems certain that economies of scale would continue to be operable beyond that size.[33]

The present plant is a surface operation using German-made digging machines. The machines actually are capable of extracting enough sands to supply a 90,000 barrel/day plant. The lease encompasses 4,000 acres with reserves of 600 million barrels—approximately a 30-year supply (if account is taken of the fact that recoverable reserves may

[22] L. A. Bellows, Jr., *et al.*, *Athabasca Oil Sands* (AIME, Society of Petroleum Engineers preprint No. 517, February 1968), p. 1.

[23] U. S. Department of the Interior, *United States Petroleum Through 1980, op. cit.* (n. 16), p. 78.

[24] W. A. Bachman, *et al.*, "Plant Starts, Athabasca Now Yielding Its Hydrocarbons," *Oil and Gas Journal*, October 23, 1967, pp. 69–88.

[25] F. K. Spragins, "Athabasca Tar Sands Place in a Total Energy Concept" (paper presented at the American Mining Congress Coal Convention, Pittsburgh, May 5-8, 1968).

[26] William H. Davis, "The Great Canadian Oil Sands Project," *Proceedings of the Council of Economics, AIME,* (New York: AIME, 1968), p. 111.

[27] Ball Associates Ltd., *Surface and Shallow Oil-Impregnated Rocks and Shallow Oil Fields in the United States* (U. S. Bureau of Mines Monograph 12, [Oklahoma City: Interstate Oil Compact Commission, 1965]), p.7.

Not included among these rocks are certain solid bitumens such as the asphaltic material, gilsonite, a rather extensive deposit of which is currently being mined in Utah. There are 10 million tons of the material in this occurrence in veins from a few inches wide to 22 feet thick. The veins go to at least 2,000 feet with mining to a depth of 1,500 feet. Part of the output has been converted to 1,300 barrels/day of gasoline near Grand Junction, Colorado. The remainder is used in the manufacture of ink, phonograph records, roofing, paints, battery cases, and similar products. (*Iron Age*, August 1, 1957, pp. 54–55.)

[28] Ball Associates, *Surface and Shallow Oil-Impregnated Rocks and Shallow Oil Fields in the United States, op. cit.*, p. 1.

[29] *Ibid.*, p. 6.

[30] D. Ball, "United States Tar Sands as a Petroleum Source" (paper presented at the U. S. Department of the Interior Symposium, Washington, D. C., March 10, 1967).

[31] U. S. Department of the Interior, *United States Petroleum Through 1980, op. cit.* (n. 16), p. 78.

[32] *Oilweek*, February 26, 1968, p. 16.

[33] Another firm has received approval for an 80,000 barrel/day plant, but with a startup of operations delayed until 1976. (*Oil and Gas Journal*, September 22, 1969, p. 121.)

be well below estimated reserves).[34] Each ton of sand contains about 20 gallons of oil. Since its opening the operation has been harassed by start-up problems. Unexpected boulders in the sand have been a problem in mining, natural gas has had to be piped in to supply additional heat, and still other problems have been encountered. If a second plant were to be built along similar lines, these problems could probably be avoided or minimized. It is also likely that a second plant would be less costly as a result of the operating and design experience gained in the first plant.

Considerable hope is held for the possibility that *in situ* methods may be possible in tar sands production. This might improve the outlook and have an important effect on cost. Pilot plant experiments using a variety of thermal methods have been technically successful. These have included both steam-drive and *in situ* combustion. These thermal processes could very well open up the deeper tar sands in Athabasca and might be applicable in the United States.[35]

Outlook for Commercial Development

The decision to build the Great Canadian Oil Sands, Limited (GCOS) plant in Athabasca was prompted by The Sun Oil Company's need for reserves for its Canadian refinery. Sun had not been successful in locating oil by conventional drilling in Canada. The company had been dependent in part on imported Venezuelan crude to supply its Canadian refining requirements, but a Canadian oil policy directed toward linking inland refineries to domestic sources forced Sun to seek a new pattern of supply.[36]

The plant, located in the Canadian wilderness, was built at a reported $235 million and is designed to produce 45,000 barrels of crude, 3,000 tons of coke, and 300 tons of sulfur daily. The cost of around $5,000 per daily barrel makes it a more expensive facility than those estimated for liquids from shale or coal. The necessity for extensive support facilities in its out-of-the-way location must have added appreciably to the capital requirement. The synthetic crude is shipped by pipeline 266 miles to Edmonton and then moved nearly 2,000 miles to Sarnia, Ontario, by the Interprovincial Pipe Line.[37]

Even though the technical problems of designing and building a plant that can produce a useable product have been overcome, operating difficulties still pose serious problems; after the first six months the venture was reported to be operating at a substantial loss.[38] Despite its present stage of development, not much has been stated publicly about the cost of syncrude from the tar sands; and there has been less published speculation about tar sands costs than about shale oil and liquids from coal. At the opening ceremonies, Sun officials in response to a question about cost replied that their official answer was $2.75 a barrel.[39] Another published report quoted $2.50 a barrel with a contracted price for delivery at Edmonton of $2.75 a barrel.[40]

In attempting to estimate a schedule for the development of commercial facilities in Canada, questions of public policy must be given considerable weight. Alberta, under pressure from the existence of shut-in oil well production capacity, initially only permitted a single plant to be developed in the province so as to keep tar sands output at 5 percent of domestic crude oil production. GCOS, in receiving the permit, was restricted as to plant size and as to completion within a specified period of time. A number of other interested firms also applied unsuccessfully.

Early in 1968, Alberta announced a change in policy designed to anticipate the need for additional output in the future. The new policy allows 150,000 barrels per day of capacity to be constructed to serve markets normally within reach of conventional crude oil output.[41] It is expected that this policy would allow tar sands to get 50 percent of "new markets" (as defined by the Alberta Oil and Gas Conservation Board) not now supplied by conventional sources in Alberta. The new total would include the existing GCOS plant. The new policy also specifies that there shall be no limits on additional capacity built to supply markets beyond reach of conventional Alberta crude because of its composition or location.

Another policy question involves the United States' limit on oil imports. The tar sands oil must be included among the exempt overland imports from Canada, and fall under such informal restrictions as this entails. Although there is some feeling that overland imports may become a larger share of U. S. oil consumption in the future, this will depend upon U. S. oil import policy and its implementation.

Thus, despite the technical head start of tar sands, the potential size of Canadian operations seems to be rather limited for the period from now until 1980. Their distant location and the necessary expenditures on pipelines, towns, and other facilities would entail large capital outlays. In addition, Alberta's restrictive (although now somewhat relaxed) policy on growth; the present barrier to free access to U. S. markets; and the difficulties encountered with the first plant will also tend to inhibit growth. Under these circumstances, it would seem reasonable to assume that Canadian operations will probably reach the current permissible 150,000 barrels/day, but would not exceed 300,000 barrels/day by 1980.

As for the United States, there is a question regarding the

[34] R. G. Dunlop, American Mining Congress Coal Convention, Pittsburgh, May 5-8, 1968.

[35] U. S. Department of the Interior, *United States Petroleum Through 1980, op. cit.* (n. 16), p. 78.

[36] Canada, Royal Commission on Energy, *Canadian Energy Prospects–Second Report* (Ottawa: Queen's Printer, 1959). Even though the first experimental shipment from Athabasca went to Sun's Toledo refinery, the motivation for building the plant was the Canadian government's decision not to permit foreign oil to move west of the Ottawa River line; this cut off Sun's Sarnia plant from Venezuelan crude.

[37] William H. Davis, "The Great Canadian Oil Sands Project" (n. 26), p. 110; and R. J. Cameron, "A Comparative Study of Oil Shale, Coal, Tar Sands As Sources of Oil," (n. 18).

[38] New York *Times*, May 12, 1969.

[39] Denver *Post*, October 8, 1967, p. E56.

[40] *Oil and Gas Journal*, October 9, 1967, p. 77.

[41] *Oilweek*, February 26, 1968, p. 15.

number of locations where there are deposits containing the several hundred million barrels of tar sands oil required for an economically-sized plant. However, if the GCOS project is a success and the current U. S. experimental work is encouraging, there is some possibility that one or two tar sands plants might exist in the United States by 1980. This would appear to be the maximum potential, however.

Outlook for Commercial Development of Liquid Fuels from Unconventional Sources: A Summary View

Despite a widespread acceptance of the inevitability of synthetic liquid fuels becoming a major supply source within North America at some future date, the timetable cannot be foreseen with any degree of clarity because of uncertainty regarding the technical processes to be employed and their costs, as well as unresolved problems in the field of government policy in the United States and Canada. In addition, major uncertainties exist because of unforeseeable developments in the future domestic supply of conventional crude oil. A rash of discoveries offshore, the opening of major new oil regions in the Arctic, or a technical breakthrough in recovery from old reservoirs could divert the flow of private capital away from synthetics for many years, and could affect the incentives for undertaking governmental programs. Despite the risks and high costs of petroleum exploration, there would appear to be an inclination to produce from the source that is known best so long as it is judged to be adequate.

An attempt has been made, nevertheless, to estimate the quantity of liquid fuels from unconventional sources that would be forthcoming from the facilities that might be built in the next decade or so, assuming that research and development yields favorable results and that the competitive threshold with crude oil is crossed in the meantime. It is assumed also that governmental and commercial incentives for pushing ahead do not change radically toward either a more or less forceful program of development. In the light of developments in the North American Arctic it is likely that these incentives will grow weaker. Summarizing now from the preceding description and analysis, which was itself based upon the integration of a great number of different sources of information (both published and unpublished), the overall assessment for 1980 is as follows:

Shale Oil

The capacity that may exist in 1980 ranges between zero and 350,000 barrels/day. The most likely expectation, perhaps, falls within the middle of the range, say, three plants with a combined capacity of some 200,000 barrels/day.

Coal Hydrogenation

The capacity that may exist in 1980 ranges between zero and 500,000 barrels/day. The most likely situation, perhaps, falls within the middle of the range, say, three plants with a combined capacity of 300,000 barrels/day.

Tar Sands

Canada may have a range of capacity anywhere from 45,000 to 300,000 barrels/day with an amount approaching current permissible capacity of 150,000 barrels/day most likely. In view of the uncertainty of U. S. import policy we attempt no estimate of the amount that might be shipped to the United States. For production plants within the United States, no estimate was made.

Combining the individual "most likely" estimates for the unconventional sources indicates that they may produce about 500,000 barrels/day in the United States under technical circumstances favorable to their development, but restrained by the economic uncertainties created by the Arctic developments as they can presently be taken into account. In Canada, an additional 150,000 barrels/day of capacity is the "most likely" estimate.

APPENDIX: ACCELERATING THE DEVELOPMENT OF LIQUID FUELS FROM UNCONVENTIONAL SOURCES

The estimated total resource base within North America of oil shale, tar sands, and coal which looms so large in strictly physical terms is, upon closer scrutiny, not to be counted on for more than a marginal increment to liquid fuel supplies in the next decade or so. Suppose, though, that it were deemed necessary to accelerate the pace of development, for whatever reason, and an early decision were made to proceed with the forced development of these sources. How much larger a flow of liquid fuels from these sources could there be; and what actions would be required to bring about such an enlarged output in an orderly fashion within a peacetime economy?

Technological Uncertainties

A first full-scale plant necessarily carries peculiarly large risks just because it is first in meeting the problems that are usually encountered in scaling-up from pilot plants or semiworks–i.e., the satisfactory functioning of each piece of equipment; the meshing of all phases of the operation, etc. These risks would remain even if all prudent commercial safeguards had been taken to minimize the impact of potential adverse market developments (e.g., by guaranteeing sales of main and by-product output under long-term contractual arrangements). But the experience gained in the first venture would be available to subsequent operators. They could be expected to avoid some of the technical difficulties and reduce lead times required before their facilities could become fully operative.

Although technological uncertainties apply to each of the three main sources of synthetics, they would appear to be smallest for tar sands because the first commercial-sized plant has already been in operation since 1967. The greatest technical uncertainties may be said to exist for coal liquefaction because development had only reached the stage at which a pilot plant for testing one of several promising processes began production in 1969. Several years of experimentation would normally occur before a decision to scale-up could be reached, and even then the next size might be a semiworks rather than a full-fledged commercial plant. (Coal liquefaction plants operated in Germany during the war, but different processes are involved in the plants now under consideration.) Shale oil falls between the other two cases. Testing of a 1,000 ton/day semiworks has been undertaken by TOSCO, and extensive research on another process has been carried on at the U. S. Bureau of Mines Rifle, Colorado facility.

Maximum Feasible Output, Limited Primarily by Physical Factors

Let us put the following question: What would be the maximum feasible output from synthetic fuel plants over the next decade if the technical feasibility of processes judged to be approximately within economic range, although not necessarily competitive, were to be taken as a justification for the construction of full-scale facilities, *and the only limiting factors were those set by physical requirements and sequential aspects that translate into time requirements*? Both a peacetime economy and an orderly process of expansion are assumed, although it is also assumed that unusual measures would be taken to accelerate the pace of development.

1. *Shale Oil*

For shale oil, where extensive testing on the pilot plant and semiworks levels has already taken place, it is generally believed that a full-scale plant could be constructed and begin operations in two to three years from the time that orders are placed with engineering and construction firms. Although it would be possible to order a number of such plants simultaneously, and to prepare the plant sites, mining locations, housing and other supporting facilities, and transportation links simultaneously, such a decision could have serious disadvantages: the first facility may well encounter serious problems which tend to raise costs and which could be designed out of subsequent plants. Hence, we will assume the construction of a single plant and its full-scale testing, followed by a decision to build additional units. For shale oil, the total time required before any additional plants could come on-stream would thus be of the order of five years–i.e., by about 1975, assuming an early decision.

How rapidly could the additional capacity be developed? At least initially, mining techniques would draw heavily on those of other industries. Retorting equipment is another matter, because even though much is known, there is no experience in building retorts of the scale that would be required for commercial operations. Upgrading would involve essentially the same type of processing units used in oil refineries; these have been installed at annual rates in excess of 800,000 barrels.[42] No insuperable manpower problems should be expected, either, since personnel with normal mining, engineering, and construction skills should be obtainable, assuming the absence of a major national emergency and the ability to offer sufficient incentives to draw such labor into the oil shale areas.

The principal obstacles to rapid large-scale growth of shale oil would appear to arise from the limited number of suitable sites for shale oil plants on private lands; the adequacy of water supplies; the problems associated with spent shale disposal and related environmental aspects; and the build-up of supporting community facilities, both public and private. What all this seems to imply, in terms of maximum speed of oil-shale development, is a rather modest scale and pace for some years even after the first plant has proven full-scale operations to be technically feasible and either economically acceptable or on the margin.

The limit under these circumstances is frequently placed at about 2 million barrels/day by 1980.[43] We judge that this figure is unlikely to be reached, even under the assumptions here adopted, and that during the next decade a shale oil industry should, at the maximum, not be expected to exceed one million barrels/day. Expansion on a larger scale would appear to require the solution to at least two problems which have as yet not been tackled: (1) mining methods adequate for recovery of the thick, deep deposits typically found on the public lands (or, alternatively, successful development of an economic *in situ* technique–neither of which can be counted upon for the next decade); and (2) improvements in recovery and processing techniques which would lower costs to the point at which operations outside Colorado (in Utah or Wyoming, where the limitations of water supply are less serious) could be contemplated.

2. *Coal Hydrogenation*

For coal hydrogenation, further research must still precede the construction of the first full-scale plant. Of the major processes under consideration, only one has arrived at the pilot plant stage, while another is just under development and has been scheduled for completion in 1970.[44] Furthermore, the research on coal gasification which may produce an answer to the hydrogen supply problem is also incomplete; and efforts to date to produce low-cost hydrogen from coal have not proved encouraging. These various processes must proceed through the pilot plant stage, the best process or combination of processes selected, and a prototype plant designed. It is unlikely that the pilot plant research period could be reduced to less than two years because operations for a shorter period probably would not yield sufficient experience to permit moving to the next stage of development. If the removal of existing restraints

[42] American Petroleum Institute, *Petroleum Facts and Figures*, Centennial Edition (New York, 1959), pp. 98–100.

[43] See, for example, Northcutt Ely, "The Oil Shale Industry's Water Problems," *Quarterly of the Colorado School of Mines, Fourth Symposium on Oil Shale,* LXII, No. 3 (July 1967), pp. 9–17.

[44] *Mining Record*, August 22, 1968.

in funding is assumed, it should be possible to have several pilot plants in operation simultaneously. These could yield results early in the 1970's which would make an initial full-size plant possible by 1974. This assumes that the pilot plant research would produce results that would justify building a full-scale plant. Allowing a three-year lead time for additional full-scale plants would mean that several commercial-sized plants could not appear on the scene before about 1977.

By that time, if all had gone well, it would appear to be physically possible to construct a million barrels/day of capacity per year. The greatest barrier to this rate of development might prove to be the coal mining phase. One million barrels/day of new synthetic fuel capacity would require between 250,000 and 600,000 tons of coal per day (on the assumption of 1½ to 4 barrels of liquids per ton of coal) or as much as 200 million tons/year. A 1980 coal liquids level of, say, 4 million barrels/day would require up to 800 million tons of coal per year—a substantially greater quantity than the entire coal industry produces at the present time.

The task involved is staggering—thirty new mines each year equivalent to the largest mines now being developed (say, 7 million tons/year). These mines would require 4 billion tons of reserves to be acquired each year, and a net addition to the mining labor force of perhaps 10,000 to 20,000 new miners would have to be found each year (at an approximate output ratio of 30 tons/man-day). Considerable doubt must be expressed as to the ability to acquire coal reserves in suitable locations at this pace and to find and train this many coal miners.

Such an expansion in coal mining might be realized by moving west to tap the large coal reserves of the mountain states. But this would give rise to some of the same problems as shale oil development would face—the need to provide transportation to markets and the great distance (and added costs) to reach those markets, limited water supplies, and the recruitment of labor supply for the plants and mines.

Another physical, as well as economic, problem connected with this level of output concerns the flood of co-products of the process. Using one design configuration, there would be 65 million tons of new char to dispose of each year either as fuel or after conversion to electric power, 700,000 tons of sulfur, and several million tons of inorganics, as well as other by-products.[45]

An overall assessment of the situation leads to the conclusion that perhaps 2 million barrels/day of capacity might be within reach by 1980. This would require an intensive search for, and development of, Midwestern and Eastern coal reserves. It is unlikely that an adequate supply of strippable coal reserves could be found, and the accompanying reclamation of stripped lands could be a problem. Therefore, underground development would be needed and perhaps too high a cost per ton of coal might be encountered.[46]

3. *Tar Sands*

The primary question with respect to tar sands is how successful the first Canadian project will be from a technical and economic standpoint, since it is a strictly commercial venture. If it is a financial failure, then the financial gap probably would have to be closed by government assistance if further development is to take place. If it is a financial success, then it becomes a case of determining if any additional incentives would be needed to accelerate the pace of subsequent development.

Owing to the remote location of the reserves, the cost of transportation to market would pose a major obstacle to rapid development of a large number of projects in Athabasca. Physically there appears to be no reason why tar sands development could not reach a level as large (1 to 2 million barrels/day) as that estimated for shale oil or coal under conditions of accelerated development, except that, so far as the U. S. market is concerned, the tar sands syncrude would probably have to seek an outlet in the same areas as shale oil and synthetic liquid fuels from Western coal.

Actions Needed to Attain Maximum Feasible Output

The actions required to attain maximum feasible output, if the controlling factors were those set by physical requirements and sequential aspects that translate into time requirements, are to be discussed under two general headings: (1) public policies and (2) investment requirements. Government funding of research and development will be included under the first topic.

1. *Public Policies*

It is obvious that the levels of output envisaged for shale oil and coal in the United States and for tar sands in Canada are not going to be realized without early decisions to proceed with great speed, involving definite actions by the respective governments. No effort will be made at this point to go beyond suggesting the nature of what appear to be the required, although not necessarily sufficient, steps to achieve the task.

A basic uncertainty that stands as an obstacle to the large-scale development of domestic sources of synthetic liquid fuels concerns U. S. policy regarding imported petro-

[45] Interview with Neil Cochran, Office of Coal Research, U. S. Department of the Interior in *Business Week*, June 1, 1968, pp. 116–118.

[46] Steam coal mined underground may run as much as $1 per ton higher than strip-mined coal in the same state. An additional $1 in the price of coal adds approximately 0.5 cents to a gallon of synthetic gasoline. Of course, a difference in cost of this order is swallowed up in the other, major uncertainties in the estimates referred to in the text. It is, nevertheless, instructive to note the effect of change in coal cost on the cost of gasoline; even such a small difference could affect the competitive position of synthetic oil from coal.

leum versus unconventional liquids as supplements to domestic crude oil. The synthetics at this stage are, at best, generally thought of as marginally competitive with high-price domestic crude oil. If a continued policy of severely limiting the country's dependence on imports, at least from overseas areas, were considered to be in the national interest and this were laid down as a long-term policy commitment, then one of the factors that has been considered a major obstacle to the development of synthetic liquids within the United States would be removed.

Federal assistance with R&D funds undoubtedly would be another requirement. This is clearly the case in coal hydrogenation, where R&D are currently being supported by the federal government. The current Office of Coal Research effort is of the order of $10 million per year, and about $2 million per year is being expended in this field by the U. S. Bureau of Mines. If the construction of pilot plants for primary processes were to be bunched, as suggested in the preceding section, this might involve spending some $40 million in two years rather than four.

However, even with successful pilot plant and semiworks experience, as in shale oil today (and as might be the case with coal hydrogenation in several years), there would still exist uncertainty as to the operational experience to be expected from the first full-scale plant, both technically and in cost terms. It is altogether likely that some form of government assistance would be needed for the first full-scale plant if quick action were wanted. There is a broad array of devices open to government. No effort is here made to suggest which, if any, should be employed, or the extent to which assistance might be needed for the first and, perhaps, subsequent full-scale plants. The government is, of course, not without considerable experience in the energy R&D field, as can be seen from its large investment in accelerating nuclear power development. It is strongly unlikely that anywhere near as difficult and costly an effort would be required for synthetic liquids.

For shale oil there is an additional area in which public policy clarification will be required if a large-scale industry is to emerge—leasing policy on the public lands. To this point only a limited beginning has been made in opening three test leases to private bids,[47] and there is every indication that the Interior Department approach has been to proceed at a very deliberate pace—not one designed to speed large-scale development on the public lands. The government, as landowner, might find itself faced with conflicting concepts as to its proper role: to behave as any other landowner with an eye to maximizing the revenues from the leasing of the land, or to behave in a way that will hasten development even at some cost in land revenues. The latter could be justified only within the framework of a well-defined and defensible conception of the broader public interest to be served.

The tar sands situation is unique in that a full-scale plant already is in existence so that R&D support may not be needed, even though its experience (as of 1968/69) had been disappointing. However, there are, as described in the body of the chapter, existing policies on the part of the province of Alberta and the U. S. government that discouraged Canadian tar sands development. It may be necessary to adopt some form of North American energy policy if tar sands are to be a significant addition to liquid fuels by 1980. The first step might be to lift the restraints now in effect on Canadian shipments of oil to eastern Canada and the United States. Then, as the shut-in Canadian crude oil capacity would have an opportunity to seek a wider North American market, Alberta might be in a position to eliminate the restriction on tar sands development, letting it seek its natural level in free competition with Canadian crude reserves for investment dollars. Tar sands growth would also have to be supported by adequate investment in transport facilities (e.g., large diameter pipelines in order to move the output to United States and Canadian markets).

[47] The results were disappointing. (See p. 56.)

2. *Investment Requirements*

In the present stage of knowledge no more can be done than to pull together the rough estimates included in the foregoing discussion (both in the chapter and appendix), and to multiply the estimated quantities by the estimated investment requirements. In multiplying two sets of figures, each inherently subject to a wide range of uncertainty, the resultant product cannot avoid being subject to serious qualifications.

The calculation follows:

	Maximum quantity by 1980 with accelerated pace of development (million barrels/day)	Estimated investment required per barrel/day[a]	Total estimated investment required (billion $)
Shale oil	1	$2,500–$4,000	$2.5– $4.0
Coal hydrogenation	2	$4,000–$5,000	$8.0–$10.0
Tar sands	1–2	$5,000	$5.0–$10.0
Total	4–5		$15.5–$24.0

[a] Including all operations, mining through refinery.

Several observations may be offered. The estimated investment requirements per million barrels/day of capacity may turn out to be too high or too low, depending upon the particular raw material and the stage in the development process that has already been reached. The tar sands estimate, based upon the first full-scale plant, may be too high for subsequent plants. The others, prior to the building of full-scale plants, probably are too low for early full-scale plants, but perhaps too high for later ones. But, at best, this is all highly conjectural.[48]

[48] In addition, the estimated investment requirements for synthetics do not provide for land acquisition costs, transportation to market, or for associated community facilities. As indicated earlier, many of the locations are in remote inland areas, so that investments in transport and community facilities would be very large.

The grand total, even spread out over a period of years (mainly a five-year period beginning in the mid-1970's), represents a very large sum of money to be spent on brand-new industries, especially when their competitive position is unknown. However, the investments, if they are made at all, would proceed on a sequential basis. This would provide the opportunity for evaluation at successive future points in time of the competitive position of the various unconventional sources, and the need and desirability of pushing their maximum development. Already early indications regarding the prospects for the North American Arctic suggest that an accelerated approach to the development of liquids from unconventional sources will not be required.

CHAPTER 5

Geographic Diversification in Oil Supplies: Developments Between the 1956/57 Crisis and the 1967 Crisis

SINCE the time, beginning in the 1950's that it became evident the major consuming countries, particularly in Western Europe, were going to become increasingly dependent upon imported oil supplies, one approach to minimizing the possible impact of supply interruptions which has received much attention is that of geographic diversification in supply sources. The desirability of diversification has been recognized by the governments of the consuming countries (and the regional political groupings of which they are members) and by the major international oil companies which constitute the main commercial link between the consumer and producer countries.

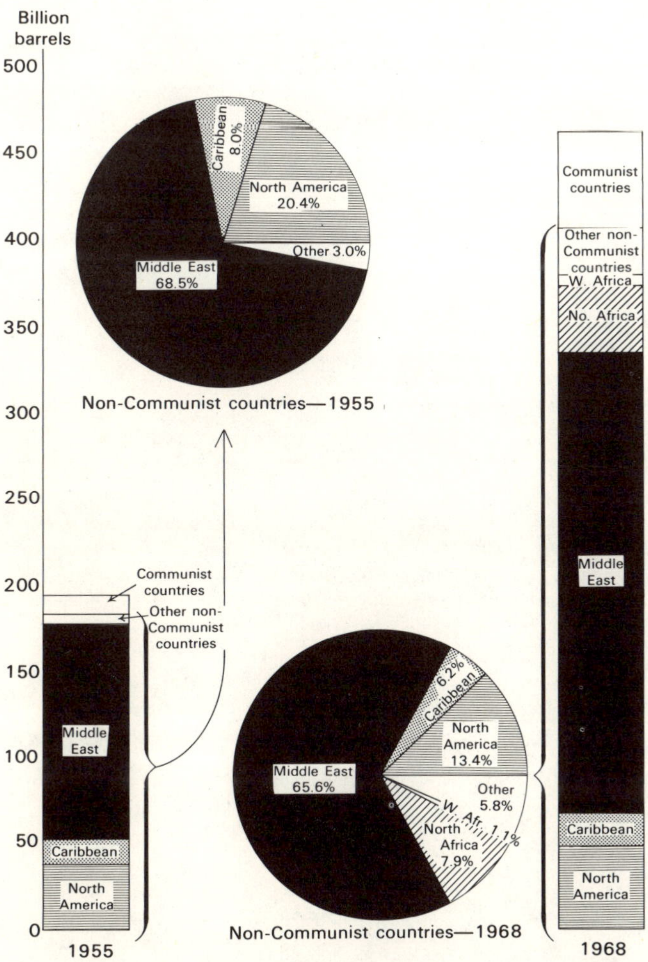

Fig. 5.1. World Proved Oil Reserves, Year-End 1955 and 1968

Source: 1955 data from Table 5.1; 1968 data from British Petroleum Co., Ltd., *Statistical Review of the World Oil Industry, 1968* (London: 1969), and *Oil and Gas Journal,* December 30, 1968.

Diversification in supply sources is made possible by the fact that although the geological circumstances favoring the existence of petroleum formations are far from ubiquitous, they are sufficiently widespread to allow oil to be produced in many parts of the world. Over time, this has made possible the emergence of major operations in new places where production either had been of small importance or completely nonexistent in the past. On a world scale it is, of course, the shift in the center of gravity of world oil industry operations made possible by the spectacular development of oil reserves in the Middle East which has resulted in the current and projected configuration of oil supply and demand patterns that are the root of the prospects and problems to which this study is directed.

The desirability of geographic diversification has received particular attention in the years since the Suez Canal crisis of 1956/57, partly as a result of that crisis and partly as a result of the general atmosphere of instability which has characterized the Middle East. It is well, therefore, to examine the actual results achieved between the Suez crisis and the most recent major supply crisis in the spring and summer of 1967. The focus of this chapter is on developments that took place during that particular time span in order to highlight the achievements of diversification during a critical decade. There have been some additional changes in the geography of world oil supply since 1967; these are taken into account in our statistical projections, or elsewhere in the book.

In assessing the diversification in the sources of oil that occurred in the decade between the two most recent major supply crises three sets of statistics bearing on geographic shifts will be examined: (1) changes in crude oil reserves; (2) changes in crude oil production; and (3) changes in the origin of imports into Western Europe and Japan.

Crude Oil Reserves

The broad geographic distribution of proved oil reserves on the eve of the 1956/57 and 1967 crises is shown in Table 5.1.[1] Between year-end 1955 and year-end 1966 the total

[1] "Proved reserves," according to U. S. practice, are the estimated quantities which geological and engineering data demonstrate with reasonable certainty to be recoverable from known resources under existing economic and operating conditions. (For definition, see *Reserves of Crude Oil, Natural Gas Liquids, and Natural Gas in the United States and Canada as of December 31, 1967* [New York: published jointly by the American Gas Association, American Petroleum Institute, and Canadian Petroleum Association, May 1968], XXII, p. 15.) The definitions used for other regions of the world, and their application in the preparation of estimates, may be more restrictive in some cases and less so in others. The standards applied undoubtedly are not uniform and country figures cannot be regarded as strictly comparable.

Table 5.1

World Proved Oil Reserves, as of end of Year, 1955 and 1966

(Quantity in Billion Barrels)

	1955		1966		Increase, 1955–66	
	Quantity	Percent of total excl. Communist countries	Quantity	Percent of total excl. Communist countries	Quantity	Percent of total excl. Communist countries
North America[a]	37.6	20.4	48.0	13.4	10.4	6.0
United States	35.4	19.2	39.8	11.1	4.4	2.5
Caribbean[b]	14.8	8.0	22.1	6.2	7.3	4.2
Middle East	126.2	68.5	234.6	65.6	108.4	62.4
North Africa	negl.	–	28.4	7.9	28.4	16.4
West Africa	negl.	–	4.0	1.1	4.0	2.3
Other non-Communist countries	5.5	3.0	20.6	5.8	15.1	8.7
Total	184.1	100.0	357.7	100.0	173.6	100.0
Communist countries	10.8		33.8		23.0	
Total world	194.9		391.5		196.6	

[a] United States and Canada; includes natural gas liquids.
[b] Mexico included with Caribbean.

Source: Basically, British Petroleum Co., *Statistical Review of the World Oil Industry, 1960* and *1966.* However, adjustments were required to accord with the geographic breakdown here employed. Data for adjustments derived from American Petroleum Institute, *Petroleum Facts and Figures, 1956* and *1967,* and from *Oil and Gas Journal,* December 26, 1966.

Table 5.2

World Oil Production, 1956 and 1966

(Quantities in Million Metric Tons and Thousand Barrels/Day)

	1956			1966			Increase, 1956-1966		
	Million metric tons	Thousand barrels/ day	Percent of total, excl. Communist countries	Million metric tons	Thousand barrels/ day	Percent of total, excl. Communist countries	Million metric tons	Thousand barrels/ day	Percent of total, excl. Communist countries
North America[a]	403.8	8,420	54.1	512.9	10,595	37.2	109.1	2,175	16.8
Caribbean[b]	151.0	2,910	18.7	214.0	4,135	14.5	63.0	1,225	9.5
Middle East	168.1	3,415	21.9	463.6	9,305	32.6	295.5	5,890	45.5
North Africa	1.8	35	0.2	113.1	2,350	8.2	111.3	2,315	17.9
West Africa	–	–	–	23.0	465	1.6	23.0	465	3.6
South East Asia	18.7	375	2.4	28.2	570	2.0	9.5	195	1.5
Other non-Communist countries	21.0	420	2.7	55.7	1,095	3.8	34.7	675	5.2
Total, above	764.4	15,575	100.0	1,410.5	28,515	100.0	646.1	12,940	100.0
Communist countries	96.4	1,965		291.8	5,870		195.4	3,905	
Total world	860.8	17,540		1,702.3	34,385		841.5	16,845	

[a] Includes United States and Canada. Natural gas liquids are included.
[b] Mexico included with Caribbean.

Source: British Petroleum Company, Ltd., *Statistical Review of the World Oil Industry, 1964* and *1967.*
Note: The conversion factors by which barrels per day are equated to metric tons per year may vary slightly among the areas shown. The percentage distribution is based on barrels per day.

reserves of the world (excluding the Communist countries) just about doubled, rising from roughly 180 billion to 360 billion barrels. In both years the Middle East provided, by far, the dominant component of the total, its share declining only slightly but remaining in the neighborhood of two-thirds during the entire period. Whatever small relative decline occurred in the Middle Eastern component of world reserves was more than made up by the growth in North African reserves from practically nothing at year-end 1955 to about 8 percent of the world's total at year-end 1966.

Measured according to the yardstick of proved reserves, North Africa was the major new region developed in the period between the two supply crises. From the standpoint of softening the impact of such crises, as they would affect Western Europe, the emergence of North Africa had the potential advantage of constituting a west-of-Suez supply source. Against the background of the 1956/57 supply crisis, resulting as it did from the closure of the Suez Canal, such a development was significant. However, during the 1967 supply crisis, which involved both the closure of the Suez Canal and concerted action by the Arab nations—including use of the selective embargo and governmental shutdown of operations—the diversification achieved through North African development was of far less value than it would have been in a crisis which involved no more than closure of the Canal. The North African producer countries, as part of the Arab world, were influenced by the same nationalistic and political sentiments that motivated the actions of other Arab countries. Libyan operations were, in fact, closed down by the government for almost a month—the lengthiest period of any of the actual closures—for fear of sabotage by strikers as well as the general attitude of the trade unions.

Crude Oil Production

Table 5.2 shows the broad geographic distribution of world crude oil production in 1956 and 1966. Annual figures for these years provide a close approximation to the pattern of international oil output as it stood prior to the Suez Canal crises that began in November of 1956, and the Arab-Israeli war of June 1967.

In broad outline these statistics tell the same story as those in Table 5.1 in that the outstanding development was the growth in output in the Middle East and North Africa. The expansion in North African output over this period was about equal to that experienced in North America, more than 100 million tons (or over 2 million barrels/day) in each case; but whereas North America began the period with production of more than 400 million tons (8.4 million barrels/day) in 1956, North African production in that year was less than 2 million tons (or 35 thousand barrels/day). Expansion in Middle Eastern and North African output taken together dominated the world scene, accounting for more than three-fifths of the growth in the world's total (excluding the U.S.S.R. and other Communist countries). On a world scale, such geographic diversification in production as took place was *toward* the Middle East and North Africa—i.e., toward greater concentration in those regions most likely to experience problems leading to concerted supply interruptions.

West Africa, another newcomer to the scene during this span of years, provided added geographic diversification and a measure of strategic diversification as well, since Nigeria, the major producer, is not a part of the Arab world. In terms of production and reserves West Africa is far from being in the same class as the Middle East or even North Africa. Unfortunately, too, a completely unrelated internal crisis disrupted supplies from this source at the same time as the supply crisis resulting from the June 1967 war. However, there appear to be large possibilities for growth in this region, and they have been taken into account in our supply projections.

The picture conveyed by Tables 5.1 and 5.2, although of major importance, must yet be filled out by examining in more detail the changing composition of output among countries within the Middle East and North Africa. This is done in Table 5.3, which includes annual production statistics by countries between 1956 and 1966 (and also carries the figures forward to 1968). Throughout this period the four major producing countries within the Middle East were Iran, Iraq, Kuwait, and Saudi Arabia. The four, including also the Neutral Zone, which is under the joint jurisdiction of Kuwait and Saudi Arabia, accounted for between 90 percent and 95 percent of all of the Middle East's production during the 1956–66 period. The Middle East, in turn, produced about 80 percent of total output within the Middle East and North Africa in 1966, declining from its 1956 position of almost complete supremacy in the two regions combined. With North Africa's emergence, production in Iraq, one of the big four of the Middle East, had, by 1966, been surpassed by that of Libya (and subsequently Kuwait was also surpassed in 1968). In 1966, five major producing countries in the Middle East and North Africa (Iran, Iraq, Kuwait, Saudi Arabia, and Libya) accounted for more than 85 percent of all the oil produced in the Middle East and North Africa.

As for the relative importance of each of the "Big Five" within the Middle Eastern and North African total over the 1956/66 decade, Table 5.3 shows the following changes: Iran was the only Middle Eastern country to increase its share, from 15.5 percent in 1956 to about 18 percent in 1966. The other three Middle Eastern countries declined relatively: Kuwait from roughly 32 percent to 20 percent, Saudi Arabia from about 28 percent to 21 percent, and Iraq from 18 percent to less than 12 percent. (Within the Middle East itself, these relative movements are explained in good part by a 1956 starting point which reflected the ground Iran had lost in production to other Persian Gulf producers as a result of the Iranian oil stoppage of 1951–54.) The most dramatic change in comparative terms was the increase experienced by Libya, from virtually no production as late as 1960 to more than 12 percent of the combined Middle Eastern and North African total in 1966 (and further relative increases in 1967 and 1968).

Even though the diversification reflected by these changes all took place within the Middle East and North Africa, they are not without significance so far as the

Table 5.3

Middle Eastern and North African Oil Production, by Country, 1956–68

Year	Iran	Iraq	Kuwait	Saudi Arabia	Neutral Zone[a]	Other Middle East[b]	Total Middle East	Algeria (incl. Sahara)	Libya	Other North Africa	Total North Africa	Total Middle East and North Africa
						Thousand barrels/day						
1968	2,835	1,505	2,420	2,830	425	1,170	11,185	920	2,600	295	3,815	15,000
1967	2,595	1,225	2,290	2,600	420	830	9,960	840	1,745	170	2,755	12,715
1966	2,110	1,390	2,275	2,395	420	715	9,305	715	1,505	130	2,350	11,655
1965	1,905	1,315	2,170	2,025	355	570	8,340	560	1,220	125	1,905	10,245
1964	1,705	1,255	2,115	1,730	360	450	7,615	560	860	125	1,545	9,160
1963	1,490	1,160	1,930	1,630	315	295	6,820	505	465	110	1,080	7,900
1962	1,330	995	1,830	1,525	245	250	6,175	435	185	90	710	6,885
1961	1,195	990	1,645	1,390	175	220	5,615	330	20	75	425	6,040
1960	1,060	955	1,620	1,245	135	220	5,235	180		65	245	5,480
1959	940	850	1,380	1,100	120	210	4,600	25		65	90	4,690
1958	825	725	1,395	1,005	80	215	4,245	10		60	70	4,315
1957	730	445	1,140	985	65	175	3,540			45	45	3,585
1956	540	630	1,085	975	30	155	3,415			35	35	3,450
						Percentages						
1968	18.9	10.0	16.1	18.9	2.8	7.8	74.6	6.1	17.3	2.0	25.4	100.0
1967	20.4	9.6	18.0	20.4	3.3	6.5	78.3	6.6	13.7	1.3	21.7	100.0
1966	18.1	11.9	19.5	20.5	3.6	6.1	79.8	6.1	12.9	1.1	20.2	100.0
1965	18.6	12.8	21.2	19.8	3.5	5.6	81.4	5.5	11.9	1.2	18.6	100.0
1964	18.6	13.7	23.1	18.9	3.9	4.9	83.1	6.1	9.4	1.4	16.9	100.0
1963	18.9	14.7	24.4	20.6	4.0	3.7	86.3	6.4	5.9	1.4	13.7	100.0
1962	19.3	14.5	26.6	22.1	3.6	3.6	89.7	6.3	2.7	1.3	10.3	100.0
1961	19.8	16.4	27.2	23.0	2.9	3.6	93.0	5.5	0.3	1.2	7.0	100.0
1960	19.3	17.4	29.6	22.7	2.5	4.0	95.5	3.3		1.2	4.5	100.0
1959	20.0	18.1	29.4	23.5	2.6	4.5	98.1	0.5		1.4	1.9	100.0
1958	19.1	16.8	32.3	23.3	1.9	5.0	98.4	0.2		1.4	1.6	100.0
1957	20.4	12.4	31.8	27.5	1.8	4.9	98.7			1.3	1.3	100.0
1956	15.7	18.3	31.4	28.3	0.9	4.5	99.0			1.0	1.0	100.0
						Increase between 1956 and 1966						
Thousand barrels/day	1,570	760	1,190	1,420	390	560	5,890	715**	1,505*	95	2,315***	8,205
Percent of increase	19.1	9.3	14.5	17.3	4.8	6.8	71.8	8.7**	18.3*	1.2	28.2***	100.0

*Between 1960 and 1966.
**Between 1957 and 1966.
***Essentially between 1960 and 1966.
[a]Neutral Zone has an ambiguous political status. Insofar as oil is concerned, it is controlled partly by Saudi Arabia and partly by Kuwait.
[b]In recent years, mainly Abu Dhabi and Qatar.

Source: British Petroleum Co., Ltd., *Statistical Review of the World Oil Industry, 1964* and *1968.*

problem of supply interruptions is concerned because the interests of the five countries are not identical. Iran, for one, is not an Arab country, so that its attitudes on certain key political issues, particularly those involving Israel, have in the past differed from those of the other Middle Eastern and North African oil-producing countries, all of which are Arab. And even within the Arab world there are differences of outlook on numerous questions which might be expected to keep the individual countries from joining together in unified action on all matters affecting oil, or from pursuing a particular course of action, even if adopted, with equal tenacity. This might be even more the case on strictly economic questions, involving bargaining between oil companies and governments, than on political questions that stir deep nationalistic feelings. And even the stand taken on political questions will involve unequal economic sacrifices to the different oil-producing countries, measured by loss of current revenues, and the ability to sustain such loss.[2] It is nonetheless true that particular countries may find themselves in a position of being forced to join in united action, although contrary to their own economic interests, by reason of popular pressure or of disruptive activities initiated from the outside.

On balance, diversification, even within the Middle East and North Africa, should be regarded as providing some measure of insurance against the threat of serious supply

[2] See Chap. 8.

Table 5.4

Origin of Net Imports of Crude Oil into Western Europe and Japan, 1956 and 1966[a]

	Western Europe imports						Japan imports					
	1956			1966			1956			1966		
Crude oil exports from:	Million metric tons	Thousand barrels/ day	Percent of crude total	Million metric tons	Thousand barrels/ day	Percent of crude total	Million metric tons	Thousand barrels/ day	Percent of crude total	Million metric tons	Thousand barrels/ day	Percent of crude total
Western Hemisphere	16.3	326	15.5	23.0	460	6.1	0.18	34	1.8	0.6	12	0.7
Middle East	86.4	1,728	82.4	219.3	4,386	58.6	7.84	157	79.7	75.0	1,500	89.8
North Africa	–	–	–	97.7	1,954	26.1	–	–	–	–	–	–
West Africa	–	–	–	15.3	306	4.1	–	–	–	–	–	–
Far East	1.3	26	1.2	0.2	4	0.1	1.81	36	18.4	5.1	10	6.1
Communist countries	0.9	18	0.9	18.2	364	4.9	–	–	–	2.8	56	3.4
Net imports of crude[b]	104.9	2,098	100.0	374.3	7,486	100.0	9.83	197	100.0	83.5	1,670	100.0
Net imports of products	17.1	342		41.0	820		3.25	65		12.6	252	
Total liquid fuel net imports	122.0	2,440		415.3	8,306		13.08	262		96.1	1,922	

[a]Since, for 1956, data on crude imports *only* were allocable by origin, the 1966 data in this table on geographic sources of imports likewise apply only to crude oil. (Note that in Table 1.3 the data on geographic flows include products as well as crude oil.) In 1966, West European product imports originated principally in the Western Hemisphere (40 percent), Middle East (20 percent), and Communist countries (30 percent); Japan's product imports came from the Western Hemisphere (25 percent), the Communist countries and Far East (15 percent each), and Middle East (45 percent).

[b]May not precisely equal sum of areas shown because of small amounts which could not be geographically allocated.

Source: United Nations, *World Energy Supplies,* Series J, Nos. 3 and 11.

Note: The United Nations data are published in terms of metric tons. The corresponding data in barrels/day shown above are based on the approximate conversion factor of 1 barrel/day = 50 metric tons/year.

interruptions, or of the intensity or endurance of oil stoppages if they were to be undertaken. Nevertheless, it is a sobering thought that concerted action by five countries would have affected about 85 percent of the oil supplies from this broad region and that concerted action by the four major Arab producing countries (or disruptive activities yielding similar results) would have affected close to 70 percent of the region's oil supplies[3]—as of the situation in 1966, after a full decade during which the *need* for diversification was stressed following the Suez Canal closure of 1956/57.

Origin of Imports: Western Europe and Japan

In Table 5.4 the origins of imports into Western Europe and Japan are compared for the years 1956 and 1966. Total imports into both increased enormously over the decade: Japan's 1966 imports were over seven times its imports in 1956, while in Western Europe the 1966 import total was about three and a half times as large as that of 1956. In absolute terms, the European increase came to roughly 300 million tons (or 6 million barrels/day), the Japanese to about 85 million tons (or 1.6 million barrels/day). These increases reflected the rise in energy consumption in Japan and the countries of Europe, and, within the energy total, the rise in the relative importance of liquid fuels from about 20 percent to 45 percent in Europe and from about 20 percent to more than 55 percent in Japan.[4]

Almost all of the liquid fuels consumed in Japan and Western Europe are imported, so that, as indicated in Chapter 3, the interruption of imports poses a particular threat to these two major consuming areas. To what extent did the broad geographic diversification of crude oil supplies between the time of the Suez crisis and the 1967 supply crisis serve to soften the impact of possible supply disruptions?

In the case of Japan, dependence upon the Middle East as a supply source of crude oil increased between 1956 and 1966. Viewed from the standpoint of a crisis of the 1956/57 type, in which the closure of the Suez Canal

[3]About 80 percent if *all* Arab producing countries were to be included.

[4]See Tables 3.1 and 3.3.

brought about the interruption in oil flows, diversification away from the Middle East was not an important objective for the Japanese. In terms of a crisis in which the Middle Eastern producing countries might shut off exports, or threaten to do so, Japan had become more vulnerable than in 1956.

Unlike the situation in Japan, the statistics for Europe do show diversification away from the Middle East in the decade following the Suez Canal crisis. Whereas the Middle East accounted for over 80 percent of 1956 crude oil imports, by 1966 it had declined to under 60 percent of the crude oil import total, or closer to half the total if imports of oil products (originating disproportionately in areas other than the Middle East) are counted in.[5] Because of its location west of Suez, this shift in supply pattern, whatever its causes, constituted a sensible protective step for a Europe that had faced a crisis of the 1956/57 type. However, the relative decrease in the Middle Eastern share was offset to a major extent by the increase in North Africa as a supply source, from practically nothing in 1956 to about 25 percent in 1966. If threatened by a crisis that involved the disruption of supplies at the point of origin in Arab countries, this substantial move toward greater geographic diversification would not add measurably to Europe's security of oil supply. In addition, reliance on Communist countries had grown from approximately 1 percent to 5 percent of total supply (8 percent if products are included), heavily concentrated in Soviet shipments to Italy; this could scarcely be looked upon as enhancing security since the position that the Communist countries would take toward making good supply deficits resulting from Arab actions would always be unknowable in advance of the event.

One move toward diversification that could be judged "politically reliable" in the event of an Arab supply threat was the growth in West Africa's import share, but this increase was modest as of 1966. And, as it turns out, supplies from this source were disrupted for another reason. However, its relative importance as a European supply source is projected to rise in the future. Also, as indicated earlier, the relative importance of Iran within the Middle East had increased. To the extent that this shift was reflected in the import pattern of European countries it should also be viewed as a move toward greater "political" diversification in supply sources.

APPENDIX: LIBYA–SOME ASPECTS OF THE DEVELOPMENT OF A NEW MAJOR OIL PROVINCE

As indicated in the chapter, the major geographic diversification in oil supplies that occurred during the decade following the Suez Canal crisis involved the emergence of North Africa, particularly Libya, as a significant source of production for export to Western Europe. The purpose of this appendix is to identify and assess certain aspects of the development of Libya that may provide some guidance in evaluating the prospects for major new provinces that look promising today. The Libyan experience cannot, of course, be looked upon as providing anything like a perfect guide of what to anticipate in other parts of the world where conditions are likely to be substantially different. Nevertheless, some valuable insights may be gained in regard to such factors as the amount of time, effort, and investment required to achieve a high level of production.

Exploration crews were at work in Libya as early as 1954, immediately following passage of the first legislation concerning oil concessions. However, it was not until 1956 and 1957, essentially at the time of the Suez Canal crisis, that the pace of the entire exploratory and development process began to be accelerated. This can be seen from the data in Table 5A.1 for the period beginning in 1954 and ending in 1961–the first year in which Libya shows up in world oil production statistics. Figures for subsequent years show a steady decline in geological and geophysical crew months (predrilling exploratory activity), and a growth in exploration and development footage drilled to peak levels of about 2,500 thousand feet in 1964–66.

Although commercial production did not begin until 1961, the announcement of discoveries began in 1958 and statistics showing moderate reserves appeared for the first time as of the beginning of 1959. Zelten, the largest field in terms of production, was found in 1959, but did not reach the stage at which all currently known reservoirs within the field had been broadly outlined by the drilling of development wells until 1964[6]. Among the other large fields, Amal-Maragh, found in 1959, did not begin to produce until 1966; Gialo, found in 1960, began production in 1965; Samah, found in 1962, began producing in 1964; for Sarir, the same dates are 1961 and 1964; for Idris A and D, 1967 and 1968; for Idris B, 1967 discovery but no production to date.[7]

All of the foregoing is by way of indicating that there are inevitable time lags in moving from exploration to development to production. Also indicated are the existence of time intervals of varying lengths between the announcement of the discovery of particular fields and the ensuing production of oil from these finds.

These relationships would need to be examined in very great detail in order to obtain a full understanding of the Libyan experience. What is wanted here, however, falls considerably short of such a detailed understanding; instead, the objective is to try to derive some very general dimensions of aggregate behavior that may be of some use in estimating the performance profile to be expected in new, potentially prolific, producing regions.

[5] See Table 5.4, n. a.

[6] Estimated reserves of this field as of 1/1/69 were 2.2 billion barrels. Production in 1961 was 60,000 barrels/day; in 1964, 388,000 barrels/day; and in 1968, 578,000 barrels/day. ("Worldwide Oil," final issue each year, *Oil and Gas Journal*; and "The World's Monster Oil Fields, and How They Rank," *Oil and Gas Journal*, January 13, 1969, pp. 43–49.)

[7] *Oil and Gas Journal*, December 30, 1968, p. 117.

Table 5A.1

Early Indicators of Petroleum Search in Libya

Year	Number of geological and geophysical crew months	Footage drilled (thousand feet)		
		Exploratory	Development	Total
	(1)	(2)	(3)	(4)
1954	10	–	–	–
1955	70	–	–	–
1956	252	9	0	9
1957	621	29	0	29
1958	989	157	0	157
1959	675	343	0	343
1960	814	142	782	924
1961	666	378	948	1,326

Source: *Bulletin,* American Association of Petroleum Geologists, August issues.

Table 5A.2.

Libya: Selected Statistics on the Development of Oil Production and Reserves, 1956–69

	Production[a]		Reserves as of Jan. 1[b] (million barrels)	Capital costs of development ($ million)[c]	Costs of exploration plus capital cost of development ($ million)[c]
	Thousand barrels/ day	Million barrels			
	(1)	(2)	(3)	(4)	(5)
1956	–	–	–	–	8.6
1957	–	–	–	–	10.9
1958	–	–	–	3.5	14.2
1959	–	–	50	16.1	38.7
1960	–	–	1,500	36.3	70.4
1961	20	7.3	2,000	83.4	138.3
1962	185	67.5	3,000	70.4	120.4
1963	465	169.7	4,500	66.4	112.3
1964	860	313.9	7,000	107.0	216.6
1965	1,220	445.3	9,000	148.6	219.8
1966	1,505	549.3	10,000	114.2	202.4
1967	1,745	636.9	20,000	109.1	188.4
1968	2,600	949.0	29,200	–	–
1969	3,500[d]	1,280.0[d]	30,000	–	–
Total (through 1967)				(755.0)	(1,341.0)

[a] Barrels per day figures through 1968 from British Petroleum Company, *Statistical Review of the World Oil Industry, 1968.*

[b] "Worldwide Oil," *Oil and Gas Journal,* final issues of each year, 1956-68.

[c] See discussion in text. Figures either taken from estimates made by M. A. Adelman, or based upon his approach.

[d] Estimated figure, as forecast in *Oil and Gas Journal*, December 12, 1968, p. 27; this daily rate was not reached until late 1969/early 1970.

The statistics of particular interest on Libyan developments are assembled in Table 5A.2. Although the figures on reserves and production are taken directly from published sources, the cost information consists wholly of estimates. It is well to begin by indicating how these estimates were derived.

Relevant cost series are not readily available for oil industry expenditures at the national level in Libya. What we have used instead are statistics published by the Chase Manhattan Bank on capital and exploration expenditures for the continent of Africa as a whole.[8] The series identified as "Capital Costs of Development" in Table 5A.2 refers to total production capital expenditures and consists essentially of drilling and other field development outlays. Another published series refers to exploration expenditures, and these consist essentially of geological and geophysical survey costs, drilling of exploratory wells, and lease rentals. In Table 5A.2, the former series appear as column (4), while the sum of the two expenditure streams is shown in column (5). Since the statistics that were used cover all of Africa, it was necessary to approximate a portion of the African total to be assigned to Libya. The available approaches involve an allocation of costs among the countries of Africa in relation to their share in total continental geological-geophysical exploratory crew months; in wildcat drilling by footage, wells, or average depth; in development drilling by the same categories; in production; in estimated reserves; and on any other suitable bases for which data are available both for Libya and for total Africa. Although whatever approach is adopted can only yield approximations at best, Libya's dominant place in the continental total should tend to result in tolerably good estimates. The method followed in obtaining the costs shown in Table 5A.2 is that devised by M. A. Adelman and explained in his article, "Oil Production Costs in Four Areas."[9]

Adelman's main purpose was to obtain per-barrel estimates of development and production costs for crude oil from several areas of the world in order to determine the minimum competitive supply prices for oil from existing producing regions under conditions of active price competition. For the purposes of this appendix, finding costs are also of concern because one of the estimates that is wanted is an *ex-ante* estimate of the total costs for all phases of the process of bringing production in a brand-new region into existence; the reason for wanting the latter figure is so as to

[8] Published in its annual pamphlet, *Capital Investments by the World Petroleum Industry* (New York: Chase Manhattan Bank, annual).

[9] *Proceedings of the Council of Economics*, American Institute of Mining, Metallurgical, and Petroleum Engineers (New York: AIME, 1966). In this article, Adelman calculated Libyan oil development capital costs for 1958–64 and total exploration and development costs for 1963. An effort has been made to use Adelman's method to compute oil development capital costs for 1965–67 and total exploration and development costs in Libya for 1956–62 and 1964–67. Although Adelman does not give enough information in his article to allow the precise computation of the data of the other years by means of all the parameters he uses (cost-depth relationships, etc.), it is believed that the results are sufficiently close to permit coupling these calculated costs with those of Adelman.

be able to apply it to wholly new provinces elsewhere in the world in which exploration may just be getting started.

Turning to the gross dimensions of the Libyan experience, column (5) of Table 5A.2 indicates estimated total exploration and development outlays over the period through 1967 of $1.341 billion. As a result of this expenditure, an average production rate of somewhat over 2½ million barrels/day was achieved during the thirteenth year following the initiation of the first sizeable investments in exploration and the eleventh year following the first sizeable investments in development. This total sum can be divided roughly into an exploration component and a development component, the latter being given in column (4). Total exploration outlays were $586 million (i.e., geological and geophysical costs, land costs, and drilling of exploratory wells) while development capital costs were $755 million.

It is important also to view the statistics shown in Table 5A.2 from another angle. Libyan oil production in 1969 is estimated to reach 3½ million barrels/day, up nearly 1 million barrels/day from 1968. According to column (4), average development investment during the most recent three-year period for which statistics are available was about $124 million. Thus, in rough terms, and *additional* expenditure of about $125 million on development in 1968 would at that stage yield another 1 million barrels/day of productive capacity.

To return to the total stream of expenditures. Assuming the expenditure of about $125 million on development in 1968, but no expenditures on exploration, a total thirteen-year outlay of about $1.5 billion would yield a productive capacity of more or less 3½ million barrels/day. Put another way, an expenditure of roughly $1.5 billion yielded about 3½ million barrels/day of productive capacity at the wellhead during the fourteenth year after the beginning of sizeable exploration and the twelfth year after the beginning of sizeable development expenditure. This implies an investment in these two categories of expenditures of about $430 for each daily barrel of productive capacity. (This is without a correction for the capitalization of past costs to their present value, and without correction for change in the price level.) However, as indicated above, to add to productive capacity, without additional exploration, would appear to require in the neighborhood of $125 per daily barrel of productive capacity.

These are low costs, comparatively (although not as low as in some locations in the Middle East), because Libya represents a prolific producing province, a fact which must be borne in mind in connection with any effort to apply this experience to assessing prospects in new regions elsewhere. Also, it must be recognized that land acquisition costs were on the order of $200 million, and that this is a category of cost that could vary widely from place to place. And only those costs are covered that carry the operation through lease and field facilities; other major costs such as pipeline and terminal facilities are omitted.

As for the period of time required to achieve the productive level reached, it is likely that a faster effort that could trim many years from that experienced in Libya is well within the realm of possibility. By how many years time requirements could be cut back is unanswerable on the basis of publicly available information, but it is relevant that Occidental Oil, over a period of less than two years, conducted exploration, discovered five fields with total recoverable reserves estimated at 2.69 billion barrels (60 percent recovery factor) and built a 135-mile, 40-inch pipeline to a new Mediterranean Sea terminal, capable of a throughput of 1 million barrels/day.[10]

On the other hand, the speed of Libyan development was aided by factors that have not been mentioned in this appendix. One factor was the market for particular crudes, Libyan crudes being of types much in demand. Another was the presence of aggressive newcomers among the oil companies and the nature of the concession contracts written. More recently the geographical advantage of nearness to the European market has been enhanced by the closure of the Suez Canal since June 1967.

---

[10] *World Petroleum*, April 1968, pp. 40–43. The two major fields are Idris A and Idris D.

Although the factors accounting for this outstanding performance may include an element of plain good fortune, credit must be assigned to advance planning of logistics, and to a determined effort to meet the problems that arose.

CHAPTER 6

Coping With Possible Supply Interruptions

THE quantitative framework dealing with the supply of liquid fuels that provides the springboard for the analysis contained in this chapter is based upon the unfolding of future supply trends, as projected in mid-1969, consistent with those governmental policies and programs which appeared to set the guidelines for the future, and consistent, in the main, with the trends in geographic origins of crude oil supplies that had been manifested up to that time (thus, excluding the North American Arctic). The next chapter will take up the prospects and implications of further geographic diversification in crude oil sources through developments in the North American Arctic that, although quite speculative in 1969, are obviously worthy of serious consideration.

Future Supply Prospects, Assuming the Continuation of Existing Trends

It is well to begin by sorting out and elucidating some of the findings of earlier chapters concerning factors that are of central importance to the analysis of the problem of supply interruptions.

Major Consumer Countries: Dependence on Oil Supplies from Middle East and North Africa

Attention has been centered on the United States, Western Europe, and Japan as the major consumer areas. (The Communist countries, taken together, now account for about 15 percent of total world consumption. However, they are essentially self-sufficient. Out of a total production of about 7 million barrels/day in 1968, they were able to export about 1 million barrels/day to the rest of the world. According to recent Western appraisals, their net export position is likely to decline. In this study it has been assumed that they will remain essentially self-sufficient, neither drawing upon nor competing with Middle Eastern and North African supplies to an extent sufficient to invalidate the broad picture here described.)

United States

The United States today accounts for about one-third of the world's total consumption of liquid fuels, and is projected to account for about 27 percent in 1975 and less than 25 percent in 1980. Throughout this period it will rank first among the oil-consuming countries of the world. Currently the importation of oil is subject to severe governmental restriction, with the net result that about 20 percent of total liquid fuels consumption consists of imports. Only a small portion of these imports (about 18 percent, or about 3½ percent of total oil consumption) currently originates in the Middle East and North Africa.

In Chapter 2 the question was posed whether the assumption employed in this study's projections—that U. S. imports would continue to hold at about 20 percent of total oil consumption over the next decade or so—can be considered realistic. To put the matter another way: Can the government of the United States continue to achieve the main quantitative results of the present-day import control program, particularly as they affect supplies from the Middle East and North Africa, at a cost penalty comparable to that now being incurred, if it should choose to do so?

The provisional conclusion reached was that the latest comprehensive analysis of U. S. oil prospects made by the U. S. Department of the Interior in 1968, prior to the North Alaskan discoveries, could not support an unqualified "yes" in reply to this crucial question. However, in light of the prospects for Northern Alaska (dealt with in the next chapter and summarized in Chapter 2), it was also concluded that the ability of the United States to continue to rely upon domestic reserves to meet the same (or even a higher) relative level of oil needs as in the past, and likely at a lower cost penalty, appeared reasonably well assured for the next decade.

Furthermore, on the basis of the analysis contained in Chapter 4, dealing with the outlook for liquid fuels from uncoventional sources in North America—i.e., more than one technology on the borderline of economic feasibility—it appeared safe to conclude that if a judgment were to be reached at an early date that crude oil from domestic reserves was going to be inadequate to the task, accelerated research and development on the unconventional sources of liquid fuels could result in commercial facilities to provide supplies, at costs comparable to domestic crude oil produced in the contiguous forty-eight states, in sufficient amounts to sustain a policy that would result in no greater comparative reliance on imports than in the late 1960's.

It is important that our conclusions on this subject be clearly understood. They are *not* intended as a recommendation that the import control program be continued unchanged. They are *not* meant as a prediction that synthetic sources will be introduced as a significant source of liquids if crude oil reserves in the contiguous forty-eight states and Alaska prove inadequate to the task of sustaining this country's present level of oil self-sufficiency. What our conclusions do say is that the outlook for domestic crude oil sources, taken together with the potentialities of liquids from unconventional sources, appears to point to the ability of the United States to remain as self-sufficient in liquid fuels supplies over the next decade or so as it is today, at a cost penalty comparable to that of the present difference between the price of imported oil and the cost of domestic supplies. Moreover, this might very well have been the case even without the North Alaskan discoveries, although at a higher cost. This option—to retain a high degree of self-sufficiency, particularly vis-à-vis the Middle East and North Africa—is an important one for the United States to have at its disposal; whether it will need (or choose) to exercise the option is not here considered.

Western Europe and Japan

Neither Western Europe nor Japan has available the option just described for the United States; there is no way, assuming the continued unfolding of present-day trends, for either to avoid an overwhelming reliance on imported oil supplies, originating mainly in North Africa and/or the Middle East.

By way of review from the preceding chapters, the present and projected situation may be briefly summarized. In 1968, Western Europe consumed roughly one-fourth of the world's total oil supply—a proportion which is not expected to shift markedly in the years ahead. Japan's share of world consumption—about 7 percent in 1968—is, however, projected as rising steadily, reaching over 10 percent by 1980. Translated into absolute terms, the 1975 amounts are 670 million tons for Western Europe (roughly 13½ million barrels/day and around 220 million tons for Japan (4.4 million barrels/day); and the quantities in 1980 are 835 million tons (16¾ million barrels/day) and more than 345 million tons (7 million barrels/day), respectively. The foregoing refer to energy uses only; including oil used for petrochemicals and in other nonenergy uses, the 1975 estimates are about 730 million tons for Western Europe (14½ million barrels/day) and about 240 million tons for Japan (4.8 million barrels/day); and the 1980 quantities are 915 million tons (18 million barrels/day) and 385 million tons (7.7 million barrels/day), respectively.

Almost all of the liquid fuels consumed in the two regions will consist of imports. The amounts originating in the Middle East and North Africa are estimated to dominate the import totals. In 1980, Western Europe's imports from the Middle East are projected at more than 400 million tons (equivalent to more than 8 million barrels/day) and from North Africa at about 350 million tons (equivalent to nearly 7 million barrels/day), for a grand total of 15 million barrels/day of oil imports from these two exporting areas. Japan in 1980 is estimated to import over 250 million tons from the Middle East, equivalent to about 5 million barrels/day. (By way of comparison, imports from the Middle East and North Africa in 1968, the year following the Arab-Israeli conflict, were at the rate of somewhat under 8½ million barrels/day for Europe and 2.6 million barrels/day in Japan.)

Efforts to maximize production from indigenous energy sources would not, as has been seen in Chapter 3, substantially alter the fact that imported oil will constitute the mainstay of European and Japanese energy consumption in 1980. In that year, Western Europe's reliance on oil imports, as a percentage of *total* energy supply, is estimated to range between 55 and 60 percent. In Japan's case, reliance on oil imports as a percentage of *total* energy supply is estimated to range between 75 and 80 percent.

The problem of supply interruptions in the future is, therefore, brought into sharp focus by examining the European and Japanese situations. We turn next to such an examination.

Measures for Dealing with Supply Interruptions, Assuming the Continuation of Existing Supply Trends

The present discussion, although concerned with measures for dealing with supply interruptions, will not attempt to specify the actual sequence of events that might give rise to such interruptions nor will it try to assess questions regarding the duration and intensity of one or more future supply crises. The underlying assumption is that supply crises may again occur, since they have occurred in the past, and that they may occur in what are peacetime circumstances in the consuming countries. As for the supplying regions, the flow of oil might be discontinued because of such eventualities as hostilities among countries in the region; civil strife within countries involving actual hostilities or acts of sabotage by dissident forces; and political factors involving the use of oil as a bargaining weapon in relationships with the world's major consuming countries.

The fact that peacetime conditions are assumed for the consuming countries needs to be underscored because this implies the ability, but for the possible interruption of energy supplies, to continue the normal operations of the economy. It also implies the absence of any patriotic motivation within consuming countries for belt tightening in the consumption of liquid fuels, or of those goods and services whose production is dependent upon the consumption of liquid fuels. Consequently, rationing, as a device for reducing liquid fuels consumption, is not considered among the measures examined below, although it would need to be kept in reserve to deal with long-lasting supply interruptions of major proportions. Instead, all of the measures here considered are concerned with meeting, so far as possible, normal demands in the face of possible shortfalls, resulting from the complete, though temporary, cutting off to supplies from all the Middle Eastern and North African countries, or all the Arab countries.

Imports into Western Europe from the Middle East and North Africa in 1980 have been estimated at 15 million barrels/day, and supplies into Japan at around 5 million barrels/day. If there were to be a complete breakdown in the flow of oil from these regions, it would be necessary, for as long as the interruption continued, and unless consumption were to be curtailed, to supply approximately 20 million barrels/day from other sources. We look at this extreme case, not because it is the most likely one, but because it represents the maximum danger against which protection would need to be provided.

What means might be available within the Western world for meeting such a supply crisis? Two of the methods for meeting major supply interruptions that could be used, and have to a degree been used in the past are (1) the activation of existing idle crude oil supply capacity, including that portion located in the Western Hemisphere; and (2) the utilization of emergency oil stocks that had been built up within Western Europe and Japan. What possibilities do these offer?

1. *Idle Crude Oil Capacity in the Western Hemisphere*

Idle productive capacity in significant amounts—in reality, excess capacity to produce at the prevailing price—exists in the United States, Canada, and Venezuela. The largest amount has been present in the United States, and its existence has been an important factor in meeting European supply crises in the past. Consequently, in Chapter 2 attention was given to the reasons for the presence of excess productive capacity in the United States, and the prospects for its continued existence, and at what level, in the future. The conclusions reached in that chapter will be briefly restated, followed by a treatment of Canada and Venezuela.

United States

At the beginning of 1968, idle crude oil productive capacity in the United States was estimated at 4 million barrels/day. This quantity, according to the American Petroleum Institute (the agency responsible for the estimate) was the amount that would be available, at the point where petroleum enters the transportation system, after ninety days of effort to increase production with existing wells, well equipment, and surface facilities.[1] Thus, as pointed out in Chapter 2, the measure refers to oil *producibility*, not to its *deliverability*; the latter, of course, is the relevant measure so far as meeting a supply crisis is concerned.

However, the experience of the June 1967 war and its aftermath raised serious doubt as to the accuracy of the measure even in terms of producibility. The maximum production achieved in the United States by August 1967 exceeded the May level by about 1 million barrels/day, and at that rate other limiting factors to a further increase in production began to be encountered.[2] In the appendix to Chapter 2 an effort was made to estimate *deliverable* capacity over a longer period of time, assuming strong motivation to expand output on the part of industry and the regulatory authorities, as well as investment in additional field and transport facilities. These estimated amounts as of the situation at year-end 1967 were 1.8 million barrels/day after six months; 2.4 million barrels/day after nine months; and 3.5 million barrels/day at the end of one year.

Even if the foregoing estimates are accepted for the present and near-term future (and serious doubt began to be raised on this score in early 1969), there remains the question of prospects for the more distant future. The Interior Department, in its 1968 study, estimated that excess productive capacity in 1980, employing the API definition of 90-day producibility, might be about 3 million barrels/day. Chapter 2 concluded that the Interior Department estimate appeared to be unrealistically high in the light of existing trends in the U. S. oil industry in the contiguous forty-eight states, but no alternative estimate was presented. It was also concluded that it was too early to assess the impact of Northern Alaskan developments on idle capacity in the United States in the future.

[1] Chap. 2, p. 35.

[2] Chap. 2, p. 33.

Canada

Alberta is the only Canadian province with significant present-day oil productive capacity. Production in 1967 was at a rate of 631,000 barrels/day. The subtraction of this level of production from the several measures of productive capacity employed by the Alberta Oil and Gas Conservation Board yields the figures on idle capacity shown in Table 6.1.

The excess capacity figure which comes closest to being relevant to the estimation of "instant" deliverable capacity that could, either directly or indirectly, contribute to meeting a supply crisis elsewhere in the world is the last one—230,000 barrels/day. Actually, even this is somewhat on the high side because the total trunk line capacity for shipping Albertan oil *outside* the province amounted to 800,000 barrels/day, as compared to the 861,000 barrels/day shown in Table 6.1. Adjustment for this factor reduces the relevant 1967 excess productive capacity figure to about 170,000 barrels/day.

Although the "instant" excess capacity is comparatively slight, there is room for increases in the portion of excess capacity that could be rendered producible and deliverable outside the province. Using an approach comparable to that employed in the appendix to Chapter 2, it could be reasoned that almost 900,000 barrels/day (current developed wellhead capacity minus current production) could be made available over a period of time extending up to a year, at the cost of additional investment, particularly

Table 6.1

Alberta: Estimated Excess Crude Oil Productive Capacity, 1967

(Thousand Barrels/Day)

	Productive capacity (12/31/67)	Production in 1967	Excess capacity 1967
Maximum efficient reservoir capacity	2,102	631	1,471
Developed wellhead capacity	1,505	631	874
Capacity adjusted for field and processing facility limitations	954	631	323
Capacity adjusted for main pipeline facility limitations	861	631	230

Source: Alberta Oil and Gas Conservation Board, *Reserves of Crude Oil, Gas, Natural Gas Liquids and Sulphur, Province of Alberta, December 31, 1967*, pp. 168-172.

in transport. However, more nearly within the realm of the reasonable would be an increase in deliverable capacity to about 650,000 barrels/day through the premature expansion of the Interprovincial Pipe Line.[3]

The foregoing may be taken to apply to the present and the near future. As for the more distant future, the only publicly available and apparently well-informed estimate of excess productive capacity in Alberta is that made by the Syncrude Company in support of a 1968 application to produce oil from Albertan tar sands. Calculations based upon the Syncrude estimates are presented in Table 6.2.

Table 6.2

Projections of Canadian Productive Capacity, Production, and Excess Capacity, for Crude Oil, 1975, and 1980

(Thousand Barrels/Day)

Year	Productive capacity (M.E.R.*)	Production	Excess capacity	Estimated deliverable excess capacity (90-day) Low	Medium	High
	(1)	(2)	(3)	(3)	(5)	(6)
1975	2,700	1,442	1,258	219	500–650	750
1980	2,900	1,620	1,280	235	a	1,000

*M.E.R. = Maximum Efficient Rate of Production. This is the rate of production for a particular reservoir which if exceeded would lead to avoidable underground waste through loss of ultimate oil recovery (see *Petroleum Conservation*, edited by Stuart E. Buckley, American Institute of Mining and Metallurgical and Petroleum Engineers [New York, 1961], pp. 151–152).

[a] No estimate made, see text.

Source: Col. (1) from transcript of Proceedings before the Alberta Oil and Gas Conservation Board at Calgary, August 8, 1968, vol. 3, pp. 260–263.

Col. (2) derived from supplementary schedules introduced by Syncrude to support material presented on page VI-4 of its proposal to produce 800,000 barrels/day of tar sands crude at the Proceedings before the Alberta Oil and Gas Conservation Board, August 8, 1968.

Cols. (3) to (6) according to calculations explained in text.

Syncrude figures indicate excess productive capacity in the neighborhood of 1,200-1,300 thousand barrels/day between 1970 and 1980, brought about by practically parallel increases in production and productive capacity. Three estimates of deliverable spare capacity were made. The low estimate multiplies total productive capacity by the 1967 ratio of deliverable excess capacity outside Alberta (170,000 barrels/day) to total productive capacity within Alberta. The medium estimate is based upon the assumption that added capacity in the Interprovincial Pipe Line will be installed so as to exceed future demand for a period of years before the full demand develops, but, as noted, this excess pipeline capacity will cease being "excess" in a matter of five years or so, hence there is no medium estimate for 1980. The high estimate assumes that the Alberta Government's new policy on tar sands development could result in an increasing level of idle capacity in crude oil production. (While the earlier policy limited tar sands oil to 5 percent of the Alberta market, the new policy contemplates an expansion of tar sands oil production by 50 percent of any "new markets" as defined by the Board. Hence, if the Board recognizes that a "new market" exists for Alberta crude oil on the order of 100,000 barrels/day, the new policy implies that it will authorize an expansion of 50,000 barrels/day in tar sands oil, to be accompanied by an equal increase in conventional crude oil production. (See Chapter 4, p. 61.)

An increasing market for oil from tar sands opens the possibility that conventional production will be prorated more strictly in the future if market demand does not live up to estimates, as it is the Board's policy not to prorate production from tar sands because its costs are far more sensitive to less-than-full capacity operation rates. Thus, if tar sands production increases, the average incidence of deliverable excess capacity in the conventional industry may very well increase. Consequently, the "high" estimate of deliverable excess capacity in column (6) of Table 6.2 is estimated to be about 3⅓ times this estimate in 1975, and about four times as large in 1980. Obviously, all of this is extremely hypothetical, and it is safe to say that these estimates are quite likely to be excessive unless large synthetics capacities (tar sands and others) are permitted to develop between now and 1980, and demand for liquid fuels proves to be less buoyant than predicted. If this were to happen, however, there is a strong likelihood that the

[3] The Interprovincial Pipe Line is the major line now carrying Albertan crude outside the province. Its capacity, as of the end of 1967, was 485,000 barrels per day. The longest crude line in existence, it extends 1,979 miles from Redwater, Alberta, to Port Credit, Ontario, near Toronto. About half of the line crosses United States territory between the North Dakota-Manitoba border and Sarnia, Ontario, going through Minnesota, Wisconsin, and Michigan. The United States stretch, known as the Lakehead line, is a 30-inch pipeline with a capacity of 490,000 barrels per day at the end of 1966. During 1967 the line added 412 miles of 34-inch and 57 miles of 20-inch loops, and 101,500 horsepower of pumping capacity, at a cost of over $70 million. (*Oil and Gas Journal,* January 16, 1967, p. 64.) During 1968 Interprovincial was in the process of laying a 34-inch crude oil line from Superior, Wisconsin, to Chicago, a distance of 430 miles, at a cost of about $75 million. Some of the oil reaching Chicago is to be shipped via the Tecumseh line to the Detroit-Toledo area. The Tecumseh line is being used as a temporary expedient, however, since the Superior-Chicago line will eventually be extended to Sarnia, Ontario, making the system a double line, in effect, between Superior and Sarnia, with one line passing around either end of Lake Michigan. (*Oil and Gas Journal,* September 4, 1967, p. 87.) The current expansion program regarding the northern part of the line calls for looping 68 percent of this 1,097-mile segment with a 34-inch pipe. Looping the remaining 32 percent would result in a capacity increase for the entire line of about 500,000 barrels per day. This might cost something like $140,000 per mile for about 352 miles of line, a total of approximately $50 million. One possible incentive to undertake this premature investment would be an interest-free loan, repayable in installments on those dates when the pipeline company would voluntarily have constructed the additional mileage. This would be a relatively inexpensive way of obtaining additional capacity, although only for about five years at most, by which time the capacity would have been required to meet the expected growth in regular demand.

policy regarding synthetics would have become more restrictive in the meantime, particularly in the light of Arctic potentials, as presently perceived.

As for the Canadian Arctic (see Chapter 7), it is, even more so than in the case of Northern Alaska, too early to estimate what impact developments there will have on idle capacity.

Venezuela

The starting point for evaluating excess productive capacity in Venezuela are the statistics for 1967, shown in Table 6.3. In April and May of 1967, the months immediately preceding the Arab-Israeli conflict, official statistics show idle capacity to have been in the neighborhood of 800,000 to 900,000 barrels/day. By July and August, the months following the war, excess capacity had declined to between 430,000 and 460,000 barrels/day. The additional production of oil from Venezuela, brought about by the supply crisis, was somewhat less than 450,000 barrels/day, or only about one-half of the excess productive capacity that the statistics showed prior to the crisis.

In analyzing the response of U. S. production to the 1967 supply crisis it was found that limitations on both producibility and deliverability resulted in an increase far below that indicated by the API measure of excess productive capacity. Detailed statistics furnished by the Venezuelan Ministry of Mines and Hydrocarbons dealing with pipeline, storage, and pumping capacity, indicate that no such limitations on either producibility or deliverability existed in Venezuela.

Under the circumstances the reason why Venezuelan output did not expand by a larger amount during the 1967 supply crisis may raise some doubt as to the validity of the official estimates of total productive capacity. On the other hand, it could be that marketing opportunities for Venezuelan crude during the crisis were not such as to stimulate the production of all the oil which could have been brought to the surface and delivered for shipment overseas. Much of the country's excess capacity seems to be in heavy crude oil, which would be less attractive in foreign trade than the lighter crudes.

Table 6.3

Venezuela: Productive Capacity and Production, by Months, 1967

(Thousand Barrels/Day)

	Production	Productive capacity	Excess capacity
January	3,549	4,153	604
February	3,553	4,188	635
March	3,395	4,156	761
April	3,266	4,145	879
May	3,328	4,139	811
June	3,593	4,133	540
July	3,750	4,181	431
August	3,717	4,181	464
September	3,638	4,192	554
October	3,558	4,210	652
November	3,565	4,236	671
December	3,596	4,238	642

Source: Venezuelan Ministry of Mines and Hydrocarbons. Data furnished by Dr. Mariano Gurfinkel, a member of the Ministry Staff.

It is, therefore, impossible to say whether the amount of excess capacity officially said to exist as of the end of 1967 of about 650,000 barrels/day is a dependable estimate. As for excess capacity in the more distant future, there are no publicly available estimates upon which to base a forecast. Clearly, much depends upon the investment climate in Venezuela compared to other areas in which the international oil companies conduct operations. A new factor entering the picture, is, of course, the increased attention expected to be directed to the North American Arctic region. This is an unpromising development so far as prospects for investment in Venezuela are concerned.

A significant piece of information is provided by the behavior of productive capacity in Venezuela in recent years. In 1956, total capacity stood at 2.7 million barrels/day. By 1965 it had increased to about 4.0 million barrels/day, and it has held constant at this level through 1968.[4] On the assumption that the static situation in the recent past is indicative of what to expect in the future, the 1967 measure of apparently deliverable excess capacity—about 650,000 barrels/day—is the best estimate that can now be made for the next decade.

Western Hemisphere: Total Idle Capacity

The estimates of idle productive capacity in the United States, Canada, and Venezuela presented in the preceding sections are summarized in Table 6.4. The United States provides the main component of the total for all of the years covered. In 1967 about 80 percent of the grand total of about 5 million barrels/day was accounted for by the United States. However, as the earlier discussion has shown, only 1 million barrels/day was effectively mobilized over a 90-day period in 1967. Given a longer period, a more serious situation, and additional investments, our estimates indicate a capability on the part of the United States, as of year-end 1967, substantially to better the 90-day performance. (See appendix to Chapter 2.)

The major concern, though, is with the future. Employing what we judge to be unrealistically high figures for both the United States and Canada, the excess productive capacity estimated to exist in 1975 and 1980 is in the neighborhood of 4½ million to 5 million barrels/day. Even so, this need only be compared to European and Japanese estimated imports from the Middle East and North Africa of around 16 million barrels/day in 1975, and 20 million barrels/day in 1980, to see that if the bulk of supplies from the Middle East and North Africa were to be shut off, idle capacity in the Western Hemisphere, even if it could be effectively and speedily mobilized, could provide, at best, only a small fraction of imports in those years.

[4] Data provided by the Venezuelan Ministry of Mines and Hydrocarbons.

Table 6.4

Idle Productive Crude Oil Capacity, Western Hemisphere, Excluding the North American Arctic

(Thousand Barrels/Day)

	1967–70	1975	1980
United States	4,000 (1,000)[a]	3,500[b]	3,000[c]
Canada	323[d]	750[e]	1,000[e]
Venezuela	650	650	650
Total	(about) 5,000	4,900	4,650

[a] Estimated total 90-day deliverable excess capacity, based upon 1967 experience.

[b] Interpolated figure which we regard as unrealistically high (see n.c.).

[c] U. S. Department of the Interior estimate made in their 1968 study, which we regard as unrealistically high.

[d] Developed capacity adjusted for facility limitations.

[e] High end of range.

Source: Preceding sections of text.

2. *Stockpiling in Western Europe and Japan*

Idle crude oil capacity in the Western Hemisphere, of some importance in meeting oil supply crises in the past, will be of greatly dwindling comparative, if not absolute, importance in the future. We turn next to another method that has been employed in meeting supply crises in the past—the stockpiling of crude oil within the major importing countries. How costly would such stockpiling efforts be in order to meet future major supply crises should they occur?

No effort will be made here to assess the duration and magnitude of a supply crisis likely to face Western Europe or Japan at some future date. This would involve an analysis of the politics of oil within the Arab world—a subject which is outside the scope of this study, though obviously of great importance to importing countries in judging the possible situations to be guarded against. Instead, using the import estimates that have been projected for future years, a schedule of costs will be derived consistent with the maintenance of stocks adequate to meet supply interruptions lasting for varying periods of time.

The most recent policy adopted by the six-nation European Economic Community calls for the maintenance, as of January 1971, of oil stocks equal to 65 days of consumption.[5] In Japan, the total oil stockpile as of the end of fiscal 1965 was said to be equivalent to 44 days of consumption.[6] This level was judged to be too low, and at the end of 1967 it was reported that Japan had decided to increase stock levels to 60 days.[7] The stock level maintained in Great Britain is reported to be higher than those planned for the Common Market countries and for Japan. Even before the June 1967 war, Britain was reported to possess 120 days worth of stocks, "not by regulation but by 'gentlemen's agreement' with oil refiners and distributors."[8]

The officially designated stock targets, as publicly (although not always officially) reported, thus range from 60 to 120 days of oil consumption. Stocks that would last for longer periods have, however, been suggested. The French Minister of Industry in November 1966 estimated six months as the required stock level, and South Africa, which is in a unique international political position, has been reported as having an 18-month supply in stock.[9] The authors of the PEP report concluded that estimating the number of months of stock that would provide an adequate safeguard "could only be a matter of guesswork."[10]

In the cost estimates presented here, a period ranging from sixty days to one year will be used, in order to indicate how costly stockpiling would be under varying assumptions concerning appropriate stock targets. A period of up to one year, it should be noted, may be excessively long if one takes into account the sensitivity of oil exporting countries to the loss of oil revenues. Moreover, concerted action by the Middle Eastern and North African Arab exporting countries can only be as strong as the weakest link among the four largest exporters. In addition, from the outset of any concerted effort, it may be surmised that Iran (a non-Arab country) would continue to try to export, and would be prepared to increase its share of the export market to the greatest extent possible. It must be recognized, though, that exports from Iran might be subject to harassment by hostile forces. However, as indicated, no effort is here being made to estimate how long a supply breakdown is likely to endure, and how thoroughgoing it is likely to be.

Let us first examine the estimated costs of building up a stockpile of oil. The basis for developing such estimates are shown in Table 6.5, which presents figures for Western Europe in 1965, 1975, and 1980 on the basis of the oil import levels applicable in each of these years (the figures for the two latter years are those projected in this study). In Table 6.5, cost estimates are developed for what are referred to as "units equal to 60 days of consumption," in other words, stocks adequate to cover estimated daily con-

[5] *Petroleum Press Service*, January 1969, p. 24.

[6] *Report on Energy Policy* by the Advisory Committee for Energy, a consultative body to Japan's Ministry of International Trade and Industry (Tokyo, February 20, 1967; translated and published by the Institute of Energy Economics), p. 21.

[7] *Petroleum Press Service*, December 1967, p. 468.

[8] Political and Economic Planning (PEP), *A Fuel Policy for Britain* (London: Political and Economic Planning, 1966), p. 183. Officially the British are reticent on the subject. The Ministry of Power's White Paper, *Fuel Policy*, states: "The companies supplying this country have agreed with the Government to hold sizeable stocks in addition to their normal commercial requirements. The stock targets, which are confidential, are designed to enable the United Kingdom to resist the denial of a large part of her supplies for a considerable period." (Great Britain, Ministry of Power, *Fuel Policy*, Presented to Parliament by the Ministry of Power by Command of Her Majesty, November 1967, Command 3848 [London: Her Majesty's Stationery Office, 1967], p. 25.)

[9] M. A. Adelman, "Security of Eastern Hemisphere Fuel Supply" (Working Paper No. 6, Department of Economics, Massachusetts Institute of Technology, December 1967), pp. 6-7.

[10] PEP, *A Fuel Policy for Britain*, *op. cit.*, p. 183.

Table 6.5

Estimated Costs of Stockpiling Oil in Western Europe: at 1965, 1975, and 1980 Levels of Consumption

(In Units Equal to 60 Days of Consumption)*

	1965		1975		1980	
	Total imports	Imports from Middle East and North Africa	Total imports	Imports from Middle East and North Africa	Total imports	Imports from Middle East and North Africa
Imports (annual averages in million barrels/day)[a]	7.4	5.7	14.4	12.3	18.1	15.1
Imports for 60 days (in million barrels)	444	342	864	738	1,086	906
Costs (in $ million, approximate)						
Stocks of oil[b]	$730	$565	$1,425	$1,220	$1,800	$1,500
Tanks for storage[c]	$550–$890	$425–$680	$1,100–$1,730	$900–$1,500	$1,350–$2,200	$1,150–$1,825
Total	$1,280–$1,620	$990–$1,245	$2,525–$3,155	$2,120–$2,720	$3,150–$4,000	$2,650–$3,325

* See text for explanation of choice of this unit.

[a] From Chap. 1, Table 1.4.

[b] At current prices, as of late 1967, of $1.65 per barrel (Adelman, *op. cit.* [text n. 9], p. 6). Adelman puts it in these terms: "Oil can be purchased f.o.b. the Persian Gulf today by big credit-worthy buyers for less than $1.25 . . . and shipped, emphasizing the cheaper summer season, at 43 cents per barrel to Rotterdam and 37 cents to Marseille or Northern Italy. (Under the usual method of rate quotation, this would be INTAscale less 65% through Suez, less 67.5% around the Cape.) An average delivered cost to Europe would then be about $1.65, North and South taken together."

[c] Assuming steel tanks such as are conventionally used. Low end of range, $1.25 per barrel, from Adelman, *op. cit.* (text n. 9), p. 6; high end of range, $2 per barrel, from Alex Hunter, "Oil Supply in Australia's Defense Strategy," *Canberra Papers on Strategy and Defense* #1 (Canberra: Australian National University Press, 1968), p. 25. The latter figure assumes what are said to be optimum-sized tanks of 500,000 bbl. capacity, each costing about $1 million to fabricate and erect. Hunter's estimate is in rough agreement with figures provided privately by a major oil company for use in this study.

sumption for 60 days. The "60-day unit" was adopted because it corresponds with the shortest stock period officially suggested and because it has the additional advantage of lending itself to easy multiplication to approximate other periods of time that will also be considered. It should be noted that the costs covered include both the oil and the necessary storage facilities. The costs of land and site preparation are not included.[11]

Using the "60-day unit" estimates derived in Table 6.5, the following schedule of costs of building stocks in Western Europe has been derived (applying not to total imports, but only to those amounts originating in the Middle East and North Africa):

Estimated costs of building stockpiles to replace Middle Eastern and North African supplies for a period of:	1965	1975	1980
	(in billions of dollars)[a]		
60 days	$1.1	$2.4	$3.0
120 days	2.2	4.8	6.0
180 days	3.3	7.2	9.0
360 days	6.6	14.4	18.0

[a] Tank storage costs roughly at midpoint of range presented in Table 6.5 (i.e., at somewhat more than $1.60 per barrel).

Let us examine the 1975 estimates. At 60 days, which approximates the ECE and Japanese targets, the investment in stocks and storage facilities in Western Europe would stand at about $2½ billion; at 120 days, the length of the period reported to have been covered by British stocks in 1966, at about $5 billion; and at 180 days, a target which has been suggested by others, at more than $7 billion. If a full year of stocks were judged to be necessary, the investment in stocks and storage facilities by the end of 1975 would have reached a grand total of about $14½ billion.

These are, of course, vast sums of money. By way of comparison it may be noted that the stockpile of strategic and critical materials in the United States at its peak level in 1964-65 was valued at about $8 billion.[12] The latter figure

[11] Information privately provided by a major oil company places land requirements at one square foot of tank base per ton of oil, with tanks taking up about a quarter of total space required—bund, pipelines, access roads, etc., accounting for the balance. No cost estimates were attempted because of wide variation in land values, site preparation required, etc.

[12] Officially designated the "Government Inventory of Strategic and Critical Materials," *Stockpile Report to the Congress*, Office of Emergency Preparedness (semiannual). Much political controversy has surrounded the U. S. stockpile. Recent efforts to draw down the amounts considered excess to national objectives have been controversial, as have the threats to sell stockpile materials to achieve other governmental objectives. Such vast amounts of stocks with a ready market are bound to engender controversy and concern in a variety of situations.

refers only to the materials contained in the stockpile (copper, lead, zinc, tin, etc.); it does not include the costs of storage facilities, which are not unique and costly as in the case of oil. Thus, six months of oil storage in Europe, at 1975 levels of supplies estimated to originate in the Middle East and North Africa, would practically equal the cost of the U.S. national stockpile of all strategic and critical materials at its peak. The cost of stocks and storage to cover longer periods, or to satisfy 1980 levels of consumption, can be read from the foregoing schedule.

The 1975 figures for Western Europe should be taken as illustrative only. (Anyone wishing to make similar calculations for Japan can easily do so beginning with the following import figures: 1965–1.6 million barrels/day [1.4 from the Middle East]; 1975–4.6 million barrels/day [3.4 from the Middle East]; 1980–7.3 million barrels/day [5.1 from the Middle East].) The estimates, although rough, indicate the very large investments that would be required to meet various target levels. Of course, these involved expenditures that would be built up year by year, beginning with the investments embodied in the stocks that already exist. An additional factor that must be kept in mind is that routine commercial stocks exist in any event. These are said to average about forty-five days of consumption.[13] However, it is quite unclear what portion of these stocks can be regarded as a security reserve in the sense that stocks could be drawn down if the regular flow of imports were to be interrupted; a certain amount (how much we have been unable to determine) would always be tied up in the sequence of operations involved in the process of converting crude oil (some of it in transit to the country of destination) into finished products for the market.[14]

Another way to look at the costs involved in maintaining stocks at levels that might be judged necessary to provide an adequate safeguard at some time in the future is to ask how much of a premium would be paid in a particular year over and above the price at which oil is purchased in order to buy the insurance provided by stocks. The following calculations provide a rough measure of this premium, as of 1975/76 in 1967 prices:

(1) Oil delivered to Europe costs, let us say, $1.65 per barrel (see Table 6.5).

(2) If 180 days is the emergency stock period chosen, the investment in the already existing inventory would be about $7 billion (see foregoing text table).

(3) Let us assume an interest rate of 8 percent and 20-year amortization on the investment in stocks and storage capacity, resulting in a capital recovery factor of somewhat over 10 percent per year.[15] Applied to the $7 billion investment, this yields an annual average capital cost of roughly $715 million, which when divided by 4½ billion barrels of then current imports from the Middle East and North Africa, works out to about 16 cents per barrel, as the annual premium (surcharge) to be added to the actual cost of oil from the Middle East and North Africa.

The illustrative and hypothetical character of the preceding calculations must be stressed. Other cost factors could be added and subtracted from those that have been presented: for example, add land costs and the maintenance costs of inventories; subtract some portion of the costs of normal commercial stocks and the cost reductions that may be realized if cheaper storage techniques prove practical.[16] (The question might also be raised whether oil purchased for stocks for national security should be priced to include or exclude company profits.) Nevertheless, it is plain that the stockpiling of oil is a costly operation, measured either in terms of (1) the investment that would be tied up in the oil and storage facilities; or (2) the per barrel costs per year that may be conceived of as the annual premium required for the maintenance of stocks at whatever is determined to be an adequate level. Despite this, Adelman has demonstrated that even including the premium involved in its stockpiling, oil is a less costly energy source than indigenous European coal in those uses in which the two are interchangeable.[17]

In the future there will, of course, be a question whether Middle Eastern and North African oil supplies, including the premium payment for security involved in their use, will be cheaper than oil from newly developed oil provinces, such as the North American Arctic.[18] Although it is premature to speculate on this matter, it will be important for the governments of the European countries and of Japan—as well as other importing nations that may engage in stockpiling—to keep this matter under close study as the prospects for the North American Arctic unfold, and as oil prices in the Middle East and North Africa change in response to political and commercial forces at work in that part of the world.[19]

---

[13] Adelman, *op. cit.* (n. 9), p. 6. Also, in private communication from a major oil company.

[14] Private communication from a major oil company.

[15] More precisely, a capital recovery factor (CRF) of 10.185 percent. The capital recovery factor defines what uniform annual stream of payments is needed to return any capital outlay, with interest.

[16] The *Oil and Gas Journal* of October 14, 1968 reported that the Japanese Institute of Ocean Development was about to test a pilot model of a subsea oil storage tank to free much needed onshore acreage for other uses. The pilot model was to be anchored on a concrete slab on the ocean floor in forty feet of water. The upper part of the tank was made of oil-proofed Vinylon and the bottom half of steel.

There may also be possibilities for storage in underground natural reservoirs, or in depleted mines after suitable treatment.

[17] Employing a different approach to the measurement of the premium than the one here used. ("Security of Eastern Hemisphere Fuel Supply," see p. 80, n. 9.)

[18] See Chap. 7. (See also Chap. 3, n. 21, dealing with newly emerging North Sea prospects.)

[19] See Part II, Chaps. 10 and 11. Commercial forces at work in the Middle East and North Africa, as represented in company and government confrontations on taxes paid, host-country participation in the ownership and management of oil operations, etc., will, in turn, be influenced by the anticipated international role of oil from newly developed oil provinces in other regions.

APPENDIX: A NOTE ON THE PURPOSEFUL CREATION OF STANDBY CAPACITY IN SECURE REGIONS

The only alternative to stockpiling examined in the chapter to guard against supply interruptions in Europe and Japan was the existence of idle productive capacity in the Western Hemisphere. However, idle capacity, as defined, was that which resulted in the normal operations of the oil industry as an unintended by-product of institutional, technological, and commercial factors. Another approach, worthy of attention, is the *purposeful* creation of standby capacity in secure regions, by the countries of Europe and Japan, as an alternative or supplement to stockpiling.

With the dwindling importance of routine idle capacity in a country such as the United States, the purposeful creation of such capacity in other "secure" regions may command increasing attention in the future. In appraising this approach, the governments of importing countries will need to be concerned, of course, with its cost, as compared to stockpiling, as well as the comparative political and strategic advantages that it appears to offer.

Let us assume that to meet possible supply interruptions, oil stockpiles equal to 180 days of consumption are judged necessary. As estimated in the chapter (p. 81), by 1975 such a stockpile already will have cost Europe a capital investment of about $7.2 billion and the ensuing 5-year buildup to 1980 would involve an additional capital outlay of about $1.8 billion (all in 1967 prices). The latter figure is based upon an estimated 3 million barrels/day increase in West European imports between 1975 and 1980 which, in accord with a 180-day reserve stock policy, would call for a 5-year increase of 540 million barrels of oil stocks.

Strictly from an investment standpoint, the question confronting European governments in 1975 would be: Can 3 million barrels/day of standby capacity, to last at least the next five years, be purchased at a lower cost than $1.8 billion? In other words, could standby capacity be made available in a *secure* region at a capital cost of less than $600 per daily barrel? Let us examine this question by looking at North America first.

Oil from unconventional sources in the United States and Canada (shale oil, coal, and tar sands) can immediately be ruled out as an economic source of standby capacity because of capital requirements running into several thousand dollars per daily barrel. (See Chapter 4.) Crude oil from reserves in the contiguous forty-eight states (see Chapter 2) can also be ruled out because of the tight squeeze that woud be involved just in meeting current U.S. domestic needs at that time even at prices within the United States well above the world oil price.[20]

These estimates would appear to rule out North America, except for the possibility of much lower costs for developing capacity on the North Slope of Alaska and in the Canadian Arctic. Let us take the year 1975 and assume that developments in the North American Arctic have unfolded sufficiently well so that large reserves are known to exist, and transport to Western Europe is feasible in the sense that tankers can deal with the ice encountered in the run through the Northwest Passage (see Chapter 7). Let us also make the assumption that the Libyan exploration and development cost derived in the appendix to Chapter 5 were to apply in the Arctic—i.e., that $430 would have been required to achieve each daily barrel of productive capacity, and that additional productive capacity, without further exploration, could then be achieved at a cost of about $125 per daily barrel (see appendix to Chapter 5). This figure is well below the $600 break-even cost derived above.

Obviously, the foregoing computations are entirely hypothetical. Nevertheless, they do illustrate the possibility that the comparative capital costs of achieving security in oil supplies could favor standby capacity in the North American Arctic in the mid-1970's as opposed to a further building up of stockpiles in Western Europe.

Clearly, though, it is not necessary to look ahead to 1975 to find oil-producing regions elsewhere in the world in which the development of standby capacity would have a capital cost of less than $600 per daily barrel. One of the major oil companies, in commenting on the comparative costs of oil stockpiles and standby capacity stated, in a letter to us, that although the investment required for spare producing capacity varies greatly from area to area, a figure in the $200–$300 barrel/day range does not seem out of line for a reasonably prolific area. Some available estimates run even lower, as does our own rough estimate for Libya, but it is not our purpose to determine the precise differences in the cost of the two approaches. The important point is that the capital costs of creating standby capacity in many proven regions outside of North America are known to be sufficiently low to provide a less costly alternative than stockpiling at the 180-day level. Conceivably, therefore, importing countries could make arrangements to buy standby capacity in secure regions, presumably through companies operating in those regions.

This fact gives rise to a number of questions, the most important of which is how to identify a secure region. There is, to begin with, the political problem that the designation of any country or region as secure (through the purposeful creation of standby capacity by an importing country) is likely to be considered offensive by those oil-producing countries not so designated. Beyond this, even though Venezuela and Iran, say, have proved dependable in past crises, it is impossible to know how they would perform in future crises. For example, company-government confrontations leading to supply interruptions could find the producing countries lining up quite differently than in the case of an Arab-Israeli crisis. And, were there to be another war between Israel and the Arab states, Iran might find its freedom of movement circumscribed by hostile forces.

---

[20] Estimates derived by us of the costs of development of standby capacity on federal offshore lands run at a minimum of about $1–$2 thousand per daily barrel. (Based upon figures in T. D. Barrow, "Economics of Offshore Development," included in *Exploration and Economics of the Petroleum Industry* [Dallas: 1967]. Mr. Barrow is Executive Vice President of the Humble Oil and Refining Company, in charge of exploration and production.)

Thus, although the capital invested in assuring the availability of oil to meet an interruption of regular supplies could be economized by a country if it paid for the development of standby capacity in "secure regions," the political and strategic hazards in so doing are great. And, in the best of circumstances, the security achieved is not comparable to that involved in having the oil supply stored on home ground. The latter consideration also indicates that the credibility of standby capacity would be weaker than the presence of adequate stocks within an importing country as a *deterrent* to actual supply interruptions, or the threat of such interruptions if brandished as a political or commercial weapon.

It is also necessary to recognize that, comparatively, the capital savings would be considerably greater than the actual savings in cost per barrel if the standby capacity, on the one hand, or the emergency stocks, on the other, should have to be used. In the event of an actual call on standby capacity, the importing countries would be faced with such cost items as tanker transportation during an emergency period, and the tax take per barrel of the exporting country, which even under ordinary circumstances is the major element in f.o.b. costs. However, these charges, although they would reduce the total cost advantage of standby capacity, would be incurred only if and when the standby needed to be activated.

We conclude that costs (clearly so for capital investment) would favor the creation of standby capacity in secure regions as opposed to the building of oil stocks within the importing country at, say, 180-day stock levels. (The capital cost advantages of standby obviously diminish with lower stock levels and increase with higher stock levels.) Even so, in our judgment, the political and strategic factors weigh heavily in favor of the stockpile approach—at least until the North American Arctic (or an equally secure region) becomes a feasible site. Nevertheless, the importing countries would be well advised periodically to evaluate the relative advantages of the two approaches (including also a combination of the two),[21] taking all relevant factors into account—economic, political, and strategic.

---

[21] For example, stocks to meet the first several months of a crisis; standby capacity to guard against a much longer, but less likely, disruption in supplies.

CHAPTER 7

Future Geographic Diversification in Crude Oil Supplies: The North American Arctic, Potentials and Implications

DURING the period that this study was under way the current oil news was dominated by the announcement of the discovery of vast new oil reserves in Northern Alaska and of associated developments in the field of transport that could lead to the eventual establishment of the Alaskan and Canadian Arctic as a major region of world oil supply. What kindled the intense commercial interest, beginning in 1968, was the size of the oilfield said to be indicated by the two discovery wells that had been drilled on acreage jointly held by the Atlantic Richfield Company and the Humble Oil and Refining Company (the U. S. arm of Standard Oil Company [New Jersey]). In the language of DeGolyer and MacNaughton, a prestigious Dallas-based firm of oil industry consultants, the discovery could develop into a field which would rate "as one of the largest petroleum accumulations known to the world today."

The field that had been discovered was estimated to contain between 5 and 10 billion barrels of recoverable oil reserves, which if eventually proved up to the higher end of the range, would make it about twice as large as the biggest oilfield ever discovered in North America. For the Alaskan Arctic as a whole, and for the Canadian Arctic, informed speculation almost immediately raised these estimates, on the assumption that the single field was not the only one that would be found, and also in the belief that the recoverable oil contained in the particular field itself may have been conservatively appraised. Estimates have since been made regarding potential discoveries in other sections of this vast region which, if validated, indicate the existence of new oil provinces rivaling those in North Africa.

As has been shown in earlier chapters, the demonstrated propensity of oil-deficient regions such as Western Europe and Japan to consume liquid fuels is based upon the known existence of enormous oil deposits in the Middle East, and more recently North Africa. Projected into the future, European and Japanese consumption patterns therefore result in an extremely heavy dependence on Middle Eastern and North African oil supplies, with the attendant problem of possible supply interruptions. For the first time, a potential source of oil supplies vast enough to make more than a slight dent in this dependence appeared to be a real possibility as a result of Arctic prospects as envisaged in 1968/69.

Geographic diversification, which produced negligible strategic benefits during the decade following the Suez Canal crisis, may, therefore, hold significant promise for the years ahead. However, at this very early stage there is almost nothing in the way of firm information regarding such vital matters as the size of the oil reserves in the region, the time scale of their development, the ability to transport the oil through the region's icebound waters, and the costs of production and transportation. Nevertheless, it is essential to sift the presently available facts and speculations in order to place in perspective, to the extent possible, the impact that developments in the North American Arctic could have over the time period with which this study deals and regarding the matters with which it is principally concerned.

The North American Arctic: Potentials

Long before the announcement in July 1968 of the first successful oil strike on leased acreage it was known that geological circumstances in the Alaskan Arctic were favorable for the accumulation of oil. Indeed, in 1923 the U. S. government set aside some 35,000 square miles of potentially oil-bearing land in this area as a U. S. Naval Petroleum Reserve.[1] The U. S. Navy conducted a program of exploration on its acreage between 1944 and 1953, at a cost of some $45 million, but did not discover reserves judged to be sufficient for economic production.[2]

Why the great delay in beginning commercial exploration of the Alaskan Arctic? As expressed in an authoritative oil industry journal: " . . . this is in the nature of oil exploration. An area is known for many decades to contain sedimentary basins of large extent and thickness but is remote and difficult; costs would be prohibitive there. A few wells are drilled for special reasons, political/nationalist perhaps, or to further academic or technical research; the geologists remain enthusiastic but management is sceptical. Then a new external factor arises, a Suez crisis, a breakthrough in techniques, a reconsideration of commercial strategy. One company goes in, makes a discovery, the industry's competitive instincts are aroused and a rush begins. . . ."[3]

The rush is under way in the Arctic. It follows upon the heels of the Arab-Israeli war of June 1967 and the continuing unrest and intermittent outbreaks of small-scale warfare in the affected region—the major oil exporting area of the world. What is more important from a commercial standpoint is that the original strike was made in Alaska. Therefore, the United States, the richest market in the

[1] Wallace E. Pratt, "Petroleum in the Polar Areas" in *World Geography of Petroleum*, ed. W. E. Pratt and Dorothy Good (Princeton: Princeton University Press, 1950), pp. 313ff.

[2] "Prospects for Alaskan Oil," *Petroleum Press Service*, January 1969, p. 9.

[3] "Treasure in the North?" *Petroleum Press Service*, December 1968, p. 444. In the same article, the Russian geologist, Prof. Mikhail Kalinko, is reported to have stated that the Arctic oil reserves may surpass the boldest predictions; that of all the world's oceans the Arctic holds the greatest prospects of oil and gas. His own main interest centered, of course, upon the Soviet Arctic region.

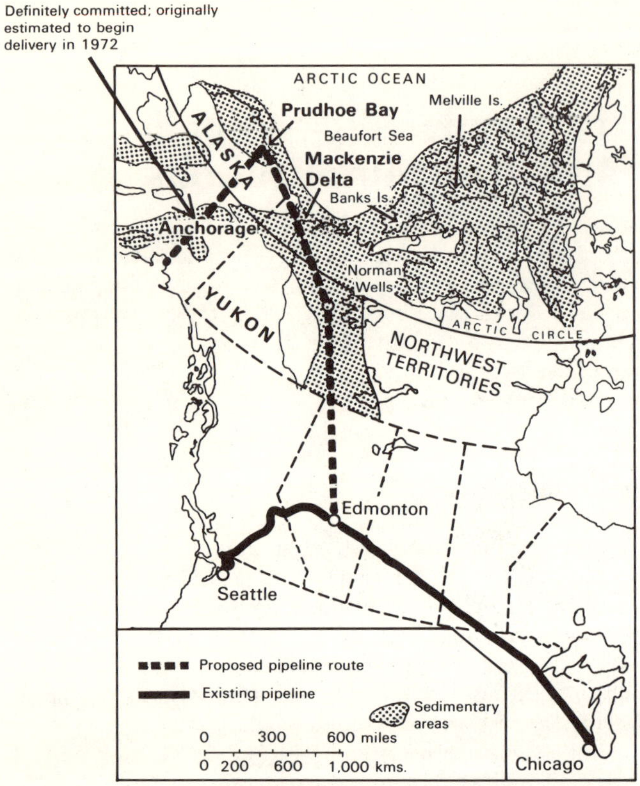

Fig. 7.1. Potential Oil and Gas Areas in Northern Canada and Northern Alaska

Source: Petroleum Press Service, February 1969, p. 64.

world, falls within the same political jurisdiction, even if thousands of miles separate the marketing centers from the site of the oil deposits. The need for supplies from outside the "lower forty-eight" states is evident from the fact that imports already are significant even though they are subject to strict quantitative restrictions; without the Alaskan discoveries, it is possible that imports would play a considerably enhanced role in the future. In addition, the inferred existence of large recoverable oil deposits in the Canadian Arctic is within a region of the world marked by great political stability and well located for shipments to major world markets if transportation problems (discussed below) can be solved.

Although many of the signs are propitious (including, in September 1969, the largest oil-lease sale ever, yielding close to $1 billion to the state of Alaska), evaluation of the prospects at this stage is based, to a considerable extent, upon informed speculation. Figure 7.1 presents a map of the broad region around which the speculation centers. It encompasses a vast, physically inhospitable area. We will confine the discussion to two broad subregions: (1) the Alaskan North Slope; and (2) the Canadian Arctic, including the Arctic Islands. In each case the discussion will consider the reserve potentials, as these have been inferred, in some cases from drilling, in others from broad geological indicators; and the prospects for the development of production. Transportation to markets will then be discussed for the entire region. Obviously all of these aspects are interdependent in that the development of reserves and production will not take place unless transportation to markets turns out to be commercially feasible.

Most of the following analysis will have to be couched in the language of what "might" or "could" happen. Although those with commercial interests at stake are in possession of more information than is available in the open literature upon which this chapter is based, what *will* happen, it may be safely said, is not known to anybody at this time.

1. *Potential Reserves and Possible Productive Capacity*

Alaskan North Slope

As indicated, DeGolyer and MacNaughton's original estimate of the strike in the Prudhoe Bay area on the North Slope (reported to Atlantic Richfield and Humble as joint leaseholders of the tracts on which the discovery wells were drilled) placed recoverable reserves in the range of 5 to 10 billion barrels. The same firm subsequently placed the recoverable reserves on British Petroleum Company leaseholds in the same general area, some within the same (Prudhoe Bay) field, at close to 5 billion barrels. To date, therefore, leases acquired during earlier sales by the state of Alaska, for which the total proceeds to the state were $12 million (as compared to $900 million in the September 1969 sale), appear to contain at least 10 billion barrels of recoverable reserves, and probably more.

In addition, in a more general evaluation of Alaska prepared for an agency of the federal government, but based to a greater extent on broad geological indicators, DeGolyer and MacNaughton has ventured an estimate of recoverable reserves for Northern Alaska of about 25 billion barrels, with the observation that "At the present stage of exploration and development in Alaska, one hazards an estimate today only with the knowledge that he may change it tomorrow."[4] Other authorities have estimated recoverable

---

[4] DeGolyer and MacNaughton, *Report on Estimates of Additional Recoverable Reserves of Oil and Gas for the United States and Canada* (Dallas, June 30, 1969), p. 17.

It is interesting to note that the late Everette DeGolyer was head of a U. S. Mission to the Middle East which in its preliminary report made public early in January 1944 stated that "the center of gravity of world oil production was shifting from the Gulf-Caribbean area to the Middle East-Persian Gulf area, and was likely to continue to shift in this direction." The preliminary estimates of Middle Eastern reserves made by the Mission showed reserves proved by developed fields and indicated by fields not yet fully explored of about 9 billion barrels in Kuwait; 6 to 7 billion barrels in Iran; 5 billion barrels in Iraq; 4 to 5 billion barrels in Saudi Arabia; and 1 billion barrels in Qatar–a grand total of about 25 billion barrels. (See *Congressional Record*, Vol. 44, Part 4, 80th Congress, 2nd session, p. 4962.)

The total estimated for the Middle East in 1944 is almost identical with the Northern Alaska estimate made in the 1969 report to the Office of Science and Technology (above). However, as indicated on p. 91, the official estimate of Middle Eastern reserves at the beginning of 1969, after very large amounts had already been produced, was 270 billion barrels; and it is generally agreed that large quantities of oil still remain to be proved up in the region. The temptation to draw an historical parallel between the preliminary 1944 report on the Middle East and its ensuing developments, and the 1969 report on the Arctic, is strong. However, only time will tell.

reserves at up to almost 50 billion barrels. For example, a figure of 45 billion barrels has been estimated by an official of an independent oil concern active in the Canadian Arctic, but not in Alaska. His estimate was derived on the basis of such indicators as cubic miles of effective marine sediments, age of sediments, structural conditions, and related factors.[5]

An effort to place the prospects for Northern Alaska into an economic framework, in order to explore some of the commercial implications of developments there, has been made by the W. J. Levy Consultants Corporation of New York. In its analysis, the highly regarded Levy organization bracketed the possibilities for the Alaskan North Slope from a low of 15–20 billion barrels to a high of 30–40 billion barrels of reserves "proved up" by 1975 to 1980. The assumed productive potential was placed at 1 million to 2½ million barrels/day by 1975 and 2½ million to 5 million barrels/day by 1980.[6] We have adopted this range of estimates as representing the best judgment now available (1969) on the economic outlook for oil production on the North Slope. Those with direct commercial interests involved are understandably reticent on the subject although it can be inferred from statements and actions by some of the companies mainly involved that shipments from the North Slope could exceed 3 million barrels/day by 1980.[7] (This may be said to be roughly consistent with the midpoint of Levy's estimated 1980 range, 3¾ million barrels/day.)

The Canadian Arctic, Including the Arctic Islands

When attention is directed to the Canadian side of the border, a region known to be rich in sedimentary rocks that are potentially productive of hydrocarbons, the basis for estimation of recoverable reserves and of possible productive capacity over the next decade becomes far weaker. Exploration activity has not progressed as far as on the Alaskan North Slope, and no significant oil discoveries have been announced. Consequently it becomes necessary to rely heavily, if not exclusively, on geological inference. By the same token, activities of the sort that have been set in motion by companies operating on the Alaskan North Slope—the start on development of pipeline transportation and the testing of an icebreaking tanker (both discussed later)—are not present as strong indicators of relatively near-term commercial promise.

Several specific subregions in the Arctic reaches of Canada have been identified as potentially productive. The two main ones are: (1) the Mackenzie Delta (more generally, the Arctic coastal plains); and (2) the Arctic Islands.

(1) *Mackenzie Delta.* The same formation that exists at Prudhoe Bay is thought to extend eastward across Alaska and through the Northwest Mackenzie Delta area. Many geologists believe it quite likely that a trend of large fields may extend all along the southern shore of the Arctic Ocean in this region. Consequently, natural conditions that would result in oil accumulation in large fields appear highly favorable, and leasing and exploration have proceeded at a fast pace. Estimates of recoverable reserves are, however, scant for this area. Employing the same indicators as he used for his Alaskan North Slope estimate, J. P. Gallagher has estimated recoverable reserves in the Mackenzie Delta and adjoining continental shelf in the Beaufort Sea at about 35 billion barrels of oil.[8]

(2) *The Arctic Islands.* Exploratory interest also runs high in the Canadian Arctic Islands. Panarctic, a consortium of private companies in partnership with the Canadian government, began an exploratory drilling program, budgeted at a minimum of $30 million, in the spring of 1969.[9] The drilling was to occur at various sites on the 50 million acres in which the consortium has interests.

The archipelago covers some 500,000 square miles and was surveyed in 1955–59. The geology of the region appeared to be ideal for the accumulation of large oil deposits, with the result that in 1960 a land rush began to obtain drilling rights. The sedimentary basin area is approximately 1,150 miles long and 450 miles wide, and about half of the area is above water. The land area is easily evaluated in surface geological terms since the islands are largely frozen deserts, devoid of vegetation, soil, and even snow due to the very dry climate (two to eight inches of precipitation per year). Surface rocks are, therefore, exposed.

The wide belt of sedimentary rocks includes a succession of folds in old geological sediments. The individual folds are up to 10 miles wide and 100 miles long, and are such that the formations have been compared favorably by geologists with those of the prolific oil-producing structures in the Middle East. North of the belt of folds there is a vast trough of downwarped sedimentary rocks with sections as much as 40,000 feet thick. Oil seeps and other surface indications are common on the islands, and there are extensive outcroppings of tar sands on Melville Island.[10] Dr. J. C. Sproule of Calgary has estimated that from 50 to 100 billion barrels of oil underlie the region.[11] Mr. Gallagher, whose other estimates have been cited, judges recoverable reserves at about 50 billion barrels.[12]

---

[5] J. P. Gallagher (President, Dome Petroleum Ltd.), "The Canadian Petroleum Industry and the New Arctic Frontier," excerpts from an address to the Oil Analysts Group of New York, November 22, 1968.

[6] "A Note on Implications of Alaskan Production" (October 1968) and "A Further Note on Implications of North Slope Potential," (May 1969) (unpublished documents, W. J. Levy Consultants Corporation, New York). See also a later estimate by Milton Lipton (Executive Vice President, W. J. Levy Consultants Corporation) of potential North Slope production of between 2 and 2½ million barrels/day by 1975, and between 3 and 5 million barrels/day by 1980 (*Oil and Gas Journal,* November 17, 1969, p. 44).

[7] *Oilweek*, September 29, 1969, p. 21, in a review article by Earle Gray mainly concerned with the test icebreaking tanker voyage through the Northwest Passage (see below).

[8] Gallagher, *op. cit.* (n. 5); *Petroleum Press Service*, February 1969, p. 63, also employs the estimate of 35 billion barrels.

[9] *Oilweek,* February 24, 1969, p. 64; and *World Petroleum,* January 1969, p. 24.

[10] "Arctic May be Canada's Next Target," *Oil and Gas Journal,* January 6, 1969, pp. 94-96.

[11] *Ibid.,* p. 96. Dr. Sproule is a consulting geologist who has done considerable work on Arctic oil prospects.

[12] Gallagher, *op. cit.* (n. 5). Also, *Petroleum Press Service*, February 1969, p. 64, refers to 50 billion barrels as the average of many reserve estimates for the Arctic Islands and adjoining offshore areas.

(*continued*)

Possible Aggregate Reserves and Productive Capacity in the Canadian Arctic

For the North Slope of Alaska it was possible to draw upon existing analyses, reflecting attention to commercial factors, to indicate a range of reserves expected to be "proved up" and of associated productive potential. Nothing of this sort was available to us for the Canadian Arctic. Consequently, the estimates to be presented involve an unusually–and undesirably–high degree of conjecture.

From the estimates referred to above, we shall use those of J. P. Gallagher because they were made within the context of an overall appraisal of Northern Alaska and the Canadian Arctic, using the same general estimating approach throughout. (His Alaskan North Slope estimate of 45 billion barrels is not far above Walter Levy's high [40 billion barrels] estimate of reserves proved up by 1980.) Gallagher's estimates of potential recoverable reserves were 35 billion barrels for the Mackenzie Delta and 50 billion barrels for the Arctic Islands–a total for the two of 85 billion barrels, approximately twice today's high estimates for the Alaskan North Slope.[13]

For lack of any better alternative we will, despite potential reserve estimates approximately twice those of the North Slope, assume the same pace of development as on the North Slope: i.e., a productive potential figure for the Canadian Arctic the same as that for Northern Alaska–a range of 1 to 2½ million barrels/day by 1975 and 2½ to 5 million barrels/day by 1980. Although arbitrary, the assumption of a similar development rate as in Northern Alaska may be judged on the high side, even in the face of higher potential reserves, because of such elements in the situation as the earlier start in Alaska; the fact that major oil companies are heavily involved in Alaska; the location of Alaska within the large, protected U. S. market; the definite plans for construction of a pipeline to southern Alaska (see below); and the fact that much of the Canadian Arctic oil is dependent for transportation to markets on a successful solution to the problem of tanker shipment (see below). On the other hand, possible success with the icebreaking tanker; and the fact that Canadian oil may be exploited by more aggressive, smaller companies are among the factors that could weigh in on the other side.

The Alaskan North Slope and the Canadian Arctic

Addition of the estimates adopted for Alaska and Canada yields the following quantitative picture of what might unfold in the North American Arctic over the next decade:

	Oil Reserves (billion bbl) 1975/80		Productive potential (million bbl/d)	
	Low	High	1975	1980
Alaskan North Slope[a]	15–20	30–40	1–2.5	2.5–5.0
Canadian Arctic[b]	43	85	1–2.5	2.5–5.0
Total	[c]	[c]	2–5.0	5.0–10.0

[a] W. J. Levy Consultants Corporation reports, cited earlier (p. 87, n. 6). Reserves are referred to as "proved up." This designation is probably meant to approximate the American Petroleum Institute "proved" category (see Chap. 5).

[b] Low figure, Canadian Petroleum Association estimate (text n. 13); high figure from text. These are estimates of recoverable reserves; it is assumed that sufficient exploratory work will have been completed to have indicated the recoverability of these quantities within reasonable limits of certainty, but that reserves will not be "proved up" to this level within the time period indicated.

[c] Not additive, owing to definitional differences. See footnotes a and b.

2. *Transportation to Markets*

No matter what the mode of transportation, formidable technical problems must be overcome if oil from the North American Arctic is to make its way into major world markets. The construction of a pipeline to move the oil to the south, already a firmly committed commercial venture, will be beset with technical difficulties because of a combination of climatic conditions and the nature of the terrain which must be traversed. Transportation by tanker through the Arctic Ocean, whether towards the east or west, must overcome what appear to be even more formidable technical difficulties because of severe ice conditions in parts of the passage for all but a few weeks of the year. The technical problems connected with both pipeline and ocean transport are now under intensive study.

Uncertainty regarding transportation prospects, even allowing for the definite decision to construct a trans-Alaskan pipeline to the south, thus serves to compound the existing uncertainty regarding recoverable reserves and productive capacity. And without a successful solution to the problem of transportation by sea the potential importance of the North American Arctic would dwindle considerably for major world oil markets other than those within the United States and Canada.

We will sketch out some of the transportation alternatives that have been suggested, along with the problems– and potentials–they present. It would be premature to attempt more at this early stage.

To Markets in the United States

Three companies (Atlantic Richfield, Humble, and British Petroleum) are the principal financial backers of a pipeline to be built at an estimated cost of about $1 billion across Alaska's rugged terrain with the hopes of having it completed during 1972. (Following the Alaskan lease sale of

See also judgments to the effect that the Arctic Islands hold reserves which are as large or larger than those on the Alaskan North Slope. (*Oil and Gas Journal*, October 13, 1969, p. 73; *Oilweek*, September 29, 1969, p. 21.)

[13] By way of comparison the Canadian Petroleum Association has placed the potential crude oil reserves of the Arctic Islands and the Arctic Coastal Plains (including the continental shelf), which constitutes essentially the same area as the two subregions identified, at 43½ billion barrels. In commenting on this estimate DeGolyer and MacNaughton, an organization not generally given to overestimation, states that it "could easily prove to be conservative" (DeGolyer and MacNaughton, *op. cit.* [n. 4], p, 27). DeGolyer and MacNaughton's estimate for the North Slope of Alaska in the same report is, as noted, about 25 billion barrels.

September 1969 the three prime backers were joined by five other companies in this venture.) The planned 48-inch diameter would be the largest of any pipeline in the Western Hemisphere, and has been described as the largest single private construction project in history.[14] It is being designed so that the initial capacity of 500,000 barrels/day could be increased to 2 million barrels/day with the installation of additional compressor horsepower. Although the venture abounds with technical uncertainties, and is faced by watchful criticism, if not opposition, from those concerned with its possible upsetting effects on the delicate ecological balance of Northern Alaska, its successful completion appears to be assured, although probably at a higher cost and a later date than was originally contemplated.[15]

The West Coast market is likely to be the first destination for output from this pipeline. From its southern terminus (Valdez) the crude would be tankered to refineries on the West Coast.

To reach the mid-continent markets in the United States from the North Slope, the easiest and most economical route would appear to be a pipeline from Prudhoe Bay east to the Mackenzie Delta and then south to Edmonton, from where the crude could then move into mid-continent U. S., making use of an already existing pipeline.[16] However, in return for permitting a pipeline to be constructed across its territory Canada might attach conditions regarding the shipment of Canadian crude from its Arctic and other producing regions to U. S. markets. If no satisfactory solution could be worked out, it is possible that oil tankered from southern Alaska to Seattle might then be shipped by pipeline, located entirely within the United States, from that point eastward as far as Chicago.

The rich East Coast markets of the United States might then be reached by pipeline from the mid-continent, or conceivably by tanker from southern Alaska, through the Panama Canal, or using mammoth tankers around Cape Horn. A far more favorable possibility, involving considerably lower estimated costs, would involve tankering through the Arctic Ocean. The latter possibility would require success with icebreaking tankers, even though more exotic means of transport such as nuclear-powered submarine tankers have also been mentioned.[17] It will be discussed next since reaching foreign markets poses the same type of problem.

To Foreign Markets

To reach major foreign markets, particularly those in Western Europe, at a competitive price, it appears certain that the use of tankers through the icebound waters of the Arctic will be required. Japan is a possible exception in that once crude oil supplies are available at a port in southern Alaska, they could be tankered to Japan by conventional means. Whether the delivered price, under these circumstances, would be competitive with alternative sources depends not only on production and transportation costs, but on the factors affecting prices in Northern Alaska and in existing export centers such as the Persian Gulf.[18]

The major hope for reaching Western European markets, and the lowest costs for reaching the eastern seaboard of the United States, rests on success with icebreaking tankers (London and New York are about the same distance from Prudhoe Bay via the Northwest Passage). The first experimental run, to test the feasibility of a voyage via the Northwest Passage between the East Coast of the United States and Northern Alaska, took place in the late summer and early fall of 1969.

The test run was made by the S.S. Manhattan, the largest commercial tanker (115,000 dwt.) in the U. S. merchant marine, and one with many more horsepower than other tankers of comparable size, or even those of larger size. This vessel was converted into the world's largest icebreaker by being equipped with a newly designed icebreaking bow, strengthened structurally, and modified in other ways. It was intended to break its way through the icebound waters of the Arctic in a test voyage, originally scheduled to begin in May or June, from the Philadelphia area to Prudhoe Bay, the approximate point at which oil destined for tanker voyage would need to be loaded. Preparing the vessel for this trip involved a race against time, which despite such extraordinary measures as slicing the vessel in several parts for simultaneous work in separate shipyards and subsequent reassembly in Chester, failed to make an early enough start (it departed Philadelphia in late August) to encounter ice under conditions as severe as those the sponsors had wanted to test.

With the aid of conventional icebreakers, including in particular the flagship of the Canadian icebreaker fleet, the S.S. Manhattan successfully completed the voyage to Prudhoe Bay following the route indicated in Figure 7.2. In reviewing the test voyage, the editor of *Oilweek,* who went along on the voyage, reported the following tentative conclusions: (1) that "ships much larger and more powerful than the Manhattan, with special design features to improve performance under adverse ice conditions, will be required"; and (2) that "conventional icebreaker ships, larger and more powerful than those now in operation, would be required for standby service at strategic positions through the Passage, to render assistance to the tankers as needed." His most important conclusion was that *there is little doubt*

[14] "Moving Arctic Oil to Market," Jack F. Bennett, Humble Oil and Refining Co. (press briefing, June 3, 1969).

[15] On the ecological aspects, see, for example, an editorial entitled "Slow Down the Oil Rush" in the New York *Times*, November 10, 1969; and an article, "North Slope: Oil Rush" in *Science*, October 3, 1969, pp. 85-89.

[16] "A North American Oil Policy?," *Petroleum Press Service,* February 1969, p. 46.

[17] In a statement released after the icebreaking tanker had made its way through the Northwest Passage, the Electric Boat Division of General Dynamics Corporation in conjunction with various oil companies was reported to be investigating the feasibility of a submarine tanker (*Wall Street Journal,* October 16, 1969, p. 16).

[18] To be more specific: at least for some time, prices in Northern Alaska probably will be determined by delivered prices of crude oil from other points of production in the lower forty-eight states, even though, after allowing for transportation costs, this level may be well above costs of production at the Alaskan source. Similarly, Persian Gulf prices include a heavy tax component (see Chap. 10) which, although it is not a true element in the costs of producing oil, must be covered by the prices oil companies charge for Middle Eastern exports.

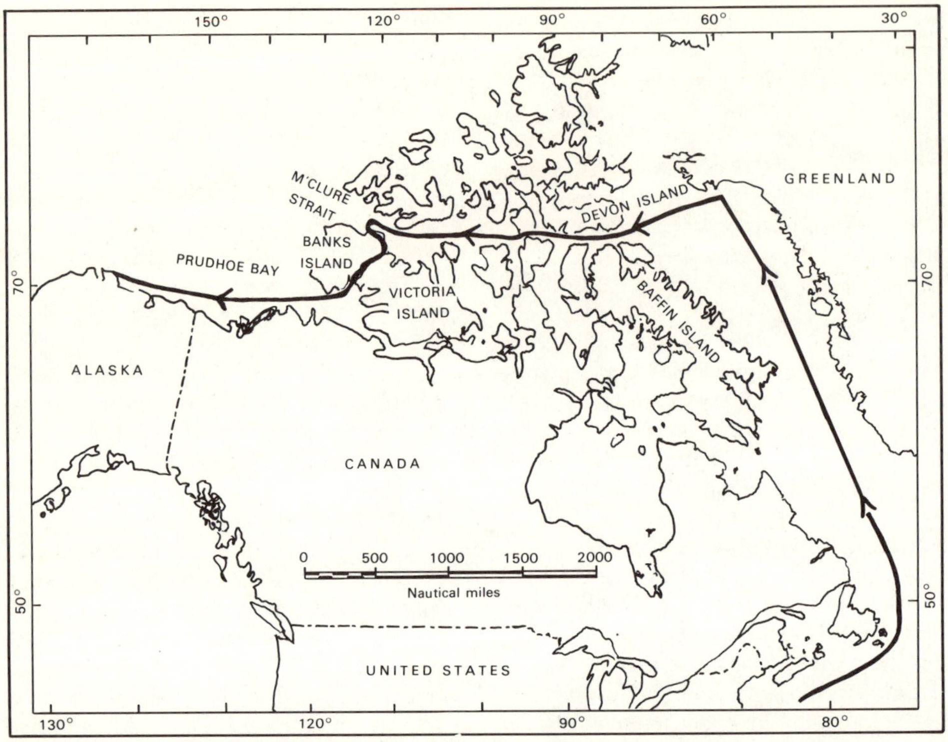

Fig. 7.2. S.S. Manhattan's Route Through Northwest Passage

Source: Oilweek, September 29, 1969, p. 20.

that ships could be built which would carry oil through the Northwest Passage, but that the voyage time involved would be the critical factor affecting economics.[19]

The critical importance of the time required for a voyage through the Northwest Passage on a routine year-round basis was also indicated following the test voyage, by Stanley B. Haas, manager of the icebreaking tanker project: "Our calculation [to be based on the data assembled] of the delays which might be encountered in shipping caused by ice conditions . . . will be a primary factor in determining the economic feasibility."[20] Nevertheless, Mr. Haas also stated: "We have seen nothing which indicates that it will not be possible to operate commercial shipping through the Northwest Passage. Our objective is still commercial shipping on a year-round basis."[21] And Humble Oil, chief financial backer of the $40 million S.S. Manhattan experiment, was reported to be planning another test run in early spring 1970. It should be added that the Bering Strait, which would have to be traversed for direct tanker shipments to Japan, presents ice problems even worse than those of the Northwest Passage.[22] It was not tested in the Manhattan's first voyage, although included in the original plan.

[19] "Oil Can Move Via NW Passage," *Oilweek,* September 29, 1969, p. 20.

[20] *Ibid.*

[21] *Ibid.*, p. 28.

[22] See *Oil and Gas Journal*, February 17, 1969, pp. 40–43. This article also reported that Humble had not completely discarded the idea of using submarines to move crude oil under the ice to open waters.

The North American Arctic: Implications

The foregoing discussion has described and assessed potentials as they are currently perceived, for reserves, productive capacity, and transportation routes for oil that has been or may be discovered in Northern Alaska and the Canadian Arctic as a result of the intensive search under way in that region. The relatively firm information that was presented regarding commercial prospects was confined to several successful strikes in the Prudhoe Bay region of Northern Alaska; the definite plans for building a pipeline to carry the oil to an all-weather port in southern Alaska; and the test voyage of an icebreaking tanker in the Northwest Passage in the late summer and fall of 1969. Beyond that, everything else was in the category of speculation, particularly regarding the possible discovery of other oil deposits, and the development of productive potential over the next decade, as well as the possible commercial feasibility of transportation of the oil to market by way of icebreaking tankers.

Cost and price analysis was omitted completely. There has been speculation regarding costs of production and transportation and some preliminary analyses have appeared publicly; but this involves piling a set of economic assumptions on top of geological and engineering calculations which are themselves highly uncertain at the present time.[23] However, despite their speculative character, the available analyses indicate that oil from the Alaskan North Slope could be sold profitably in various regions of the United States.[24] Conceivably it could also be sold profitably in Japan and Europe (if transport problems are solved), but not as profitably as in the United States.

In the remainder of this chapter, analysis of the costs of production and transport to market and of prices at the points of production and delivery will be bypassed. Instead the question will be asked: What are the implications of developments in the North American Arctic over the next decade, particularly from the standpoint of geographic patterns of supply for the United States, Western Europe, and Japan on the assumptions (1) that the amount of reserves and productive capacity might fall within the range shown earlier, and (2) that oil supplies from this region might be price-competitive in the United States, Japan, and Western Europe, the world's major oil markets?

1. *Reserves: North American Arctic, Middle East, and North Africa*

Official estimates of Middle Eastern and North African reserves as of the beginning of 1969 were:

Middle East	270 billion barrels
North Africa	40 billion barrels

These figures are undoubtedly conservative, generally conforming, although not with the same degree of strictness, to the proved reserves concept in the United States. Within industry circles underestimation of between 50 and 100 percent of the amounts that could be proved up is not considered unlikely. If a 50 percent underestimation is assumed, a more realistic figure for potential recoverable reserves in the Middle East would be on the order of 400 billion barrels, and for North Africa, 60 billion barrels.

In comparative terms, therefore, estimated recoverable reserves in the North American Arctic, Middle East, and North Africa, could shape up as follows:

	Billion barrels
Middle East	270–400 (as of January 1, 1969)
North Africa	40–60 (as of January 1, 1969)
North American Arctic	
Northern Alaska[a]	30–40 (as of 1975–80)
Canadian Arctic[b]	45–85 (as of 1975–80)

[a] Referred to in the underlying source as reserves estimated to be "proved up" by 1975–80.

[b] Referred to in the underlying sources as estimated recoverable reserves, with no time schedule provided as to the rate at which the reserves might be "proved up." We venture the guess that some figure within the range shown, more likely toward the lower end, will be proved up within the decade of the 1970's.

Although not nearly as large as the Middle East, Northern Alaska alone might, according to these estimates, rival North Africa's 1969 proved reserves level by the middle or late 1970's; and, together with the Canadian Arctic, might substantially exceed North Africa's 1969 level by 1980. But this is of course counting unhatched chickens.

Even if it is granted that the North American Arctic *might* reach these levels, the question still remains: *Will* it? There is, to begin with, the question of incentives for exploration and development of reserves. Activity is great in the Arctic, but to date, much of this activity is in *anticipation* of further favorable discovery experience; solutions to transportation problems; and profitable delivered prices in major markets, particularly the United States (where restrictive import policies, undergoing Cabinet-level review in 1969 by order of the President, have served to keep prices well above the world market). Activity is, no doubt, also motivated by the potitical-strategic considerations that favor the development of major new producing provinces outside the troubled regions of the Middle East and North Africa. Whether activity will be sustained at a high-enough rate to achieve the reserve levels indicated above must remain a moot question.

An important fact to be borne in mind is that the average reserves to production ratio in the Middle East in 1967, on the basis of official reserves figures, was almost 70:1.[25] The Middle East, therefore, contains enough reserves to satisfy major markets for a long period of years into the future.[26] For those international companies which have a strong position in Middle Eastern reserves the strongest economic incentive for investing the large sums of money required to establish productive capacity and deliverability in the remote reaches of the Arctic would probably arise out of the ability to reach the rich and relatively high-priced U. S. oil market with supplies that can be produced and delivered

[23] For example, in a working paper prepared by the staff of the Cabinet Task Force on Oil Import Control (released August 5, 1969), wellhead costs in the Prudhoe Bay field were estimated at 36 cents per barrel. This was, however, an estimate of cost in one highly prolific field, not all of Northern Alaska or the North American Arctic. Moreover, it did not take into account the full burden of state taxes and royalties that might need to be met when commercial production begins to flow. It needs to be emphasized, too, that price could differ substantially from cost, depending upon the marginal costs of less prolific fields in the "lower forty-eight," whose production will be needed in large amounts to satisfy consumption in continental United States. The latter situation is clearly to be expected for a long period of time if the United States continues a policy that has the effect of severely restricting oil imports from the lowest cost foreign supply sources. In this connection, it is instructive to compare the Task Force staff's estimated cost of 36 cents per barrel with an earlier Atlantic Richfield statement to the Task Force that North Slope oil would be priced "in the neighborhood" of \$1.65 to \$2.10 per barrel to meet delivered prices at Los Angeles (*Oil and Gas Journal*, July 28, 1969, p. 103).

[24] See, for example, remarks attributed to Walter J. Levy, and to other well-informed specialists in an article entitled "Alaska Strikes It Rich," *Business Week*, February 1, 1969, pp. 49-53.

[25] *World Oil*, August 15, 1968, p. 45.

[26] Ahmad Zaki Yamani, Saudi Arabia's Minister of Petroleum and Mineral Resources, has been quoted as saying that Saudi Arabia, Libya, and Kuwait could double their output in a fairly short time (*Middle East Economic Survey*, December 27, 1968, p. 4).

at a lesser cost than in the "lower forty-eight" and that are not subject to import restrictions.

In addition, there are the strategic-political factors already referred to: the fact that the various oil companies are differently situated so far as the size of their reserve holdings in the Middle East are concerned; that oil companies are being subjected to increasing economic pressure, resulting in profit erosion, by governments in the Middle East and North Africa (see Part II, Chapter 10); and that supplies from the North American Arctic are also well located to serve foreign markets, if tanker transport problems can be solved. The development of large-scale alternative reserves and productive capability could, therefore, be an element in straightforward business strategy for many companies, even aside from supplying the U. S. market.

2. *Time Scale of Development in the North American Arctic*

Even if one accepts the assumption that the incentives will exist for development on the Alaskan North Slope and in the Canadian Arctic, it is necessary to consider the question of the physical practicability of achieving the productive capacity levels indicated above. Within a period of seven years from the first major discovery, a range of productive capacity of 1 to 2½ million barrels/day has been projected as a possibility for Northern Alaska, and within a period of twelve years, the possible productive capacity was placed at 2½ to 5 million barrels/day. Inclusion of the Canadian Arctic resulted in a doubling of these estimates.

Although the environmental factors are widely different, the Libyan experience, briefly described in the appendix to Chapter 5, may be of some help in evaluating physical practicability. Neither region is a physically hospitable one—but for almost opposite climatic reasons—and Libya enjoyed the advantage of prior development experience of a similar sort in the Persian Gulf region. The largest Libyan field in terms of production (Zelten) was found in 1959, and began to approach mature development by about 1964—a waiting period of about five years. (See appendix to Chapter 5, n. 6.) If major discoveries are dated from 1959, the level of reserves achieved in seven years, as of the end of 1966, was 20 billion barrels with production at the rate of more than 1½ million barrels/day. As of 1969, ten years after the first major discovery, official reserves stood at 30 billion barrels, and it was estimated that by the end of 1969 production would reach a level of about 3½ million barrels/day.[27]

Without pushing the comparison too far, it may be noted that the midpoints between the high and low estimates of productive capacity for Northern Alaska in 1975 and 1980 are roughly comparable with the levels achieved in Libya over similar periods of time. Adding the Canadian Arctic estimates into the picture implies the equivalent of the development of two Libyas over the next decade in the North American Arctic.

Such an accomplishment does not seem unreasonable. It would, however, require strong motivation. Strategic reasons for diversification away from the Middle East and North Africa probably are stronger now than after the first Suez Canal crisis. And, as for the North Slope of Alaska (the rough equivalent of one Libya in our estimates), there is, of course, the powerful attraction of the rich, protected U. S. market. But, of course, achievement of the projected levels will, to repeat, be dependent upon further rich discoveries in the Arctic and the solution to some essentially new problems in which the earlier technical experience of the industry would be of limited use.

[27] *World Oil*, August 1, 1969, p. 55.

3. *Coping with Supply Interruptions: A Reevaluation of the Problem*[28]

At this point the problem considered in Chapter 6 will be reexamined in the light of the potentials for the North American Arctic as discussed in this chapter. As in Chapter 6, the problem will be considered for the major consumer countries: the United States, Japan, and those in Western Europe. And, again as in Chapter 6, measures for dealing with supply interruptions will be examined.

United States

Having taken into account both the U. S. Department of the Interior 1968 study (covered in Chapter 2) and the possibilities for liquid fuels from unconventional sources (covered in Chapter 4), we have concluded that, if it were to choose this course, the United States could manage to hold the line on imports of foreign crude at a relative level consistent with that of the late 1960's. This was a crucial conclusion, because the United States, among the major consumer countries, is not now dependent to any significant extent on Middle Eastern or North African oil supplies.[29]

At this point it is necessary only to add that this conclusion is further bolstered by the potential envisaged for the North American Arctic, particularly on the Alaskan North Slope. In fact, as a result of the outlook for Northern Alaska, an altogether new policy problem may be in the making for the United States: whether, and for how long, the U. S. market would be able to absorb the relatively low-cost supplies that might be forthcoming from Alaska without significant changes in the operations of existing state regulatory systems in the "lower forty-eight" which, among their other results, serve to limit statewide production to

[28] In the remainder of this chapter, some of the subjects covered in Chap. 6 will be reexamined in the light of the potentials that may exist in the North American Arctic. The reader should refer back to the discussion in Chap. 6, as an effort will be made to keep repetition to a minimum.

[29] Even if policy were to be relaxed in such a way as to permit larger relative imports of supplies originating mainly in the Western Hemisphere (particularly in light of prospects in the Canadian Arctic, and in recognition of the comparative security of supplies from both Canada and Venezuela in the event of disruption of oil flows from the Middle East and North Africa), the force of this conclusion would not be affected insofar as imports from the Middle East and North Africa are concerned.

"market demand" and to allocate shares of the total to all producers located within the affected states.[30]

Western Europe and Japan

Estimated imports into Western Europe and Japan from the Middle East and North Africa in 1975 and 1980, and actual amounts imported in 1965, are as follows:

	Million barrels/day		
	Western Europe	Japan	Total
1965	5.7	1.4	7.1
1975	12.3	3.4	15.7
1980	15.1	5.1	20.2

Source: Chapter 1, Table 1.4.

As the earliest production is not expected, at best, to begin to flow from the North American Arctic before 1972, it is obvious that in the early 1970's methods of coping with supply crises would need to deal with the situation essentially as depicted in Chapter 6.

By 1975 and 1980, sizeable amounts of production might be available from Northern Alaska and the Canadian Arctic. The possible range of productive capacity in 1975 was placed at 2 to 5 million barrels/day, and in 1980 the possible range was placed at 5 to 10 million barrels/day.

Compared to estimated Japanese imports from the Middle East and North Africa, the possible 1975 flows from the North American Arctic are of significance. Looking ahead to 1980, the high estimate of productive capacity in the North American Arctic is significant for both Japan and Western Europe, equaling about half of their combined projected imports from the Middle East and North Africa. Much of this total, certainly of that portion originating in Alaska, seems sure to be consumed in North American markets, particularly the United States, and this fact alone introduces a major uncertainty in estimating to what extent the North American Arctic might serve as a future source of imports for Western Europe and Japan.

It must be stressed, though, that all of this is in terms of hypothetical developments. Although a major new element in the world oil scene appears to be in the making that could affect European and Japanese supply sources over the long run, it is, as yet, too early to evaluate its impact. Policies regarding security of supply dare not be altered in anticipation of what *might* occur in the Arctic. The new prospects should, however, be kept under close and constant study for they may open up possibilities for adding another region as a low-cost source of vital energy supplies to Europe and Japan and thereby reduce an otherwise overwhelming reliance upon the Middle East and North Africa. Until considerably more dependable information is available regarding the supplies that will be forthcoming and their prices, stockpiling of oil to cope with possible supply interruptions (as indicated in Chapter 6) appears to be the prudent course.

[30]See Chap. 2, pp. 34 to 36. It is possible, of course, that the State of Alaska might itself prefer to limit production and engage in proration, rather than move into a lower-price market that would result from breaking the existing regulatory structure in other producing states. However, the high capital investment required for development and transport would appear to require that large volumes be brought to market, thus militating against severe restriction of output. By the same token, as noted in Chap. 2, the cost penalty to the United States of maintaining its current degree of self-sufficiency in oil would be lessened by the flow of oil from Northern Alaska.

PART II

The Economic Importance Of Oil To The Exporting Countries Of The Middle East And North Africa

CHAPTER 8

Some Measures of the Economic Dependence of the Exporting Countries on Oil-Industry Operations

IN Part II we turn from the problems of oil supply of Western countries and Japan to the position of the countries of the Middle East and North Africa, the principal sources of supply. It is the interdependence of these two ends of the oil axis which provides the central theme of the present study. In this chapter we shall present certain measures of the degree of dependence of the producing countries upon the operations of the oil industry. In later chapters we shall examine the structure of relationships between the countries and the oil companies which conduct most of the operations within these countries and constitute their main commercial link with the world's oil consumption centers.

Our analysis will be mainly limited to five countries–Iran, Iraq, Kuwait, Saudi Arabia, and Libya–which among them accounted in 1968 for 85 percent of oil production in the Middle East and North Africa (see Table 8.1) and hold a major fraction of the world's known oil reserves.[1] In the four Middle Eastern countries, oil operations are, for the most part, conducted by eight foreign-owned parent companies which, in various joint ownership patterns, own a single large concessionaire company in each country.[2] In Libya, by way of contrast, there are numerous concessionaires.

Oil provides the principal foundation for the economic life of all these countries. In the first instance, practically all the revenues from oil accrue to the foreign companies through export and sale of crude oil or refined products in markets around the world. The revenues to the countries arise essentially in two ways: (1) through company tax and royalty payments to governments and (2) payments by the companies to local factors of production. The first of these is by far the larger amount, which gives to the economies of the countries a unique characteristic–the fact that the stream of payments which supports most of the economic life originates in the public treasury in the form of foreign exchange. From there a portion passes into local currency to support the various forms of public expenditure, including much direct expenditure on economic development projects. The flow then fans out into private hands and, through the multiplier effect, supports the bulk of the indigenous economic operations of the nongovernmental, nonoil sector. A portion of the exchange earnings, as will be seen, also finds its way immediately into reserves held in foreign accounts.

[1] In quantitative terms, the main exclusions are Abu Dhabi and Oman in the Middle East and Algeria in North Africa.

[2] This does not accurately describe the situation in Iran, where the properties are technically owned by the government. See Chap. 9.

This chain of relations makes it necessary to approach an analysis of the dependence of the countries upon oil operations through a study of oil exports and foreign-exchange earnings based on these exports. The statistical tables in this chapter, which illustrate the economic dependence of the countries upon the ongoing operations of the foreign-owned companies within their borders, will be confined, in the main, to measures of these export-and-foreign-exchange aspects. This emphasis excludes from consideration the broader developmental impact of oil operations. Also excluded is the relative importance of oil operations in the country's Gross National Product. Quite apart from the rudimentary character of GNP accounting in these countries, there is the problem that the pro forma measures of the contribution of oil operations yield a result that is obviously too low. In Kuwait, for example, the pro forma calculation indicates that almost 60 percent of the GNP was generated outside the oil sector. Yet intuitively one knows that the real contribution of the oil sector must be very much greater since Kuwait is devoid of practically all other sources of productive income, except those following in the wake of oil in the form of various services rendered within the country. The reason for the anomaly is plain. These other services, although they would eventually be choked off without a continuing flow of oil income, are not attributed to oil in national income accounting. Kuwait, and the other exporting countries, profit from a sustained multiplier effect of the oil revenues and local factor payments as they circulate through the economy before finally being leaked off in the form of savings abroad or imports of goods and services. It was not possible for us to undertake the intensive study that would be required to measure this multiplier effect.

Oil Exports in Total Merchandise Exports

The first measure to come to mind in assessing the importance of oil exports for the producing countries is one which sets them off against total merchandise exports. This relationship is shown in Table 8.2; but, for reasons explained later, the figures overstate the real importance of oil earnings to the national economy. The percentage share of petroleum in their exports varies from very high to "near total." In the case of Saudi Arabia, oil makes up more than 99 percent of the total value of visible exports in recent years, while Iran, for which the percentage dependence is, relatively speaking, the least of the five, the oil share nonetheless has run to about 86–88 percent over the last five years shown. For Iraq the oil sector's share in total exports hovers around 92 percent, for Kuwait 98 percent,

Table 8.1

Oil Production: Total Middle East and North Africa and Five-Country Survey Group,[a] 1961–68

(Million Metric Tons)[b]

	1961	1962	1963	1964	1965	1966	1967	1968
Total Middle East and North Africa	300.9	342.2	391.6	455.0	507.3	576.7	626.3	741.5
Middle East	280.3	308.0	339.6	380.5	415.5	463.6	496.6	557.7
Iran	59.4	66.0	73.8	84.8	94.8	105.1	129.3	141.5
Iraq	49.0	49.2	56.7	61.6	64.4	68.0	60.1	74.0
Kuwait[c]	87.5	98.7	105.5	116.4	118.6	125.6	126.3	133.0
Saudi Arabia[c]	74.0	82.3	89.4	96.1	110.4	130.5	140.8	152.2
Above four countries	269.9	296.2	325.4	358.9	388.2	429.2	456.6	500.7
As percent of Middle East	(96.3%)	(96.2%)	(95.8%)	(94.3%)	(93.4%)	(92.6%)	(91.9%)	(89.8%)
North Africa	20.6	34.2	52.0	74.5	91.8	113.1	129.7	183.8
Libya	0.9	8.7	22.3	41.4	58.7	72.3	83.5	126.0
As percent of North Africa	(4.4%)	(25.4%)	(42.9%)	(55.6%)	(63.9%)	(63.9%)	(64.4%)	(68.6%)
Five-country survey group's share in Middle East and North Africa production	(90.0%)	(89.1%)	(88.8%)	(88.0%)	(88.1%)	(87.1%)	(86.2%)	(84.5%)

[a] Refers to the five leading area producers around which discussion in this chapter centers.

[b] A metric ton may be converted to barrels, on the average, at 7.2 barrels per ton.

[c] Including, throughout this chapter, one-half of the Neutral Zone.

Source: British Petroleum Company, *Statistical Review of the World Oil Industry,* various issues.

and in the Libyan case the share of petroleum rose from 94.3 percent to 99.3 percent between 1962 and 1966. However, in the latter instance, since appreciable volumes of oil exports began only in 1962, the percentage was very much lower in earlier years.

The measurement of the gross export values in Table 8.2 requires careful interpretation because the valuation of petroleum exports is ambiguous. The major fraction of such exports of oil moves between subsidiaries or affiliates of geographically diversified, vertically integrated oil companies, so that for them there does not exist a real open market price for crude oil. Thus the values entered for most of the exports are, according to the terminology ordinarily applied, "notional." For practically all of the exports for the countries concerned these notional values, based upon posted prices, exceed values based on open market prices by perhaps 25 percent.[3] In Libya, where a hybrid system for invoicing oil exports was in operation, the overstatement is probably closer to 15 percent.[4]

The discrepancy between official export values and open market valuations affects the absolute value of the exports as shown. However, the percentage figures given for the share of petroleum exports in total exports would be quite insensitive to revaluations in oil exports, particularly when oil constitutes practically the whole of merchandise exports. Even in Iran, where the oil share is least, a reduction of 25 percent in the invoiced value of the oil exports would reduce the oil sector's share in gross merchandise exports by only three percentage points; for example, the correction would have caused a drop from 88.3 percent to 85.0 percent in 1966.[5]

Retained Value from Oil Operations

A substantial fraction of the value of oil exports is retained or disbursed abroad by the companies. What matters to the exporting countries is the remainder, the amount of foreign-exchange receipts payable to the government and the amount earmarked for expenditure in the host country. These make up the "retained value" generated by the export of oil from the foreign-owned oil sector, which is defined as the sum of two distinct components: (1) the income taxes, royalties, and miscellaneous fees which are paid to the host government; and (2) the payments made by the operating companies or its contractors to local factors of production. The latter consist chiefly of wages and salaries (including that part of the salaries of expatriate personnel which is disbursed in local currency), but they also cover any materials or services which might be locally purchased. Fortunately, the oil revenues which accrue to the host government (the major component, by far), and the other elements in the retained value which remained within the domestic economy of the host country can both

[3] For example, Persian Gulf prices on open market transactions were, in early 1969, said to be in the range of $1.30–$1.40 per barrel, whereas comparable posted prices were around $1.80. (Statement by Helmut J. Frank in *Governmental Intervention in the Market Mechanism*, Hearings before the Subcommittee on Antitrust and Monopoly of the Committee on the Judiciary, U. S. Senate, 91st Congress, 1st sess., *The Petroleum Industry,* Part 1, April 2, 1969, p. 348.) See Chap. 10 for discussion of "realized prices" and "posted prices."

[4] In Libya, the export values consist of a mixture of transactions at market prices and others at posted prices between affiliates. Hence, its figures overstate the "market" value by relatively less.

[5] If the 1966 figure for total oil exports is reduced by 25 percent, the value of the oil drops from $1,229 million to $922 million, and total "revalued" exports become 163 + 922 = 1,085; 922 ÷ 1,085 = 85 percent.

Table 8.2

Share of Oil in Total Merchandise Exports,[a] 1961–67[b]

(Values in Million $U. S.)

	1962	1963	1964	1965	1966	1967
Iran[c]						
Oil share (percent)	(88.0)	(88.5)	(87.2)	(86.6)	(88.3)	n.a.
Oil exports	870	940	992	1,085	1,229	
Total exports	989	1,062	1,137	1,253	1,392	
Iraq[d]						
Oil share (percent)	(90.5)	(92.9)	(94.0)	(93.3)	(92.0)	(91.1)
Oil exports	626	725	789	823	860	761
Total exports	692	781	840	882	934	835
Kuwait[e]						
Oil share (percent)	(98.1)	(97.8)	(97.5)	(98.0)	(97.6)	n.a.
Oil exports	1,160	1,215	1,310	1,386	1,461	
Total exports	1,182	1,243	1,344	1,414	1,498	
Libya[c]						
Oil share (percent)	(94.3)	(98.1)	(98.6)	(98.9)	(99.3)	n.a.
Oil exports	133	329	611	788	995	
Total exports	141	335	620	797	1,002	
Saudi Arabia[c]						
Oil share (percent)	.... (99.0+)					n.a.
Oil exports	992	1,054	1,035	1,190	1,536	
Total exports	997	1,059	1,045	1,197	1,543	
Total						
Oil share (percent)	(96.9)	(95.2)	(95.0)	(95.1)	(95.5)	n.a.
Oil exports	3,781	4,263	4,737	5,272	6,081	
Total exports	3,901	4,480	4,986	5,543	6,369	

n.a.–When shown for most recent years should be taken to mean that the statistics were not available when research on this chapter was done.

[a] There is no consistent relationship between the nominal official valuation of oil exports and the discernible market price for such oil. Moreover, even within the same source, as the International Monetary Fund, *Balance of Payments Yearbook*, the methods are not mutually consistent; e.g., the Saudi figures are essentially posted prices, whereas the Libyan data represent a variable weighted average of posted prices and actual market realizations.

[b]Throughout this chapter, data obtained from national sources (such as Kuwait, *Statistical Abstract* or the Saudi Arabia, *Statistical Yearbook*) frequently refer to these countries' fiscal and/or lunar years. Such data have been entered in the tables under the closest corresponding year in the Gregorian calendar.

[c]International Monetary Fund, *Balance of Payments Yearbook*, Vol. 19; oil exports and total exports taken from lines (A.l.l) and (A.l), respectively, as amended in Notes to country pages in the *Balance of Payments Yearbook*.

[d] International Monetary Fund, *International Financial Statistics*, current issues. Oil exports and total exports drawn from lines (70a) and (70), respectively.

[e] Government of Kuwait, *Statistical Abstract* (1968); exports include some re-exports.

be determined independently of the figures presented in Table 8.2.

1. *Payments to Governments*

The oil revenues themselves arise in the forms of a royalty per barrel and a tax levied upon income ascribed to the companies on their oil exports, both of which are computed upon the basis of "posted" or "tax reference" prices. (The mechanism is described in Chapter 10.) The actual payments to the governments are made by the companies almost without exception in foreign exchange, the latter, in turn, being either dollars or pounds sterling, except for a very small fraction of the transfers conducted in other currencies. The cash receipts of the governments for the years 1961-67 are recorded in Table 8.3, which shows that the aggregate revenues of this group of countries stood at more than $3 billion in 1966, having more than doubled since 1961.[6] This rapid increase in cash receipts of over 16 percent per annum is the result of two separate trends: (1) the increase in production levels, and (2) a rise

[6] As indicated in the footnotes to Table 8.3, the government's revenues are shown on the basis of cash payments received during the year. This results in differences for the several countries from the statistics of government oil revenues published by *Petroleum Press Service*. For the five countries taken together the differences from *PPS* are not too great; the 1966 *PPS* total for the five is $2.96 billion compared with $3.04 billion in Table 8.3. *PPS* also provides a total for the five countries for 1967, a year not completely covered in Table 8.3, of $3.3 billion; and a provisional total for 1968 for the same group of countries of about $4.0 billion (*Petroleum Press Service*, August 1969, p. 297).

Table 8.3

Government Receipts of Oil Revenues (Cash Basis)[a] 1961–67

(Million $U.S.)

	1961	1962	1963	1964	1965	1966	1967
Iran[b]	299	342	388	665[f]	519	652	742
Iraq[d]	265	266	308	353	368	392	n.a.
Kuwait[c]	467	484	533	724	781	818	n.a.
Libya[d]	0.3	15	57	144	227	390[g]	n.a.
Saudi Arabia[e]	378	410	608	523	663	790	909
Total	1,409	1,517	1,894	2,409	2,558	3,042	n.a.

n.a.–When shown for most recent years should be taken to mean that the statistics were not available when research on this chapter was done.

[a]These figures are not entirely comparable with those published periodically by *Petroleum Press Service* (hereafter *PPS*). The data in the table above are on a cash basis with retroactive settlements credited to the year in which cash disbursements were made, and are the figures most relevant to balance of payments analyses. The revenue figures cited in *PPS* (see n. 6, p. 99) are based upon the accrued tax liabilities of the companies, which can be appreciably more than the cash tax payments where production has been growing rapidly or where tax settlements significantly lagged exports. In *PPS* statistics, retroactive settlements also are carried back and entered for the year in which the retroactive liability had been incurred.

[b]Iran, Bank Markazi, *Annual Reports.*

[c]Includes nonrecurring tax acceleration payments for 1964/65–1966/67.

[d] *Balance of Payments Yearbook,* Vol. 19–Analytical Presentation.

[e] Saudi Arabian Monetary Agency (SAMA), *Annual Reports.*

[f] Includes nonrecurrent bonus of $185 million received in connection with the joint-ventures in the offshore areas.

[g] Provisional estimate.

in the per-barrel tax payments over the same period (see Table 10.1, p. 122).

The pattern of growth in oil revenues varied among the several countries. In Libya production began only late in 1961, so that the volume of exports was increasing rapidly while unit costs were falling sharply (and the tax base consequently rising) over the same period. As a result, per-barrel payments to the government climbed from about 65 cents in 1962 to close to 90 cents in 1966 at the same time as the level of exports soared from a trickle in 1961 to some 1.5 million barrels/day in 1966. The revenues of Iran and Saudi Arabia slightly more than doubled in those five years, while those of Kuwait rose by some 75 percent. Iraq has lagged in growth of tax revenues for two reasons. Because of the unresolved political confrontation with the government, the Iraq Petroleum Company (IPC) has undertaken no further development investment, and offtake has consequently risen but little since the surge which followed the opening of the marine terminal at Khor al-Amaya. Secondly, Iraq did not sign the royalty-expensing agreement (see Chapter 10), and hence has not benefited from the related increase in per-barrel revenues which the other Middle Eastern OPEC members have since enjoyed (though adjustments for Mediterranean supplies after the 1967 Arab-Israeli war have somewhat offset this loss).

2. *Payments to Local Factors of Production*

The oil revenues, which are collected directly by the governments concerned, understate the foreign exchange receipts to the country from the foreign-owned oil sector by the amount of the payments to local factors of production. These are not large relative to the flow of tax revenues, but the aggregate amount for the area is nonetheless a quite significant sum. Thus, in 1966 the total retained value generated by the foreign oil sector was almost $3.5 billion, of which over $3 billion was in payments to governments, and over $400 million in local expenditures.

The additional foreign-exchange receipts disbursed in local expenditures by the oil companies are summarized in Table 8.4. This table was computed from the data on local currency purchased by the companies as required to cover local outlays. These figures understate the companies' actual payments to local factors by the amount of any local currency income which they may collect–from direct marketing, as in Libya, or sales of refined products to the national company, as in Iran. On the other hand, the data also include an unknown amount of imports which have been purchased via local merchants and paid for in local currency; the benefit to the national economy of the latter category of transactions consists only of the merchant's or agent's markup. In the case of Saudi Arabia, a correction was explicitly estimated for this effect.[7]

Despite the measurement problems, it is clear that net local outlays did not increase by much between 1961 and 1967. They may, in fact, have been essentially stationary, if one allows for the possibility that much of the ostensible increase from a total of $323 million to $401 million in-

[7]ARAMCO's local outlays rose from $72 million to $122 million between 1961 and 1967. A substantial portion of the increase reflected purchases of "materials"; these, in turn, consisted, in good part, of imported goods. It was assumed that 85 percent of the sum expended locally for "materials" covered imports which had been purchased through local agents. For recent years this calculation yields an estimate of about $45 million worth of imported goods thus obtained. (See Table 8.5, n.d.)

cludes an increasing volume of imported goods which the companies order through local firms as part of an overall effort to foster host country sources of supply. Both in Kuwait and in Libya, for example, the reported local purchases of materials and supplies when compared with data from the industrial censuses strongly imply a significant import content in those figures. As the total retained value generated by the oil sector has been growing, the relatively constant level of local expenditures is reflected in their steadily diminishing comparative importance, from 18.6 percent of retained value in 1961 to 12.1 percent in 1966.

Retained Value and Total Export Earnings

In contrast to Table 8.2, which relates the value of oil exports to total merchandise exports, another and far more relevant measure is shown in Table 8.5: retained value of foreign exchange earned in oil operations in comparison with total foreign-exchange earnings on merchandise and nonmerchandise account. Retained value has been defined and measured in the preceding section, and the total of the two components is shown in Table 8.5. Gross earnings of foreign exchange, as shown in the table, are defined to equal:

(1) Retained value of the foreign oil sector, *plus*
(2) Exports of merchandise, excluding oil and re-exports, *plus*
(3) Receipts on current services account, such as tourism, portfolio income on overseas investments, etc.

This definition embraces only those foreign-exchange receipts which are earned, as distinguished from inflows of foreign exchange arising from investment by foreigners, long-term borrowings by the governments, or other forms of credit. Only in Iran is there even a trickle of foreign equity investment, but the stream of net foreign borrowings of one sort or another would significantly enhance the flow of foreign resources available to the Iranian and Iraqi economies in certain of the years. Such borrowings have been explicitly excluded in this context in order to emphasize as clearly as possible the contribution of the oil sector to the current flows of earned receipts. Borrowings and transfer payments (such as foreign aid or rentals on military installations) are, of course, responsive to an additional set of considerations.

In all cases, the real dependence of the country upon foreign exchange earned by the oil sector is less than that indicated by the position of oil in total merchandise exports shown in Table 8.2. For Iran and Iraq, some 25 percent of the country's foreign-exchange earnings are, in general, derived outside of the oil sector, in both cases due to significant receipts on services account, as well as to the exports of nonoil merchandise. Saudi Arabia earns some 10 percent of its receipts of foreign exchange from nonoil sources, split roughly equally between the gross proceeds of the annual pilgrimage (*Hadj*) and income on its sizeable investment portfolio. In the case of Kuwait, the nonoil earnings generate about 15 percent of the State's foreign exchange receipts. There, the degree of dependence on oil is mitigated chiefly by the quite substantial income which the government and the commercial banking system enjoy on

Table 8.4

Local Currency Expenditures, 1961–67

(Million $U.S.)

	1961	1962	1963	1964	1965	1966	1967
Iran[a]	100	95	83	89	94	107	106
Iraq[b]	54	48	44	38	37	38	n.a.
Kuwait[c]	28	35	38	38	43	49	53
Libya[d]	77	90	111	131	153	131	117
Saudi Arabia[e]	64	65	69	70	78	92	87
Total	323	333	345	366	405	417	401[f]
Local currency expenditures as a share of oil sector total retained value[g]	18.6%	18.1%	15.4%	13.3%	13.7%	12.1%	n.a.

n.a.—When shown for most recent years should be taken to mean that the statistics were not available when research on this chapter was done.

[a]Represents purchases of rials by the Consortium and other foreign operating or exploration companies, net of any domestic income receipts. Iran, Bank Markazi, *Annual Reports.*

[b] *Balance of Payments Yearbook*, Vol. 19—Analytical Presentation.

[c] Private communication; data from Ministry of Finance and Oil.

[d]Bank of Libya figures except 1961, which is from *Balance of Payments Yearbook*, Vol. 19.

[e] Estimated from data in ARAMCO *Annual Reports*; estimated indirect imports—i.e., purchases from abroad via local firms—have been subtracted from the reported annual figures (see n. d, Table 8.5).

[f] Incorporates an estimated figure for Iraq of $38 million.

[g] Oil sector retained value taken from Table 8.5.

Note: (1) Owing to some differences in sources, the figures in this table may differ very slightly from local currency expenditures derived as the difference between the retained-value figures of Table 8.5 and the oil revenue figures of Table 8.3. (2) In addition to the footnotes below, see the accompanying text discussion regarding the derivation of local currency expenditures.

Table 8.5

Oil Sector Retained Value Compared to Gross Foreign-Exchange Earnings,[a] 1961–67

(Million $U.S. and Percentages)

	1961	1962	1963	1964	1965	1966	1967
Iran[b]							
RV	399	437	471	754	613	757	848
GFEX	n.a.	570	619	900	822	986	1,139
Percent RV of GFEX	n.a.	(77%)	(76%)	(84%)	(75%)	(77%)	(74%)
Iraq[c]							
RV	319	314	352	391	405	430	n.a.
GFEX	n.a.	454	454	520	558	601	n.a.
Percent RV of GFEX	n.a.	(75%)	(78%)	(75%)	(73%)	(71%)	n.a.
Kuwait[e]							
RV	501	521	571	767	832	871	n.a.
GFEX	n.a.	608	672	888	983	1,053	n.a.
Percent RV of GFEX	n.a.	(86%)	(85%)	(86%)	(85%)	(83%)	n.a.
Libya[c]							
RV	77	94	164	250	358	504	n.a.
GFEX	150	176	258	324	424	593	n.a.
Percent RV of GFEX	(51%)	(54%)	(64%)	(77%)	(84%)	(85%)	n.a.
Saudi Arabia[d]							
RV	442	475	677	593	741	882	996
GFEX	514	533	739	675	837	987	1,123
Percent RV of GFEX	(86%)	(89%)	(92%)	(88%)	(89%)	(89%)	(89%)
Total							
RV	1,738	1,841	2,235	2,755	2,949	3,444	n.a.
GFEX	n.a.	2,341	2,742	3,307	3,624	4,280	n.a.
Percent RV of GFEX	n.a.	(79%)	(82%)	(83%)	(81%)	(80%)	n.a.

RV = oil sector retained value

GFEX = gross foreign-exchange earnings

n.a.—When shown for most recent years should be taken to mean that the statistics were not available when research on this chapter was done.

[a] Retained value equals government oil revenues plus local currency expenditures by the companies; gross foreign exchange earnings are equal to the retained value defined above *plus* receipts on current account (merchandise and other receipts) from nonoil sources, specifically excluding any long- or short-term borrowings by the country.

[b] From Iran, Bank Markazi, *Annual Reports*. The local currency expenditure component is based upon purchases of rials by the foreign oil companies; local outlays by the Consortium recently have exceeded its rial purchases by as much as $20–$25 million, reflecting rial income earned by the companies.

[c] *Balance of Payments Yearbook*, Vol. 19—Analytical Presentations for Iraq and Libya.

[d] Derived from data in Saudi Arabian Monetary Agency publications and ARAMCO *Annual Reports*; the published figures have been revised by reducing local expenditures by the amount of the estimated direct and indirect imports of ARAMCO via local firms. The approximate magnitude of the correction in recent years was some $45 million.

[e] Private communication; data from Ministry of Finance and Oil.

their overseas portfolio investment. For fiscal year 1966–67 this flow was estimated to aggregate approximately $175 million.[8]

For the period since the early 1960's, no significant trend in relative dependence upon the oil sector can be discerned in any of the countries except Libya, where the oil sector's share in effective foreign-exchange receipts has risen from 51 percent to 85 percent. In view of the projections of a sustained high rate of increase in Libyan oil exports, one may expect Libya's percentage dependence upon foreign exchange earnings in oil to rise still further, reaching a plateau in the vicinity of perhaps 95 percent unless there were to be concomitant appreciable increases in foreign savings.[9]

[8] Computed from Kuwait, *Statistical Abstract, 1968*.

Oil Revenues and the Government Budget

Since oil revenues themselves accrue directly to the government exchequers, their most immediate impact upon the host country is via the government's own budget. Therefore, it is important to measure the share of oil revenues in the disposable funds available to the governments.

[9] In the absence of increased foreign saving, no rise in investment income receipts can be envisaged, which appears to be the only potential source of significant nonoil foreign-exchange earnings.

Table 8.6

Oil Revenues as a Percentage Share of Total Government Revenues, 1961–67

	1961	1962	1963	1964	1965	1966	1967
Iran[a]	n.a.	n.a.	56%	61%	62%	58%	57%
Iraq[b]	n.a.	n.a.	n.a.	68	68	63	n.a.
Kuwait[c]	92%	91%	92	93	87	83	n.a.
Libya[d]	n.a.	n.a.	40	57	66	73	84
Saudi Arabia[e]	86	88	90	89	89	89	90

[a] The government revenue has been defined as the composite receipts of Ordinary and Plan Organization budgets, excluding borrowings and revenues of public enterprises. The entry therefore does not conform with the definition used in the Iran, Bank Markazi, *Annual Reports.*

[b] Government revenues = consolidation of Ordinary Budget and Economic Plan, but after deducting receipts of "Government Administration." All autonomous budgets are excluded; source: Central Bank of Iraq, *Bulletin*.

[c] Calculated from *Kuwait al-Youm*, 1968; oil revenues exclude the tax acceleration payments, which were nonrecurrent.

[d] Computed from Bank of Libya, *Bulletin,* various years. In the absence of the necessary data, the gross government revenues have *not* been corrected for entreprenuerial receipts, such as those from the tobacco monopoly. The correction would have been small.

[e] Computed from Saudi Arabia, *Statistical Yearbook, 1967.* "Departmental Revenues," such as the Railway revenues, have been excluded.

Note: The n.a.'s arise primarily from the nonavailability of the government-revenue figures needed for this calculation. See text for discussion of budgetary concepts employed.

However, before proceeding to measure the contribution of oil earnings to the government's budget, it is necessary to delineate that budget. In the case of Iran and Iraq, a primary distinction is drawn for official purposes between the ordinary and development budgets; in Iran the latter is administered by the Plan Organization, whereas in Iraq, its counterpart is the "Economic Plan," which is funded separately but otherwise has been administered through the Ministries. In both instances a specified share of the oil revenues is said to be allocated to the development budgets, the splits being 70 percent and 50 percent, respectively. That allocation constitutes the overwhelming source of funds on current account for the development budgets.

The ordinary budget receives the remainder of the oil revenues in Iran and Iraq, plus whatever taxes, excises, or customs tariffs the government collects. Both in Iran and Iraq there have been recurrent transfers from the development budget into the ordinary budget, while, conversely, it has become a practice to allocate to the development budget certain categories of recurrent expenditures which can be plausibly depicted as "developmental," such as infrastructural maintenance costs. In addition, in Iraq, parallel with the increased number of expropriations and nationalizations since 1961, the government budget, including the appended and autonomous budgets, has come to encompass a steadily increasing fraction of the nation's economy, and its limits cannot be precisely defined.

Accordingly, the following approach has been adopted in order to establish a reasonably specific compass for the "budgets" which would permit comparable presentations among countries:

(1) The ordinary and development budgets have been consolidated and treated as a composite, single budget. This bypasses the problem of interbudgetary transfers.

(2) Only governmental revenues on current account, as narrowly defined, have been included. Specifically, revenues from government enterprises, whether gross or net, have been excluded, together with any net borrowings from abroad.

The excluded transactions constitute a large element in some of the official budget presentations.

The share of oil revenues in government budgets, as just defined, is shown in Table 8.6. The relationship between oil revenues and government revenues seems to be essentially stable, no trend being discernible except for Libya. In Iran oil revenues have made up some 60 percent of composite government revenues in recent years, while customs tariffs on imports contributed another 20 percent of the total.[10] (However, the imports are in large measure made possible by the earnings available from oil revenues.) Iraq's budgetary dependence upon its oil revenues is much the same as Iran's, the share being about two-thirds; and, as in the case of Iran, tariffs on imports are in large measure an indirect product of oil earnings. The role of oil revenues in the fiscal balance in Saudi Arabia is also essentially constant, having hovered in the vicinity of 90 percent since the early 1960's. In Kuwait, to the contrary, the relative dependence upon oil revenues actually declined between 1961 and 1966, chiefly as a result of greater receipts of portfolio income from abroad.

Unlike the others, Libya's fiscal dependence upon oil has risen sharply from 40 percent, based upon a low level of government revenues in 1963, to 84 percent as of 1967, with a further rise quite likely. In Libya tariffs are relatively low, control over imports being loosely exercised via exchange licensing, and the recently imposed income

[10] Iran, Bank Markazi, *Annual Report.*

Table 8.7

Direct Employment[a] by Foreign-owned Oil Companies, 1961–67

	1961	1962	1963	1964	1965	1966	1967
Iran[b]	38,398	33,282	31,369	30,708	29,657	28,246	25,918
Iraq	12,870	12,157	11,553	11,255	11,076	10,954	10,765
Kuwait[d]	6,487	6,075	5,908	5,672	5,443	5,520[c]	5,560[c]
Libya[e]	n.a.	n.a.	n.a.	5,510	5,872	5,669	5,378
Saudi Arabia[f]	n.a.	13,573	12,988	12,880	12,783	12,664	12,073
Total	77,255[g]	70,587[g]	67,318[g]	66,025	64,831	63,053	59,694

[a] Including expatriates.

[b] Consortium only, but including the nonbasic personnel who are officially employed by National Iranian Oil Company. Aggregate direct employment by the joint-venture operations was about 500 late in 1968, exclusive of contractors.

[c] In absence of available data for 1966–67, Arabian Oil Company employment estimated at 900.

[d] Covering Kuwait Oil Company and one-half of the Arabian Oil Company in the Neutral Zone; data for Aminoil (in the Neutral Zone) not publicly available.

[e] Exclusive of contractors' employment, which might amount to an additional 50 to 75 percent of the basic figures carried here.

[f] ARAMCO only; employment by Getty Oil and the one-half share of Arabian Oil Company's work force in the Neutral Zone might aggregate another 1,000 persons.

[g] Including estimates for Libya of 5,500 for the years 1961–63 and 14,000 for ARAMCO in 1961.

Source: Reports of the companies, except for Libya, where the annual industrial census data were used.

tax law affects a very small portion of the population. Consequently, there is no reason to expect nonoil revenues to rise at a rate comparable with those received from oil operations in the future.

Employment in the Oil Sector

Employment in the five countries by foreign-owned oil companies has been falling steadily in the years shown, despite the substantial increases in output (see Table 8.7). Specifically, gross employment by the industry has fallen by roughly 25 percent, while the total production of crude oil in the same countries has about doubled during the same period. Although we were unable to obtain the necessary statistics in these countries, it appears that direct oil industry employment accounts for a small fraction of the total work force in the countries here covered.[11] (Bahrain, Qatar, and Abu Dhabi, not covered in this discussion, seem to be an exception to the rule.)

There are, nevertheless, substantial numbers directly employed in oil industry operations, and their earnings constitute the largest identifiable component in the local currency expenditures shown in Table 8.4. Because of its economic importance, and also owing to its high visibility, employment policy has been the source of much difficulty between companies and governments, and of frictions with the local population.

One source of difficulty has involved the employment of more labor than is actually required. In spite of the decline in employment that has taken place (partly as a result of programs which were consciously implemented to accelerate voluntary separations), as much as one-third of the remaining labor force is said to be superfluous in some countries. The workers are retained because of the potentially adverse political repercussions of any further, extensive layoffs. Since the annual wages and salaries are a 100 percent deduction against taxable income in the countries here covered, the financial burden of the supernumeraries is borne by both the governments and the concessionaires, each acquiescing for its own reasons. Each dollar of unnecessary labor cost, on the usual 50 percent profit tax rate, reduces company profits by 50 cents and government revenue by the same amount. From the standpoint of the host government, however, each such dollar expended on local labor represents 50 cents direct benefit to the economy, except to whatever extent the labor employed in the petroleum sector might have been equally productively employed elsewhere within the domestic economy. The latter eventuality is remote.

The size of the expatriate Western staff has also proved to be a sensitive subject in that the presence of sizeable blocks of expatriate technicians and management personnel acts as a continual irritant, an obtrusive reminder to the host country of the enclave character of the oil sector. All of the producing countries have worked for greater employment of their nationals and especially for

[11] Although no figures on total work force were available to us, some notion of the comparative importance of oil industry employment is conveyed by the following comparisons between oil company employment (from Table 8.7) and total male population (from United Nations, *Statistical Yearbook, 1967*):

	Male population (thousands)	*Oil employment (thousands)*
Libya (1964)	800	6
Iran (1966)	13,000	28
Iraq (1965)	4,000	11

In the case of Kuwait, oil company employment of 5½ thousand in 1967 compares with a *total* estimated population of 520,000 (from *International Financial Statistics*).

accelerated promotion of more local personnel into staff and management positions. The greatest reduction is by the Iraq Petroleum Company Group, which had reduced its expatriate staff within Iraq to a mere 65 persons by 1967, due in part no doubt to cessation of exploration and development activities after 1961 (see Chapter 9).

	Western expatriate personnel					
	Iran	Iraq	Kuwait	Saudi Arabia	Total 4 Middle East countries	Libya
1961	635	379	850	2,180[a]	4,044[b]	n.a.
1967	328	65	530	1,320[a]	2,243	1,653

[a] Computed from tabulated percentages of total work force; 1961 figure is actually for 1960.

Source: Same as Table 8.7.

The total number of Western expatriate staff has been declining. With an allowance for the unknown number of expatriates active in Libya in 1961, the decline may have been from about 5,500 total in 1961 to some 4,000 in 1967. However, the percentage share has scarcely fallen at all. Western expatriates still comprise some 7 percent of total direct employment. The development of indigenous staff has gone furthest in Iran and Iraq, where the process started earliest, and where the domestic education system, supplemented by the companies' own in-house and scholarship programs, was adequate to the task. This is reflected directly in the relatively low incidence of expatriate staff–less than 1½ percent of total employment in Iran and less than 1 percent in Iraq. By way of comparison, Western expatriate staff is approximately 10 percent in Kuwait and Saudi Arabia. It comes to 25 percent in Libya, where no doubt the large number of concessionaire companies and the early stage of development affect the numbers.

In addition to Western expatriates, there exist fairly large numbers of third-country nationals working in the industry, chiefly in Kuwait and Saudi Arabia, but also to a lesser extent in Libya. These are mainly Pakistanis, Indians, and Palestinians who migrated into the oil-exporting states in order to fill posts requiring clerical and technical skills which did not exist locally in the numbers required by the burgeoning oil industry. The long tradition of English-language education in the subcontinent and in Palestine under the Mandate was a principal factor in that influx. In Kuwait, for example, nationals still make up only about one-quarter of the industry's direct work force, most of the remainder being Palestinians or Pakistanis.

Foreign-Exchange Earnings and Reserves Compared with Nonoil Foreign-Exchange Expenditures

Here we ask the question: How does the foreign exchange spent by the oil-exporting countries, on their own account, compare with their foreign-exchange earnings and reserves? The following three critical elements are involved:

(1) How much foreign exchange does the country consume on its own account, exclusive of any transactions by the foreign oil companies?
(2) What foreign exchange does the country earn annually besides that generated by the oil companies?
(3) What is the size of the country's foreign-exchange reserves?

With this information it is possible to determine a particular country's degree of dependence on the *current* flow of foreign-exchange earnings from oil operations in supporting its *current* level of imports on nonoil account. To assess dependence on continuity in the current flow of foreign-exchange earnings it is necessary, however, to have some measure of the size of the foreign-exchange reserves that have been accumulated. The latter are summarized in Table 8.8, based upon officially recognized balances as defined by the International Monetary Fund.

Table 8.8

Gross Holdings of Foreign-Exchange Reserves,[a] 1961–68

(Million $U.S.)

	1961	1962	1963	1964	1965	1966	1967	1968[b]
Iran	n.a.	229	253	207	284	263	289	292
Iraq	222	195	296	251	237	326	368	438
Kuwait	n.a.	n.a.	1,253	1,479	1,465	1,834	1,754	n.a.
Libya	87	96	122	172	246	339	385	539
Saudi Arabia	238	268	522	596	755	813	944	n.a.
Total	n.a.	n.a.	2,446	2,705	2,987	3,575	3,740	n.a.

n.a.–When shown for most recent years should be taken to mean that the statistics were not available when research on this chapter was done.

[a]Includes gold. Gross foreign-exchange reserves are essentially those reserves declared in *International Financial Statistics,* before any deduction of short-term foreign liabilities of either the Government or the Central Bank, and specifically including the assets assigned to the currency cover. The entries in *International Financial Statistics,* however, are not fully comparable, from one country to another.

[b] May not be completely comparable with prior-year data because of differences in treatment of the commercial banks' foreign holdings.

Source: International Financial Statistics and various Central Bank reports. Entries are on an end-of-calendar year basis.

Table 8.9

Disposable Foreign-Exchange Reserves,[a] 1961–67

(Million $U.S.)

	1961	1962	1963	1964	1965	1966	1967
Iran[b]	n.a.	118	125	65	130	102	108
Iraq[c]	66	28	95	46	8	79	105
Kuwait[d]	n.a.	n.a.	1,143	1,369	1,341	1,670	1,570
Libya[e]	n.a.	47	65	96	142	194	200
Saudi Arabia[f]	86	92	306	370	503	532	625
Total	n.a.	n.a.	1,734	1,946	2,124	2,577	2,608

[a] "Disposable foreign-exchange reserves" are obtained after deducting those foreign assets statutorily designated as part of the currency cover from gross foreign-exchange holdings. The latter conforms definitionally to the data in Table 8.8, although because of problems arising from noncomparability of reporting periods, different sources were to some extent relied upon in constructing the present table.

[b] Adapted from *International Financial Statistics;* debts of the government and short-term borrowings by the Iran, Bank Markazi, have *not* been deducted.

[c]From Central Bank of Iraq, *Bulletin*; Central Bank's foreign liabilities have *not* been deducted.

[d]Sum of holdings by the commercial banks and the liquid foreign assets of the government; *International Financial Statistics,* lines (3) plus (7a). The foreign assets of the Kuwait Currency Board are in effect those assigned to the currency cover.

[e] Bank of Libya, *Bulletin.*

[f]*International Financial Statistics*; net foreign assets of Saudi Arabia Monetary Agency (line 11) less "currency outside banks" (line 14a).

From Table 8.8, it is immediately apparent that very considerable differences exist among the several countries. Kuwait's balances, which include quite sizeable deposits held abroad by the four Kuwaiti commercial banks, totaled $1.75 billion as of the end of calendar year 1967, whereas Iran held only some $289 million. Indeed, the group can be divided into two distinct subgroups—Kuwait and Saudi Arabia, whose holdings approximate a billion dollars or more, on one hand, and Iran and Iraq, whose balances as of 1967 clustered about the level of $300-plus millions. Libya is a special case by virtue of its relative newness as a major producer.

The trends of gross holdings since 1961 are equally diverse; Iran's position as far as official holdings are concerned has changed but little, while Saudi Arabia has managed to almost quadruple its foreign-exchange balances. Libya has increased its balances from a paltry $87 million (1961) to $539 million in 1968, which may reflect the fact that Libya's recent oil revenues have outstripped the government's ability to spend. Total holdings of the five countries rose by more than 50 percent in the five years for which consistent statistical coverage is available; the aggregate figure for 1967—$3,740 million—represented approximately 30 percent of all foreign-exchange reserves (including gold) held by the nonindustrial countries, which is indicative of the financial resources concentrated in the oil-exporting countries of the Middle East.[12]

Three supplementary considerations must be introduced at this point which affect the interpretation of these reserve figures: (1) statutory requirements for the gold and/or foreign-exchange backing of the country's currency which could restrict its power of disposition over part of its gross foreign-exchange holdings; (2) short-term debts of the Central Bank; and (3) long-term foreign debt of the country.

[12] *International Financial Statistics*, February 1969, p. 14.

Each of the countries is obligated by domestic legislation to maintain a specified level of reserves of gold or foreign exchange as backing for its currency issue, as follows: Iran (40 percent); Iraq (70 percent); Kuwait (100 percent); Libya (100 percent); and Saudi Arabia (100 percent). This requirement freezes a certain component of its gross holdings of foreign exchange as far as commercial transactions are concerned. It is accordingly more appropriate to define "disposable foreign-exchange reserves" as those over and above the amounts specifically required as currency cover. These "free" balances are in fact those which can be deployed or drawn upon to meet the short-term requirements of the economy and to buffer any shortfalls between revenues and foreign-exchange outflows, unless currency cover rules are relaxed.

The disposable foreign assets, thus defined, are presented in Table 8.9. On this basis, Iran and Iraq recently have been operating with balances rather close to the margin. As of the end of calendar year 1965, Iraq had at its disposal only $8 million in free reserves, while Iran's balance of disposable reserves has averaged only just over $100 million since 1964. Kuwait's and Saudi Arabia's positions, to the contrary, are not strongly affected by subtraction of the currency cover, and their positive balances remain high.

If the short- and long-term foreign debts of the countries are taken into account, the foreign-exchange reserves of Iran and Iraq are even lower than is indicated in Table 8.9. For example, the short-term foreign borrowings and/or liabilities of the Iran Bank Markazi amounted to almost $110 million as of the close of the Iranian year 1346 (March 20, 1968).[13] If these outstanding, short-term liabilities are subtracted from the disposable foreign

[13] The increase in "short-term foreign exchange obligations" during the year was reported to be $91.1 million exclusive of the International Monetary Fund drawing of $17.2 million; Iran, Bank Markazi, *Annual Report* (1968), p. 353.

Table 8.10

Foreign-Exchange Expenditures on Current, Nonoil Account,[a] 1962–67

(Million $U.S.)

	1962	1963	1964	1965	1966	1967
Iran[b]	528	548	759	923	1,092	1,344
Iraq[c]	434	400	526	575	619	n.a.
Kuwait[d]	342	395	395	493	582	n.a.
Libya[e]	194	234	291	357	513	n.a.
Saudi Arabia[f]	349	384	484	645	861	950
Total	1,487	1,961	2,455	2,993	3,667	n.a.

n.a.–When shown for most recent years should be taken to mean that the statistics were not available when research on this chapter was done.

[a]Because of the effects of borrowings, unofficial capital transfers, etc., annual totals from this table, if subtracted from the total foreign-exchange receipts shown in Table 8.5, do not measure changes in foreign-exchange holdings, as shown, e.g., in Table 8.8.

[b] Iran, Bank Markazi, *Annual Reports*.

[c] *Balance of Payments Yearbook*, Vol. 19–Analytical Presentation.

[d] *International Financial Statistics*, current issues; the small amount of direct imports by the oil companies have not been subtracted.

[e]*Balance of Payments Yearbook,* Vol. 19–Analytical Presentation.

[f]Adapted from Saudi Arabia Monetary Agency, *Annual Report,* various years. The foreign exchange payments declared officially were reduced by (1) Tapline outlays; (2) the entire entry for each year, which is designated "factor income repatriations"; and (3) the sum of $100 million, which represents an estimate of salary remittances by expatriate personnel, direct and indirect imports by ARAMCO, and other overseas outlays by the Company.

exchange the residual balance appears to be slightly negative in some months[14]–i.e., Iran maintained no liquidity whatsoever in the technical sense. In the case of Iraq, the Central Bank had outstanding short-term foreign liabilities of at least $55 million at year-end 1967, which if subtracted, would have just about halved its ostensible level of disposable foreign-exchange reserves. Closer examination of the Iraqi figures on a monthly basis, deducting the short-term liabilities of the Central Bank of Iraq as well, would reveal that the country has only barely maintained positive liquidity in recent years. Thus, for Iran and Iraq, the estimates of disposable foreign exchange as computed are somewhat generous. With long-term debt taken into account, the apparent foreign-exchange reserves position of the two countries is further weakened.

So far, attention has been directed to foreign-exchange earnings and reserves. The other half of this comparison involves the annual outlays of foreign exchange on current account of the countries here covered. These are shown in Table 8.10, in which the figures cover the imports of merchandise and services *exclusive* of those on behalf of the foreign-owned oil companies. Thus, the entries reflect the current foreign-exchange requirements of the domestic nonoil economy. In the case of Kuwait it was not possible to extract the oil company transactions, but these are understood to be quite small compared to the total flows during the years in question. In the case of Saudi Arabia it proved necessary to restructure the country's balance of payments presentation quite substantially in order to identify those transactions which pertained to the Saudi economy itself. The Saudi figures, hence, are subject to some error, which is believed not to be serious. The gross foreign-exchange outlays as reported by Saudi Arabia Monetary Agency (SAMA) include both ARAMCO's transactions and those of the domestic economy, and these were revised downwards to include only those of the domestic economy.[15]

Once the foreign-exchange requirements of the domestic economy have been determined, these can be compared with the annual earnings of foreign exchange on nonoil account from exports of other goods and services or receipts from portfolio income. The excess of the first category over the second may be referred to as the "uncovered foreign-exchange requirement"–uncovered, that is to say, in the absence of current foreign earnings from oil. This is a measure of the amount of the foreign-exchange shortfall in the event that the oil revenues

[14]Iran, Bank Markazi, *Annual Reports* and the *International Financial Statistics* figures are not directly comparable, partly due to timing differences and partly due to different statistical treatment of the settlement accounts under bilateral trade agreements and short-term foreign liabilities.

[15]The following items were deducted: (1) factor income repatriations abroad, which subsume both ARAMCO's real "profit" as well as the putative profits which are a balancing item required by the "notional" valuation of the oil exports in the official balance of payments data (see pp. 97 to 98); (2) Tapline's expenditures outside of Saudi Arabia; and (3) the constant figure of $100 million, which represents an estimate of the remittances of salaries by expatriate personnel, direct and indirect imports by ARAMCO, and other overseas outlays by the Company. These corrections are quite substantial, aggregating 65 percent of the total outflows in the official data in the early 1960's and about 45 percent more recently.

are not sustained for some reason. The "import coverage coefficient" (Table 8.11) expresses the number of months during which each of the countries could continue its current nonoil industry imports in the absence of oil revenues. It is the ratio of disposable foreign-exchange reserves (Table 8.9) to the uncovered foreign-exchange requirements as defined above.

In calculating the extent to which each of the countries is covered by foreign-exchange reserve holdings in the event of a crisis affecting its oil revenues, the balances of long-term debt have not been subtracted from the disposable foreign-exchange reserves, although correction was made in the case of Iran and Iraq for the short-term borrowings by the Central Bank. Nor have contractual repayments of principal been included in the foreign-exchange outlays for each year.

The results are quite diverse. As of 1966, Kuwait could have sustained its current level of imports for 31 months based only on its official reserves, without recourse to those of the commercial banking system which would have added about 18 months. Iran, however, maintained disposable balances adequate for only about 1 month as of 1967, while the Iraqi figure gyrated between zero and about 3½ months over the years shown. Both Libya and Saudi Arabia are seen to hold comfortable balances; Libya's coverage remained in the vicinity of 5 months in spite of its steadily rising level of imports, while Saudi Arabia's coverage fell slightly from its high of 11 months during the years 1963 through 1965 to a level of 8 to 9 months in 1966/67. Thus, as of 1966/67, the members of the "creditor" subgroup would have been in a position to sustain an undiminished flow of imports for 6 months or more in the event of any serious disruption in their oil exports, whereas both of the "debtor" countries would have experienced acute economic consequences if the revenue flows were interrupted even for very brief periods, and an attempt were made to sustain imports at normal levels without resort to foreign aid or borrowing. Of course, the foreign-exchange reserve positions are rather more flexible than the above figures suggest, since the local currency cover could be relaxed and current imports reduced in case of need.

Table 8.11

Import Coverage Coefficients,[a] 1962–67

(Months)

	1962	1963	1964	1965	1966	1967
Iran[b]	n.a.	3	4	2.2	1	0.9
Iraq[c]	0.6	3.3	1.0	negative	1.5	n.a.
Kuwait[d]	n.a.	35	42	32	31	n.a.
Libya	4	4	5	6	5	n.a.
Saudi Arabia	4	11	11	11	8	9

n.a.–When shown for most recent years should be taken to mean that the statistics were not available when research on this chapter was done.

[a] Except for the derivation of the Iranian and Iraqi figures (described in footnotes b and c), the "uncovered foreign-exchange requirement" is computed as the difference between the foreign-exchange outlays in any year excluding oil company transactions (Table 8.10), and the nonoil foreign-exchange receipts in that year (Table 8.5: difference between "oil sector retained value" and "gross foreign-exchange earnings"). The import coverage coefficient then equals the "disposable foreign-exchange reserves" (Table 8.9) divided by the "uncovered foreign-exchange requirement" as defined above, multiplied by twelve to give the coverage in months.

[b] Based upon Iranian fiscal year figures and obtained from Iran, Bank Markazi, *Annual Reports*; short-term foreign liabilities of the Bank Markazi have been deducted. The reserve data are *not* directly comparable with those used in Tables 8.8 and 8.9, which were developed from Gregorian year data.

[c] Short-term foreign liabilities of the Central Bank of Iraq have been deducted from disposable foreign-exchange reserves, as obtained from Bank of Iraq *Bulletins*.

[d] This figure is conservative since it is based on only the disposable liquid foreign assets of the government, as recorded in line (3d) of *International Financial Statistics*. The net foreign assets of the commercial banks in Kuwait are also quite sizeable; as of December 1968 their holdings totaled $834 million.

Any discussion of the comparative effectiveness with which the several oil-producing countries utilized their oil revenues was beyond the scope of this study, since this would have entailed intensive analyses of the domestic economies. Yet it should be pointed out that to accumulate extensive foreign reserves is not necessarily economically meritorious; it may, indeed, be economically unwise if worthy developmental possibilities within the country are being neglected. Nor does the failure to accumulate reserves, or the action of borrowing abroad, imply profligacy on the part of a government; it may rather be a case of taking advantage of domestic developmental opportunities. Iran and Iraq, as noted above, the two countries with the largest domestic economies, and hence the greatest absorptive capacity, are precisely those of the group which have the smallest foreign-exchange reserves. Conversely, the three countries where productive investment opportunities are least have placed notable fractions of their oil sector earnings into foreign savings. Moreover, the disparity between the two subgroups in terms of per capita receipts of oil revenues is very wide. They range from over $1,500 per year in Kuwait to $30 in Iran, and about $50 in Iraq. It is obviously much easier for a country to accumulate foreign exchange where receipts are $1,500 per capita than in the case where the per capita oil earnings are substantially less.

Projections of Future Foreign Earnings

As has been seen, the foreign-exchange earnings of the exporting countries, around which the preceding discussion has centered, provide the economic foundation on which practically everything else rests. What will happen to these earnings in the future? The remaining chapters of this book, particularly Chapters 10 and 11, are concerned with the institutional and economic forces, now in a state of flux, that might seriously affect future earnings. Here we wish only to present a projection of these country earnings on the assumption that the forces now at work do not bring about fundamental changes. (This is, of course, not to be

Table 8.12

Government Oil Revenues: 1965 and Projected 1975 and 1980

	Actual		Projected			
	1965		1975		1980	
	Production (mill. bbl.)	Revenues (mill.$)	Production (mill.bbl.)	Revenues (mill.$)	Production (mill.bbl.)	Revenues (mill.$)
Libya	445	$ 371	1,700	$1,935	2,180	$2,480
Middle East	3,048	2,354	5,550	4,980	7,120	6,395
4-country subgroup[a]	2,836	2,235	4,350	3,900	5,270	4,730
Other Middle East	212	119	1,200	1,080	1,850	1,665
Total Libya plus Middle East	3,493	2,725	7,250	6,915	9,300	8,875

[a] Iran, Iraq, Kuwait, and Saudi Arabia.

Source: The 1965 production data are taken from British Petroleum Company, *Statistical Review of the World Oil Industry, 1967.* The 1965 revenue data are taken from *Petroleum Press Service,* August 1968. Note that the *PPS* data, being based on tax accrual accounts, differ from the revenue figures presented in Table 8.3, as explained in the footnote to the latter table.

The derivation of the projections is explained in the accompanying text.

taken as a prediction that basic changes will not take place, as a reading of Chapters 10 and 11 will indicate.) The assumption is adopted in order: (1) to make the projections exercise possible, and (2) to provide some measure of the huge financial stakes involved in the confrontations that may occur in company-government relationships.

Projections of the estimated oil revenues in the Middle East and Libya for 1975 and 1980, together with the actual figures for 1965, are given in Table 8.12. On the assumption that the existing institutional and economic arrangements, which ensure the current high levels of per-barrel taxation, are perpetuated, the estimated flow of oil revenues to producer country governments in 1975 will come to slightly under $7 billion, and by 1980 they will have risen to nearly $9 billion. These dollar amounts are respectively about 2½ to 3¼ times receipts of oil revenues by these countries in 1965.[16] No separate projection of local currency expenditures has been made. These remained virtually static in the vicinity of $300–$400 million per annum during the 1960's (see Table 8.4). On the assumption that they will continue to remain almost stationary in the years ahead, despite substantial increases in output, the amount of such payments has been arbitrarily set at $500 million. Consequently, retained value in foreign-exchange earnings generated by the oil companies (see Table 8.5) is estimated at something like $7.5 billion in 1975 and $9.4 billion in 1980 for the countries covered in Table 8.12.

In order to provide a more complete picture of the financial flows which accompany the export of Middle Eastern oil, Table 8.12 has been extended to include the secondary producing countries as well; the latter accounted for a small portion of aggregate revenues in 1965, but their share is projected to rise to almost 20 percent by 1980, based upon optimistic forecasts for Abu Dhabi, Oman, and Dubai.

The revenue projections which are shown in Table 8.12 are the product of two separate estimates. First, the projected levels of interregional exports (as adapted from Statistical Appendix Table VI) were scaled up to provide estimates of production levels for each of the two subregions, Libya and the Middle Eastern producers east of Suez.[17] These quantity estimates were then multiplied by estimated levels of the per-barrel revenues for each of the producing areas.

As for the quantities, the allocation of production between, say, Libya and the Middle Eastern countries is highly conjectural, because the actual distribution of the increased supplies among various potential sources will depend upon numerous economic factors which cannot be predicted, on top of which will enter political factors. Thus the split between Libya and the Middle East is appreciably less reliable than the estimates for total supply from the entire area, one factor being the uncertain availability of reserves in Libya to support the projected production levels.

The estimates of the per-barrel revenue receipts, as compared with those in 1965, are composed of two distinct elements: the increases that would result from (1) the phase-out of existing discounts[18] and (2) the decline in

[16] It should be noted that the 1965 figures are those provided by *Petroleum Press Service.* As explained earlier, the *PPS* figures were not used in Table 8.3, which contains the historical counterpart to the figures in Table 8.12. The reason for the difference is that in Table 8.3 the measure wanted was receipts on a cash basis during the year in question for comparison with other financial data for the same year. Here we are concerned with the dollar values earned by the countries as a result of exports in the particular year whether or not payment is received in that year.

[17] In using production as the quantitative base of the estimates, foreign-exchange revenues, which are based upon exports, will be overstated by any domestic consumption of petroleum products. In practice, any error on this account would be more than outweighed by others inherent in the projections.

[18] Chap. 10, p. 122.

Table 8.13

Estimated Unit Government Revenues per Barrel of Oil, 1975–80[a]

(Cents Per Barrel)

	Base year unit revenues[c]	Revenue increments resulting from:[b] Phasing out of discounts	Accounting cost reductions	Estimated 1975-1980 unit revenues
Iran	83.2	6.3	1.6	91.1
Iraq	81.7	7.9[d]	2.5	92.1[d]
Kuwait	78.5	5.7	negligible	84.2
Libya	99.7	5.2	8.9	113.8
Saudi Arabia	84.2	6.4	1.5	92.1

[a] It should be carefully noted that these projections are essentially predicated upon continuation of existing concessionary arrangements.

[b] See discussion in text. The estimated increments in revenues resulting from accounting cost reductions are very rough approximations.

[c] Source is *Petroleum Press Service,* August 1968, based upon the tax accrual accounts. The base year is 1967, except for Iraq, where 1965 was used, owing to possible distortions–the extent of which are unknown, because of closures in the pipeline to the Mediterranean in 1966 and 1967.

[d] Assuming full expensing of royalties. Up to the present a royalty-expensing agreement has not been reached.

accounting costs of production. Both the preceding items affect the per-barrel revenues in that the negotiated discounts and the accounting costs are direct deductions which are introduced in the calculation of the taxable income imputed to the foreign oil companies. One-half of any reduction in either component accrues directly to the exchequer of the host government.[19] The increments to each country's unit revenues per barrel which are estimated to arise from the two sources are shown in Table 8.13.[20]

The "discount" against the posted price of the exported crude oil which is at present allowed for purposes of computing taxable income consists of a "percentage allowance" and a "gravity allowance," both of which were introduced as of 1964 to reduce the impact of the royalty-expensing agreement upon the companies by providing a form of transitional calculation.[21]

As for the accounting costs of production, these have shown a decline for almost a decade, and it is foreseen that they will fall still further, partly due to increased economies in the development of new fields or in terminaling facilities and partly due to the completed write-off of the heavy initial and infrastructural investments still carried on the books of some of the companies as of 1967. In addition, the total bill for wages and salaries in the major producing areas has been seen to stabilize.[22] (The effects of higher levels of salaries and wages have been offset by compensatory reductions in aggregate employment.) Consequently, in the more mature producing areas, the per-barrel burden of the current operating costs is estimated to decline in rough proportion to the rise in offtake levels, whereas the accounting costs of capital are estimated to decline by a lesser percentage on a per-barrel basis.

The cost reductions play a secondary role, compared to the effect of the disappearance of the discount, in the more mature producing areas (see Table 8.13). Libya is, of course, different by virtue of its newness. The table shows that the projected decline in the accounting cost in Libya is estimated to yield an additional 9 cents per barrel by 1975, whereas elsewhere the additional yield from estimated cost reductions is at most 2.5 cents per barrel.

In closing, it is important to refer again to Table 8.12, which presents the results of the calculations summarized in the preceding paragraphs. By 1980, the government receipts from oil revenues are projected at close to $9 billion. As indicated, this is a measure of the financial stake involved for the countries of the region in the continued growth in the flow of oil exports to world markets. It is also a measure of the financial returns provided by way of the *present* structure of institutional arrangements. The evolution of this structure and the outlook for its future constitute the subject matter of the following chapters.

[19] See Chapter 10, *passim.*

[20] As for the secondary Middle East producing countries of Qatar, Abu Dhabi, Bahrain, Dubai, and Oman, average per barrel revenues for this group for 1975 and 1980 are projected at 90 cents compared to 82 cents in 1967, which was roughly the average per-barrel revenue for the entire Middle East in the latter year.

[21] The former is a fixed percentage of the posted price which falls from 6.5 percent in 1967 to 2 percent in 1971 and vanishes thereafter. The "gravity allowance" was contrived as a partial compensation for the relative overvaluation of the lighter crudes; it actually rises slightly through 1972, but then it also is phased out rapidly and vanishes by 1975. Only a small "marketing allowance" of one-half cent per barrel will be admissible after 1975. (See *Petroleum Press Service*, February 1968, pp. 44–47.)

[22] Indicative of this is the fact that local expenditures have remained practically stationary (see Table 8.4).

CHAPTER 9

The Major Oil Concessions: A Brief Review

PRIOR to 1950 the oil resources of the Middle East were almost entirely developed under exclusive concessions granted by countries bordering the Persian Gulf to companies owned, separately or jointly, by seven large international oil companies—American, British, and Dutch—and one French company in which the government had a substantial share. The same companies still occupy a dominant place in the region, accounting for over 90 percent of production in 1968. Five of the parent companies were American-owned. They were:

The Standard Oil Company (New Jersey)
The Socony-Vacuum Company (now Mobil Oil Corporation)
The Standard Oil Company of California
The Gulf Oil Corporation
The Texas Company (now Texaco, Inc.)

The others were:

The Anglo-Persian Oil Company (later Anglo-Iranian, now British Petroleum)
The Royal Dutch/Shell Group
Compagnie Francaise des Petroles

Prior to World War II, the general practice was for countries to grant exclusive concessions covering either the whole, or a very large portion, of the territories of the countries. In the four countries that still account for most of the oil production in the Middle East, the original Persian concession covered the whole country except five northern provinces; the Iraq concession, as later extended by affiliated concessions, covered the whole country except a narrow strip on the eastern border that was formerly a part of Persia; the Saudi Arabia concession covered an immense territory mainly in central and eastern Saudi Arabia; the Kuwait concession covered the whole of the small sheikhdom. In later years large areas of these concessions were relinquished; but all the territory upon which the companies' producing properties were located has been retained.

It is pertinent to mention that the United States government played an active part in securing participation of American companies in some of the concessions, as against British predominance in the region. This was especially true after World War I, when there were apprehensions concerning the running out of U.S. oil reserves; the government desired replenished sources of supply under American control, in addition to supporting the principle of the "Open Door" for American enterprise in foreign investment. The breakthrough came with participation in the Iraq concession after 1925. The support of government was also of importance in the case of the Kuwait concession and in other connections.

A Summary Picture of Early Major Concessions

In the chronological order of the granting of the more important concessions, the countries, the concessionary companies, and their owning parent companies were as follows:

Persia, 1901

The Anglo-Persian Oil Company in 1909 secured sole control of a concession acquired in 1901 by W. K. D'Arcy, a substantial portion of the area being later relinquished. In 1951 the properties were taken over by the government; and in 1954 their operation was entrusted to a Consortium, originally consisting of the eight companies named above, later expanded to include a group of nine smaller American companies collectively organized as the Iricon Agency.

Iraq, 1925

The Iraq Petroleum Company, which in 1925 acquired rights based on a prewar German concession, was the first concessionaire company to be jointly owned by a group of Western companies. After some shift of membership, the permanent interests in the group became:

Anglo-Persian (British Petroleum)	23.75%
Royal Dutch/Shell	23.75%
Compagnie Francaise des Petroles	23.75%
Standard Oil (New Jersey)	11.875%
Socony-Vacuum (Mobil)	11.875%
Partex (Gulbenkian Estate)	5.00%

In the course of time, Iraq Petroleum Company with its two affiliates, Mosul Petroleum Company and Basrah Petroleum Company, acquired rights over almost the whole of Iraq. In 1961 the government by unilateral action deprived the three companies of their rights over 99.5 percent of the concession territory; but they retained the producing area which accounts for almost the whole of Iraq's production.

Saudi Arabia, 1933

The Arabian American Oil Company (ARAMCO) operates in a concession area originally granted in 1933 to Standard Oil Company of California, portions of which were later relinquished. In 1936 the Texas Company acquired a half interest. In 1947 Jersey Standard and Socony-Vacuum (Mobil) each acquired an interest in the company. Since then participation in the ownership of ARAMCO has been:

Standard Oil of California	30%
Texas Company	30%
Standard Oil (New Jersey)	30%
Socony-Vaccum (Mobil)	10%

Worldwide oil production outside of North America and Communist area

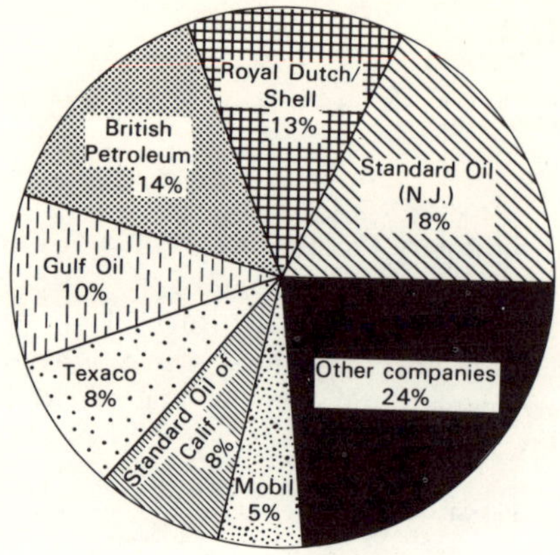

Middle East

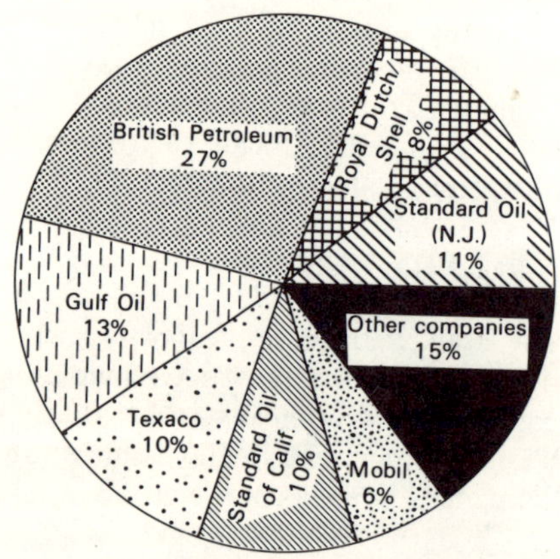

Fig. 9.1. Share of Seven Major Companies in Oil Production of Two Broad Geographic Areas

Notes and sources: Worldwide data, referring to 1966, were obtained from U.S. Congress, Senate Committee on the Judiciary, *The Petroleum Industry*, Hearings Before the Subcommittee on Antitrust and Monopoly (Washington: Government Printing Office, 1969), Part I, p. 159. (Table presented by Edith Penrose.)

Middle East data, referring to 1967, are based on statistics of source cited above, p. 447, and *World Oil*, August 15, 1968, and August 15, 1969.

Kuwait, 1934

The concession in Kuwait was granted to the Kuwait Oil Company which was, and still is, jointly owned:

Gulf	50%
Anglo-Persian (British Petroleum)	50%

A portion of the area was later relinquished.

Lesser Concessionary Areas

The four concessions named above have accounted for a large fraction of the oil produced in the Middle East, and in 1968 still produced 85 percent of the total. Several lesser concessions were granted to the members of the same group of eight companies. The following concessions have proved relatively successful:

Bahrain, 1930

The Bahrain Petroleum Company (BAPCO) was organized under a concession to Standard Oil of California. In 1935 the Texas Company acquired a half interest. The original concession covered only a part of the island; but in 1940, after failure of efforts of the Iraq Petroleum Company to secure a foothold, the concession was extended to cover the entire sheikhdom.

Qatar, 1935

The Qatar Petroleum Company (formerly Petroleum Development [Qatar]) which held the original onshore concession, now partly relinquished, is owned by the same companies as the Iraq Petroleum Company. A later offshore concession is held by the Shell Company of Qatar, and another was granted in 1969 to a Japanese group, organized as Qatar Oil Company, Limited (Japan).

Abu Dhabi, 1939

The Abu Dhabi Petroleum Company is likewise held by the same companies that own the Iraq Petroleum Company. Along with Qatar, it represents the successful element in a series of concessions negotiated by Iraq Petroleum Company during the 1930's with rulers of several small sheikhdoms around the perimeter of the Arabian peninsula. In 1950 Anglo-Iranian (British Petroleum) and the French CFP secured a concession for offshore rights, and hold shares of two-thirds and one-third, respectively, in the Abu Dhabi Marine Areas, Ltd.

Oman, 1937

Oman was another area for which an affiliate of Iraq Petroleum Company received an early concession. Under unfavorable political conditions, little exploration was done until the late 1950's. Disappointing results led the other IPC members to withdraw in 1960, leaving an 85 percent interest in the hands of Shell, with 15 percent for Partex (Gulbenkian), from whom Compagnie Francaise des Petroles later took 10 percent after oil had been discovered. Production began in 1967.

The Concessionary Position of Individual Companies, 1950

In the preceding telescoped accounts, we have summarized the Middle East concessions participated in by the seven major international companies and CFP. In the present section we shall rearrange the same data to show the position of the various companies in relation to the concessions in the several countries.

Anglo-Persian Oil Company (later Anglo-Iranian; now British Petroleum)

Originally 100% of an exclusive concession covering most of Persia. Now 40% of the successor Iranian Consortium.

23.75% ownership of the Iraq Petroleum Company and its affiliates.

23.75% interest in the Abu Dhabi Petroleum Company, an associate of Iraq Petroleum Company.

23.75% interest in the Qatar Petroleum Company, an associate of Iraq Petroleum Company.
50% ownership in the Kuwait Oil Company.
66⅔% ownership in the Abu Dhabi Marine Areas, Ltd.

Royal Dutch/Shell Group
23.75% ownership in Iraq Petroleum Company and its affiliates, and associated companies in Qatar and Abu Dhabi.
14% ownership in the Iranian Consortium.
85% ownership in the Oman concession.

Compagnie Francaise des Petroles
23.75% ownership in Iraq Petroleum Company and its affiliates, and associated companies in Qatar and Abu Dhabi.
33⅓% ownership in Abu Dhabi Marine Areas, Ltd.
6% ownership in the Iranian Consortium.

Standard Oil Company (New Jersey)
11.875% ownership in Iraq Petroleum Company and its affiliates, and associated companies in Qatar and Abu Dhabi.
30% ownership in Arabian American Oil Company (ARAMCO).
7% ownership in the Iranian Consortium.

Standard Oil Company of California
30% ownership in ARAMCO.
50% ownership in Bahrain Petroleum Company.
7% ownership in the Iranian Consortium.

Texaco, Inc.
30% ownership in ARAMCO.
50% ownership in Bahrain Petroleum Company.
7% ownership in the Iranian Consortium.

Gulf Oil Corporation
50% ownership in Kuwait Oil Company.
7% ownership in the Iranian Consortium.

Mobil Oil Corporation
11.875% ownership in Iraq Petroleum Company and its affiliates, and associated companies in Qatar and Abu Dhabi.
10% ownership in ARAMCO.
7% ownership in the Iranian Consortium.

Looking at joint company participation in concessions, the following combinations appear:

(1) Eight-company participation in the Iranian Consortium, plus the Iricon group.
(2) Standard Oil of California and Texaco in Bahrain.
(3) Standard Oil of California, Texaco, Standard Oil (New Jersey), and Mobil in Saudi Arabia.
(4) British Petroleum and Gulf Oil in Kuwait.
(5) British Petroleum, Royal Dutch/Shell, Compagnie Francaise des Petroles, Standard Oil (New Jersey), and Mobil in Iraq, and associated companies in Qatar and Abu Dhabi.

There were originally single parent companies in Persia and Saudi Arabia, but none of the major concessions has been held by single parent companies since the cancellation of the Iranian concession in 1951.

New Postwar Concessions

All the concessions reviewed above, except Abu Dhabi offshore, had been granted before 1939; by that date Iran and Iraq had achieved significant production and Saudi Arabia was under way. During and after the war, the great productive potential of Saudi Arabia and Kuwait began to be demonstrated. With respect to existing concessions, a process was under way to revise their terms to the greater advantage of the host countries. At the same time there were still openings for new concessions or other types of agreement (1) in territories not yet covered by concessions, (2) in areas relinquished under old concessions, and (3) in offshore areas of the Continental Shelf beyond the conventional territorial waters.

The two concessions of importance first to be granted after the war were located in the so-called Neutral Zone, some 2,000 square miles of uninhabited desert lying on the Persian Gulf Coast between Kuwait and Saudi Arabia, with powers over it divided between the rulers of the two countries. The Sheikh of Kuwait offered a concession on the basis of free auction. Attracting wide interest, the concession was awarded in 1948 to the American Independent Oil Company (Aminoil), organized by an ad hoc group of American companies without prior experience in the Middle East.[1] In 1949 the Saudi Arabia concession was awarded to Pacific Western Oil Company (later Getty Oil Company). In 1957 and 1958, two concessions to the Arabian Oil Company (Japanese) were awarded by Kuwait and Saudi Arabia for offshore operations in the Continental Shelf.

Beginning in 1957, Iran adopted a policy of entering into new agreements in relinquished territory and offshore. By 1965 agreements had been made with nine companies, most of them jointly organized by two or more foreign companies who were newcomers to Middle East oil operations. Agreements with new companies have also been entered into by Iraq, Saudi Arabia, Kuwait, and Abu Dhabi. These agreements are mainly of a new type which give the host country a participating share and are dealt with in Chapter 10.

Stages of Change in Early Major Concessions

Early concession agreements were in general relatively simple documents. They usually gave exclusive rights of

[1] The larger participants were Phillips Petroleum, Signal Oil and Gas, and Ashland Oil and Refining. There were five other participants.

exploration and development over large areas—in some cases the whole area of a country. The common form of payment was a royalty on the oil produced, which might be combined with a bonus payment upon signing and some form of deadrent independent of royalty, either before production started or permanently. The company membership in the concessionary groups was in part the result of strong Western government influences and support in some of the early negotiations. The important concessions were, except for Saudi Arabia, in countries where the British government held a position of special political influence. The road to the eventual compositions was paved by diplomatic negotiations between the British and American, and in one case French, governments—the American government insisting upon opportunities for American companies to enter a region which the British, by reason of their established political protectorate, had hoped to reserve for themselves, in addition to barring Russian advance in the region. The history of the concessions has been marked at times by disagreements between host governments and companies leading to great changes in the financial terms and geographical coverage, and in one case, Iran, to actual revocation. Rather than attempt to generalize, we will give brief accounts of the original terms and stages in the development of the more important concessions.

1. *Persia (Iran)*

The concession granted to W.K.D'Arcy in 1901, covering the whole of Persia except the five northern provinces, provided for a payment of £20,000 in cash and £20,000 in shares in the company set up to develop the concession. The royalty was to be 16 percent of net profits. Exemptions from customs and taxes were provided. After oil was discovered in paying amounts in 1908, the syndicate operating the concession transferred its rights to the newly formed Anglo-Persian Oil Company. By 1914 the Company's operations in southwestern Persia had established this country as a major producer, and a large-scale refinery had been erected in Abadan. In that year, the British government acquired a 51 percent ownership in the shares of the company. In the early years, cash payments to local rulers, outside the concession terms, were a necessary condition of peaceable operations. Various claims and counterclaims between the Company and the government arising out of World War I were settled without basic change in the agreement. An effort of Anglo-Persian to secure a foothold in the northern provinces failed due to Russian influence, as did also a concession arranged with American interests.

After World War I, while physical expansion proceeded rapidly, the Company was under constant attack in a highly nationalistic political situation. And when, in the early depression years, payments by the Company to the government under the profit-sharing plan fell drastically, the government at the end of 1932 announced the cancellation of the concession. The case was referred by the British government to the League of Nations through whose good offices a new agreement was signed in 1933. The length of the concession was extended from 1961 to 1993; but its area was cut to about one-quarter of the original grant, to be further reduced in 1938, but still amounting to 100,000 square miles. The royalty basis was changed from a percentage of profits to a fixed payment of 4 shillings per ton, protected against currency depreciation, with a guaranteed minimum. Further payments of 9 pence per ton were to be made as a commutation of taxes; certain payments were to be made out of profits when dividends exceeded a certain amount. A large immediate cash payment was made to the government by applying the new royalty formula to the preceding two years and in settlement of other claims. A further provision was the development of a more northern field, previous activities having been mainly confined to the region near the head of the Persian Gulf. Under the new agreement the Company for several years enjoyed a fairly quiet existence in the turbulent political milieu.

Out of the area excised from Anglo-Persian in the northeast, a large new concession, ultimately to cover 100,000 square miles, was made in 1937 to an American company controlled by Seaboard Oil Company on terms similar to the revised Anglo-Persian agreement. But the concession was abandoned by the company after two years because of unsatisfactory exploratory results.

During World War II, Persia (now renamed Iran) was subject to foreign occupation by the Russians in the north and the British in the south. After the war an agreement for Russian oil rights in the north was rejected by the Iranian Parliament in 1947. Meantime, negotiations between the government and the Company led in 1949 to a Supplemental Agreement to the concession of 1933. The royalty rate was increased by 50 percent and other financial benefits, including the sharing of profits, increased. This agreement was rejected by the Parliament of Iran in the midst of an unfavorable political atmosphere in late 1950. A bill committing the government to nationalize the Company's properties was passed in March 1951, and on May 1, 1951, a definitive Nationalization Law went into effect. The National Iranian Oil Company was set up to take over administrative responsibility. The British staff of Anglo-Iranian left the country in October 1951. Efforts to effect a new modus vivendi, with due regard to compensation of the confiscated Company interests, were made through intervention by the U. S. President and by resort to international agencies, but came to nothing. Faced by the inability of the Iranian government to sell oil to the international companies who could distribute it in world markets, the Iranian industry went into a debacle and the loss of revenue brought the government close to the point of bankruptcy.

The Consortium

So it went for two years up to mid-1953, when a palace revolution in Teheran opened the way for negotiations to find a basis for effective resumption of the former Anglo-Iranian operations. The solution arrived at, after complicated diplomatic and intercompany negotiations, was the Agreement of October 1954, investing control of producing and refining operations within the geographical area of the Anglo-Iranian Oil Company in a so-called Consortium of oil companies. The original membership of the Consortium was Anglo-Iranian (later British Petroleum), 40 percent; Standard Oil (New Jersey), Socony-Vacuum (Mobil), Standard Oil of California, Gulf, and Texaco, 8 percent each; Royal Dutch/Shell, 14 percent; and Compagnie Francaise des Petroles, 6 percent. The membership was thus identical with the companies already holding all the other major concessions in the Persian Gulf region. In 1955 each of the five American companies gave up one-eighth of its holdings to give a 5 percent share to a group of nine American companies organized as the Iricon Agency.[2]

Under the Consortium agreement, titular ownership of the former properties of Anglo-Iranian was retained by the government-owned National Iranian Oil Company. Control over their use and development was assigned by the Consortium to two operating companies, one for exploration and production, one for refining, formed by a holding company, Iranian Oil Participants, Ltd., in which the member companies of the Consortium held pro rata shares. The operating companies were to operate on the basis of cost plus a service fee met by the Consortium companies, making a profit of 1 shilling per ton taxed at the 50 percent rate. The crude oil and products were assigned to trading companies set up by each of the Consortium members. Payments for the account of the government were made by each of these companies. Export prices were to be set independently by each of the trading companies.

There was a "stated payment" (equivalent to royalty) of 12.5 percent of the value of crude oil produced, based on posted price; or NIOC had the option of taking the payment in the form of crude oil. The tax rate was 50 percent of the profits of the companies, calculated on the basis of posted prices, with the "royalty" credited against the tax. Refined products for distribution within Iran were made available to National Iranian Oil Company. Minimum levels of crude production were designated.

A primary feature of the Consortium was the process by which the financial interests of Anglo-Iranian were cared for. Its properties had in effect been confiscated without compensation. The terms of the Consortium agreement promised over time to restore its financial interests in large degree. This was to be accomplished in the following ways: (1) a payment of £25 million by the Iranian government in tax-free oil in ten annual installments; (2) retention by the Company of £51 million in accounts which would have been due the government under the old concession; (3) payment by other Consortium members of an immediate £32.4 million and an overriding royalty of 10 cents per barrel on their exports up to a final aggregate sum of $510 million; and (4) retention of a 40 percent interest in the future earnings of the Consortium.

Under the agreement the National Iranian Oil Company was gradually to take over administrative responsibility for the so-called "nonbasic" functions in the Consortium area formerly performed by the Company, such as housing, roads, welfare activities, and transport—though mainly at the expense of the Consortium.

The Consortium agreement covered solely successor arrangements to the former activities of Anglo-Iranian Oil Company in its defined territory of 100,000 square miles, to be further reduced later. The agreement was to run for twenty-five years until 1979 with an option on three further extensions of five years each under stated conditions. The state company, National Iranian Oil Company, was left with the responsibility for exploratory and producing activities in all the rest of the country. This it has carried out in two ways: (1) by direct operations on its own account, and (2) by special forms of partnership or contract arrangement with other companies. These will be described in Chapter 10.

[2] These companies were Richfield, Standard Oil (Ohio), Signal, Atlantic, Hancock, Tidewater, San Jacinto, Getty, and American Independent. The last two of these were already holders of concessions in the Neutral Zone. The others entered the field of Middle East oil for the first time.

Iricon now has six members with the following shares in the total Consortium: Atlantic Richfield (1.667%), American Independent (0.833%), Signal (0.833%), Getty (0.833%), Continental (0.417%), Standard Oil (Ohio) (0.417%). American Independent is itself made up of eight members.

2. *Iraq*

The Iraq Petroleum Company, which ultimately came to control rights for almost the whole of Iraq through three concessions, was spawned after World War I out of a complex set of diplomatic and company maneuvers to acquire the prewar rights of the Turkish Petroleum Company—a company with large German shareholding interests. Holdings in this company (renamed Iraq Petroleum Company in 1929) were first acquired by British, French, and Dutch interests, with the Anglo-Persian Oil Company holding 47.5 percent of the shares. The concession covered most of the country and was to run for seventy-five years from 1925.

Meantime, however, American companies were demanding access to the country, strongly backed by diplomatic pressure since after the war the government, apprehensive concerning the state of domestic reserves, desired assured new sources. In 1928 the Near East Development Company, jointly owned by five companies, was assigned one-half the Anglo-Iranian interest, the latter company being compensated by a 10 percent overriding royalty. Three of the companies (Gulf, Atlantic, Mexican) later sold out to the remaining two, Standard Oil (New Jersey) and Socony-Vacuum (Mobil). With these and other adjustments the present participating shares in Iraq Petroleum Company were established.

By the so-called Red Line Agreement[3] the constituent members of IPC agreed not to seek other concessions in most other parts of the Middle East except in association with the other members. On this basis, IPC made a number of agreements with rulers of the small Arab sheikhdoms, two of which, Qatar and Abu Dhabi, later proved productive.

[3] The Red Line Agreement got its name from a line on the map enclosing most of the territory of the former Turkish Empire.

The concession of 1925 was revised in 1931 limiting the area to 32,000 square miles east of the Tigris in the northern region of the country. Until exports started an annual payment of £400,000 was to be made to the government, one-half being deadrent, the remainder recoverable from future royalties. Royalties were set at 4 shillings (gold) per metric ton, with a minimum annual payment of £400,000 (gold). A substantial yearly payment was to be made in lieu of taxes. The Company was to build a pipeline to the Mediterranean within four years. There were no specific obligations with respect to drilling.

In 1932 a 75-year concession was granted to the British Oil Development Company in the north and west of the Tigris covering some 46,000 square miles. The royalty was fixed, as in the case of Iraq Petroleum Company, at 4 shillings (gold) per ton. Deadrent, pending production, was to start at £100,000 (gold), rising to £200,000. A pipeline to the Mediterranean was specified. An added feature, unlike the Iraq Petroleum Company concession, was that the government was entitled to 20 percent of oil produced. After the company ran into financial difficulties, Iraq Petroleum Company bought up shares and by 1937 was in effective control. The concession was assigned to an IPC affiliate, Mosul Petroleum Company.

In a 1938 agreement, Iraq Petroleum Company secured a 75-year concession for the southern portion of Iraq, covering all territory not included in earlier concessions. For this purpose it organized an affiliate, Basrah Petroleum Company, whose concession was substantially similar to that of the Mosul Company.

By these transactions, Iraq Petroleum Company and affiliates gained control over the whole area of Iraq except for a narrow eastern strip controlled by Anglo-Iranian Oil Company. Its later relations with the government were somewhat complicated by the fact that the three concessions contained nonidentical terms.

After the war, the diverse interests of the members of IPC led to disagreements among them, especially concerning the Red Line agreement which was distasteful to the American members. In addition, a series of disputes began to exacerbate its country relationships, partly resolved over time, but in the end revolutionizing the status of the concession. There were issues concerning the raising of the royalty rate, the "gold" valuation of royalties in devalued currencies, the initiation of a system of income taxation based on a 50-50 sharing in profits, government shareholding in the company, increasing the rate of output, placing Iraqis in high administrative positions, and other matters. Most of these issues were at least temporarily resolved by an agreement reached in 1952. But political sentiments highly hostile to the company remained active and made the outlook for any further settlement highly uncertain.

Conditions took a turn for the worse for the companies after the Qasim dictatorship succeeded the monarchy in 1958. Increasingly severe demands were made for larger revenues and relinquishment of territory. The companies expressed willingness to engage in a program of relinquishment, but agreement could not be reached on delineation of the areas. Under great pressure on all its concessionary privileges, the companies had reached the point by 1961 of offering to surrender 75 percent of the territory covered by existing concessions, but the offer was rejected. In December 1961, under the new Law 80, the government unilaterally voided the companies' rights over 99.5 percent of the existing concession area of 435,000 square kilometers. The areas reserved to them, less than 2,000 square kilometers, contained all the active producing properties, so that the companies continued to produce and pay taxes and royalties; but all exploratory activity outside the limited area was halted. The Iraq National Oil Company was set up with somewhat uncertain powers over the future exploitation of the surrendered territories.

After a new government succeeded General Qasim in 1963, negotiations were resumed between government and companies. A draft agreement of 1965 provided that the IPC companies would relinquish 99 percent of the areas under their concessions, but would receive additional areas, amounting to about 7 percent of the original concessions, to be explored and developed in conjunction with the Iraq National Oil Company; but the agreement was not ratified. In 1967, under Law 97, all territory except that small amount reserved to the companies in 1961 was granted to Iraq National Oil Company, which may operate on its own account or enter into operations in association with foreign companies. The companies have not accepted that their rights can be forfeited by unilateral legislation; but no basis for mutual agreement has been established.

The position of the companies in Iraq is more precarious than in any other Middle East country, in an unstable political environment where vilification of the foreign companies is popular and demands for nationalization strong. However, the revenues from the Iraq Petroleum Company and its affiliates are of critical importance to the government. If the companies were to be nationalized, producing and pipeline operations might possibly be carried on effectively; but the services of the companies in the marketing field appear at present to be irreplaceable.

3. *Saudi Arabia*

After some rivalry with other interests, the Standard Oil Company of California in 1933 obtained the exclusive concession to a vast area in central and eastern Saudi Arabia for a period of sixty years. The cash bonus of £30,000 outbid what its foremost competitor, the Iraq Petroleum Company, was willing to pay to King Ibn Saud, who was in financial extremities. The royalty rate was 4 shillings (gold) per ton; later loans and benefits were promised; and the company was exempt from taxation. Later additions granted in 1939 brought the area under concession to 440,000 square miles, involving an additional cash payment of £140,000 and a deadrent of £25,000 per year. In 1936 the Texas Company acquired a half interest. Rich deposits were discovered and by 1939 exports had reached over 500,000 tons.

During World War II, after the company, in fear of loss of its concession, had urged the American government to make large payments to the King to meet demands that the company felt unable to meet, President Roosevelt approved a proposal that the American government purchase a controlling interest in the company; but the negotiations failed.

In 1947 Standard Oil (New Jersey) and Socony-Vacuum (Mobil) acquired their participation in the concessionary company, ARAMCO. By this action these two companies, who were members of the Iraq Petroleum Company, overstepped the Red Line Agreement, holding that it had been invalidated by events during World War II. Their membership in ARAMCO fulfilled the need of the two companies for new sources of supply for their widespread and expanding market outlets.

Serious disputes arose after the war concerning the financial terms of the concession. In the outcome, Saudi Arabia was the pioneer in introducing in the Middle East the income tax plan which came to be called the 50-50 principle of sharing of company profits.[4] Originally, in 1950, the profits were calculated *after* U. S. taxes; but on further demand this was changed, to *before* foreign taxes in 1952–a change that immensely increased Saudi revenues but cost the company little because the local tax could be credited against U.S. tax liability.

Apart from financial terms, another primary issue to arise was that of relinquishment of large areas covered by the original concession. The original territory granted in 1933 to Standard Oil of California and continued in ARAMCO covered the whole of eastern Saudi Arabia, an immense region. In 1947 ARAMCO surrendered its "preferential" rights[5] in a large area in the western part of the concession outside the established producing zones, and in 1949 a considerable area in the southwestern portion. These divestments left it still with a vast territory. In 1948 ARAMCO also divested itself of rights in the Neutral Zone, but received in return offshore rights beyond "territorial" waters under the new concept of the Continental Shelf. Further limited relinquishments were made in 1955 and 1960, but these still left the company with over 300,000 square miles in Eastern Arabia. In 1963 a further large-scale divestment reduced the area of the concession to 125,000 square miles. And it was agreed that this would be reduced by successive stages over thirty years to 20,000 square miles.

At the same time that these changes provided the government with vast territories which might be exploited by its own national company (Petromin, set up in 1962) or by new agreements with other foreign companies, it left ARAMCO with all its established producing properties, which contained reserves probably ample for the remaining years of its concession, together with a still large surrounding area for further exploration.

Growing out of the ARAMCO concession, but not directly a part of it, was the building of the pipeline, Tapline, over 1,000 miles long from the Persian Gulf to the Mediterranean.

[4] This designation is a misnomer, as will be explained later in Chap. 10.

[5] Preferential rights applied to certain areas where other companies might bid for concessions.

Some further issues in the relations between ARAMCO and the government have arisen in connection with the setting up of the national oil company, Petromin, to be discussed in Chapter 10.

As in all other host countries, Saudi Arabia has brought strong pressures upon the company in order to modify the terms and secure greater benefits under the concession; but, relatively speaking, the relations between country and company have been more amicable, or perhaps one should say less acrimonious, than in Iran (prior to the Consortium agreement) and Iraq; and nothing comparable to the rough treatment of the original concessionaires in those two countries has occurred.

4. *Kuwait*

The concession arose out of a rivalry between the Gulf Oil Corporation and the Anglo-Persian Oil Company. Gulf was strongly backed by the American government in its policy of insistence that American firms be given equal access in a region where British political overlordship was used to establish a dominant economic position. In the outcome, the two companies compromised their differences and in 1934 organized the Kuwait Oil Company with equal shares. The concession, granted in the same year, extended to the whole realm, 6,000 square miles, and was to run for seventy-five years. The royalty was set at 3 rupees per ton, with an initial payment, an annual deadrent, and an annual minimum payment, all of moderate amounts. In 1937 the company proved the Burgan field, which turned out to be the most prolific field in the world. After interruption during the war, other richly productive fields were discovered.

The two parent companies whose affiliates were partners in the Kuwait Oil Company were both companies with crude oil supplies largely surplus to their own integrated needs. To give them firm outlets for a large share of Kuwait oil, both companies made long-term contracts with other companies for a large offtake, Gulf with Shell in 1947, and Anglo-Iranian with Standard Oil (New Jersey) and Socony-Vacuum (Mobil).

During the 1950's and 1960's the concession underwent changes in terms similar to those in other countries, referred to above. The 50 percent income tax (to be described later) was introduced, based on posted prices (1952) and modified by expensing of royalties (1965). In carrying out these and other adjustments, the relations of the company and the government remained more amicable than in other countries, and nothing like the confrontations in Iran and Iraq occurred. In 1960 a national oil company, the Kuwait National Petroleum Company, was formed to carry the types of functions to be described later.

In 1962, 2,200 square miles were relinquished and it was expected that these would be explored and developed by the new Kuwait National Petroleum Company either directly or in association with other companies. One such joint arrangement was made in 1967 with a consortium of Spanish companies (Hispanoil). It was expected that further relinquishments would be called for, especially in the islands and offshore areas.

5. *The Special Case of Libya*

To this point we have dealt only with major concessions antedating World War II in the Persian Gulf region. The recent and sudden arrival of Libya in the front rank of producing countries is a very important event in the world of international oil and needs to be brought into the picture at this point. The first discovery was in the late 1950's, the first production in 1961; but by the end of 1969 production was approaching 3½ million barrels/day, second only to Iran in the Middle East and North Africa.

The striking fact about Libya is that its oil resources were developed on an entirely different plan from that of the older source countries. In those countries, on the pattern set in the 1920's and 1930's, there was a single large concessionaire in whose hands still lie a large major fraction of the existing production in each country. In Libya, following the Petroleum Law of 1955, oil companies were invited to bid for concessions which by 1968 were held by 37 companies of which 17 were producing and exporting oil. Many of them are American, and others are British, French, Italian, and German. Some are themselves groups representing partnerships of foreign companies, others single foreign companies or their affiliates. Some are state-controlled agencies of consuming countries.

This method of promoting exploration was supported initially by attractive inducements. Concession areas were large, the early ones running usually between 30,000 and 100,000 square kilometers. There was a 25 percent depletion allowance. Rents and drilling stipulations were not onerous. Taxation, though following the 50-50 formula, was not based on posted prices. Royalties were the usual 12.5 percent. Under these conditions, numerous applicants appeared; exploration went forward at a rapid pace; and there were a number of discoveries in the late 1950's and early 1960's. Two groups in particular rapidly became large producers, Standard Oil (New Jersey) and the Oasis Group consisting of three American companies, Amerada,[6] Marathon, and Continental.

Under these favorable circumstances, the government began to stiffen its legislation. In 1960 the depletion allowance was abolished and relinquishment requirements increased, so that in three stages of five, three, and two years, concessionaires were required within ten years to reduce their holdings to one-third or one-fourth of the original areas, providing the basis for new waves of concession granting. Taxes, however, continued to be levied on the basis of discounted prices corresponding closely to market realizations, a fact that greatly benefitted the three members of the Oasis Group, who had no established European outlets and used cut-rate crude oil prices as a means of buying their way in in competition with the major international companies.

In 1962 Libya joined OPEC[7] and was brought under great pressure by other members to adopt the OPEC standards. A law designed to meet these standards (basing taxes on posted prices with expensed royalties) was passed in 1965; and the government applied severe pressure to old concessions to secure their acceptance of this change in concession terms. The change was bitterly opposed by the independent group, as it cut into the advantage they had had of marketing their oil at cut-rate prices, but still profitably. Under various sorts of pressure from the government, they gave way. In this controversy, at least some of the major companies are supposed to have taken the side of the government as a means of limiting price competition. The circumstances of the case are too complicated to attempt to describe here. The outcome was a victory for the OPEC in its effort to secure a relatively uniform basis of taxation in all countries.

In spite of the stiffer terms on old and new concessions, the demand for concessions remained strong when new or relinquished areas were put up for bidding. This is not surprising in view of two considerations: (1) the great geographical advantage in relation to European markets, and (2) the fact that concession terms were still less onerous than those being negotiated in the Persian Gulf region for new areas. Indeed, apart from taxation and relinquishment, the terms were left somewhat open, the applicants being invited to offer special inducements to the government.

Having, by its earlier policy of inviting in many applicants, greatly speeded up the exploration process, the Libyan government began to entertain new thoughts concerning the direction of future policy. Early in 1968, a national oil company, Libyan General Petroleum Corporation (LIPETCO), was set up, charged with supervision of all new concessions and other extensive responsibilities. It is expected that it will enter into partnership arrangements under all new agreements with foreign companies. (See Chapter 10, p. 134.)

Through 1967, the great successes were in the hands of two groups, Standard Oil (New Jersey) and Oasis, which in 1967 accounted for 70 percent of production in almost equal proportions. In 1968 Occidental Petroleum also became a larger producer; and three other groups were also substantial producers. So far the big payers are all in American company hands, except for a British Petroleum partnership in one concession and a German group in another.

[6] By a long-term purchase arrangement with Amerada, Royal Dutch/Shell acquired something approaching a partnership interest.

[7] The Organization of Petroleum Exporting Countries, to be described in Chap. 10.

Bibliographical Note

The factual material in Chapters 9 and 10 has been assembled from a number of sources. Citations to sources are made only in the case of direct quotation.

The most useful books for our purposes were:

J. E. Hartshorn, *Politics and World Oil Economics,* rev. ed. (New York: Frederick A. Praeger, 1967; English title, *Oil Companies and Governments*).

S. H. Longrigg, *Oil in the Middle East,* 3rd ed. (London: Oxford University Press, 1968).

E. T. Penrose, *The Large International Firm in Developing Countries: The International Petroleum Industry* (London: George Allen and Unwin, 1968).

M. A. Adelman, *The World Petroleum Market,* unpublished draft manuscript of a study being prepared for publication by Resources for the Future, Inc.

Other books include:

Henry Cattan, *The Evolution of Oil Concessions in the Middle East and North Africa* (Dobbs Ferry, New York: Oceana Publications, 1967).

Helmut J. Frank, *Crude Oil Prices in the Middle East* (New York: Frederick A. Praeger, 1966).

Paul H. Frankel, *Oil: The Facts of Life* (London: Weidenfeld and Nicolson, 1962).

Charles Issawi and Mohammed Yeganeh, *The Economics of Middle Eastern Oil* (New York: Frederick A. Praeger, 1962).

W. A. Leeman, *The Price of Middle East Oil* (Ithaca, New York: Cornell University Press, 1962).

George Lenczowski, *Oil and State in the Middle East* (Ithaca, New York: Cornell University Press, 1969).

Harold Lubell, *Middle East Oil Crises and Western European Energy Supplies* (Baltimore: The Johns Hopkins Press, 1962).

Ashraf Lutfi, *OPEC Oil* (Beirut: The Middle East Research and Publishing Center, 1968).

Zuhayr Mikdashi, *A Financial Analysis of Middle Eastern Oil Concessions, 1901-65* (New York: Frederick A. Praeger, 1966).

K. S. Sayegh, *Oil and Arab Regional Development* (New York: Frederick A. Praeger, 1968).

Benjamin Shwadran, *The Middle East, Oil and the Great Powers* (New York: Frederick A. Praeger, 1955).

The periodical sources mainly used were:

Middle East Economic Survey (Beirut)

Petroleum Press Service (London)

Publications of the Organization of Petroleum Exporting Countries (OPEC)

The Economist (London) also includes important current information. The most significant review article which appeared in that journal is Walter J. Levy's "World Oil in Transition," August 19, 1961.

CHAPTER 10

Changing Patterns of Country-Company Relationships

IN the period since 1950, the governments of producing countries have initiated a number of important changes that not only modify the position of the old concessionaires but substantially alter the institutional environment within which countries seek to promote their national interests. The principal directions of change to which we must pay attention are:

(1) The revision of financial and geographical terms of the old concessions.
(2) The setting up of the Organization of Petroleum Exporting Countries (OPEC) with worldwide membership through which countries can act jointly to promote their interests.
(3) The entrance of new companies under new types of concession, partnership, or contract agreements.
(4) The organization of national oil companies designed to participate in oil activities both on their own and/or in various forms of relationship to companies under the old concession agreements or under the various types of agreements with newcomers.
(5) The setting up of Organization of Arab Petroleum Exporting Countries (OAPEC) to promote forms of joint action among a limited group of important oil-producing countries.
(6) The as yet unfilled demand of some countries to be admitted to a participating ownership and management status in the old concessionaire companies.

These various topics are so intimately interrelated that, as we build them up seriatim, what is said on each depends in part upon what is to be said on those later in the series. Despite this unavoidable expository difficulty, we shall attempt to build up cumulatively a fairly comprehensive account of the forces that will shape the still uncertain structures of country-company relationships in the future.

Changes in the Old Major Concessions

While the revision of early concession agreements differs greatly in detail from country to country, the principal changes have been of four sorts:

(1) the sharing in profits through direct taxation after 1950;
(2) the virtual freezing of posted prices after 1960 and the calculation of profits for purposes of taxation on this basis;
(3) changes in the position of royalties in the calculation of taxes after 1964;
(4) provisions for the relinquishment of portions of the original concession area.

To this list might be added the refusal of governments to accept other than indigenous courts to settle arbitration matters and other disputes. The changes in the financial terms of the concession agreements under the second and third of these headings were promoted by the Organization of Petroleum Exporting Countries (OPEC), to be treated in the next section of the chapter; but their substance will be presented in this section.

1. *Direct Taxation: The 50-50 Principle*

The usual basis of payment by companies to governments under the major concessions was originally, or by early revision, a royalty based on tonnage of production or exports, 4 shillings (gold) per metric ton being the common formula, later adjusted in terms of dollars or pounds to offset changes in the gold value of currencies. There might also be deadrents and other fees or bonuses, but generally other taxes and import duties were waived.

In 1950, following the Venezuela precedent (which was originally put into effect in 1948), the Saudi Arabia agreement was changed to the basis of 50–50 sharing of net profits–profits being arrived at on the basis of posted prices (with some agreed discounts) and agreed methods of calculating operating costs. Royalties paid were credited in arriving at the government's 50 percent; that is to say, the government share was inclusive of earlier royalty commitments by the company. Originally, the Saudi plan covered profits *after* income tax payments to the U. S. government; but after the setting up of a general income tax law by Saudi Arabia in 1951, payments were made on profits *before* U. S. tax. The tax paid could then be taken as a credit against U. S. tax liability, following the recognized principle of avoiding double taxation of income. The plan, while vastly increasing the revenues of the Saudi Arabia government, cost the concessionaire company, ARAMCO, little if anything because of U. S. tax credits.

The same approach was demanded by other countries, and 50–50 quickly became the standard formula, though it did not mean quite the same thing in all cases, due to differences in detailed financial terms. In the case of Iraq there were difficulties in arriving at a formula because the net earnings of parent companies of the Iraq Petroleum Company would be affected in different ways, due to different tax laws in the United States, Great Britain, France, and Holland. The American companies were in the best position because the enlarged tax payments could be wholly offset against U. S. taxes, though some advantages were available under the tax laws of other countries. In administration, the 50–50 plan at times gave rise to difficulties due to disputes over the costs to be used in arriving at taxable income, over certain discounts on prices desired by companies, and over profits of subsidiary companies.

The immediate effect of introducing the 50–50 profit sharing was a large increase in revenues to the producing countries. Later, through the decade of the 1950's, a great expansion of production of low-cost oil benefited both companies and countries. While highly acceptable to the companies as a standard which fortified their company-government relations without costing them much, the 50–50 formula came under fire after 1959 because the posted prices on which profits were calculated, subject to company control, were lowered to the disadvantage of country revenues.

The 50–50 principle of sharing profits between company and country is not as simple as it sounds because of the character of the posted prices on which it was based. As the saying goes, company profits under the formula were somewhat "notional" for reasons we must state, the effective tax rate being substantially higher than the nominal rate.

2. *The Pricing Problem*

Prior to 1950 and the introduction of the direct 50 percent taxation of profits, there had been no publicly announced, or "posted," selling price for crude oil at the Middle East points of origin. Before the war there had been a rather loose system of "basing point" pricing, under which *delivered* prices were "posted" for *refined products,* closely tied to Gulf of Mexico posted prices. During and after the war a limited system of posting f.o.b. Persian Gulf prices for *products* was initiated under pressure from British and American government purchasers, still loosely tied to U. S. Gulf or Venezuela posted prices. Most of the crude oil produced was utilized by refining affiliates of the producing companies, and simply had "transfer prices" for internal integrated accounting and tax purposes. The rest was mainly exchanged among the major international companies at negotiated prices, often on long-term contract. Very little was sold at arm's length bargaining to outsiders so that there did not exist what could be called a "market price" for crude oil. Of course, it was necessary to have prices of a sort for purposes of internal transfer and inter-company exchange and for such open-market sales as did occur, but there was no system of public announcement of such prices. For integrated companies largely utilizing their own oil, the profits calculated from transfer prices merely indicated the extent to which net integrated profits were assigned to the producing stage rather than to the later refining and marketing stages. In setting these transfer prices, various factors were considered, including market factors and the tax laws of the home countries of the parent companies and of the countries in which they operated.

When concession agreements were revised on the so-called 50–50 profit-sharing basis, it was necessary to have crude oil reference prices for purposes of computing the profit on company operations within the producing countries. Thus the practice of "posting" crude oil prices was initiated. Such postings were at the discretion of the companies operating concessions in each of the producing countries, and changes in postings were administrative decisions which might reflect a variety of considerations, such as maintaining a certain relation to product prices. It is not clear just how the prices originally posted were related to current transfer prices, but presumably they were more or less in line with each other. The posted prices served as the base for determining the income of the companies, as required by the tax laws of the United States and the producing countries. They also had a stabilizing effect on product prices by their effect on nominal refinery costs of competing companies.

During the 1950's, the countries were on the whole well satisfied with the results under the 50–50 profit sharing as calculated from posted prices. In addition to the great boost to their revenues initially, production was rising rapidly and the tendency of posted prices was upward, presumably somewhat in relation to market factors. However, after 1957, a market situation was developing that was less than satisfactory to the companies. With large reserves and excess producing capacity beginning to overhang the market, the major companies stiffened their competition in product markets; newcomers began to enter crude oil production; and the growing band of independent refiners could bargain for lower-priced crude oil. The producing affiliates of major companies were under pressure to reduce prices in true sales and exchanges of crude oil, and also to reduce transfer prices to their refining affiliates. Crude oil thus moved in recorded transactions at various discounts from posted prices. This movement had a strong economic basis in the fact that long-run supply costs, both f.o.b. and c.i.f., had been declining steadily during the 1950's.[1]

In this new situation, the major producers in 1959 and 1960 started to revise their posted prices downward to get them into closer conformity with the "realized" prices in sales and transfer transactions. This move reduced country revenues per barrel under the 50–50 profit-sharing plan, and at once met the concerted resistance of governments in combating the right of the companies unilaterally to depress the public revenues by changes in posted prices. In the earlier situations, prior to 1950, where revenues had been obtained from fixed tonnage royalties, deadrents, and other fees, the play of the companies with prices had been a matter of no concern. But where the level of stated profits, and the resulting take of the income tax, were directly affected, the countries insisted upon participating in decisions concerning posted prices. In this situation, the companies had no compelling argument to support their freedom to alter posted prices of crude oil at their sole discretion, since the posted prices had no clear rationale.

A direct result of the price reductions in 1959 and 1960 was the establishment of the Organization of Petroleum Exporting Countries (OPEC), committed to restoring the pre-1960 posted prices. Though unsuccessful in this, they have prevented any further reduction and posted prices have been largely frozen. As a result, the posted prices no longer have any relation to the realized or transfer prices which would normally serve as the basis of calculating

[1] The rapid increase of production based on earlier investment reduced capital costs per barrel and declining tanker rates reduced delivered costs.

accounting profits; their only function is to serve as a "tax reference" price for determining the gross revenue upon which, after deduction of accounting costs, a 50 percent tax is levied. The effective tax rate is thus made higher than if based on transfer prices and realized prices in actual market transactions.

The further result is to divest this levy of the normal characteristics of an income tax based on actual revenues, and to transform it in substance into a fixed charge per unit of production. Assume, for example, a posted price of $1.80 per barrel and unit cost of 20 cents; the 50 percent tax, originally including royalty, would amount to 80 cents per barrel. Assuming, further, that the prices on sales and transfers average $1.35, the company's accounting profit would be 35 cents per barrel. The company's realization on actual sales or transfers may go up or down without affecting the government's revenue. The concept of 50–50 thus becomes extremely tenuous. This analysis is elaborated at later points.

This is an oversimplified picture, in that for purposes of calculating gross income, the agreements may allow for discounts from posted prices or other deviations on special types of sales or transfer transactions. But the general situation is as stated.

3. *Expensing of Royalties*

Since the original 50–50 agreements were made and then fortified by the freezing of posted prices, OPEC has succeeded in imposing a further change that substantially increases the tax revenue of governments. Originally, the royalties paid by the companies (and in some cases rents or other fees) were included in the government's 50 percent, that is, after the government's 50 percent share was computed, the royalties paid were deducted and only the balance was paid as tax. Thus, if the 50 percent paid to the government amounted to 80 cents a barrel and the 12.5 percent royalty based on posted price amounted to 22.5 cents, the tax paid was an additional 57.5 cents.

The new arrangement is based on what is called "the expensing of royalties." In this calculation the royalty (now normally 12.5 percent of posted price) is deducted as an expense before arriving at the net "profit" before tax, of which the government receives 50 percent. In addition, the government receives the royalty. The computation of revenue per barrel to the government then runs as follows (the figures used are illustrative):

Posted price	$1.80[2]
less Cost	0.20
less Royalty (12.5 percent)	0.225
Putative company profit before tax	1.375
Tax (50 percent)	0.6875
plus Royalty	0.225
Government take	$0.9125

The effect of expensing royalties, it will be seen, is to raise government revenue per barrel from 80 cents to 91.25 cents. On an assumed realized price of $1.35, the company's accounting profit would go down from 35 cents to 23.75 cents ($1.35 realized price less production costs of 20 cents and tax and royalty of 91.25 cents).

As initially introduced in 1964–65 the full force of the new plan of expensing royalties was not made effective at once, but was modified by certain discounts from posted prices. The discounts were set at 8.5 percent for 1964, 7.5 percent for 1965, and 6.5 percent thereafter except as later modified. In 1968 a schedule was negotiated under which the allowances were to be completely phased out by 1975.

The course of country revenues per barrel from 1957 to 1967 is shown in Table 10.1. The decline from 1958 to 1961 may be attributed mainly to lower posted prices; the rise in recent years, perhaps mainly to expensed royalties.

It hardly need be said that in detail the tax systems and regulations differ from country to country.[3] But in general they conform to the pattern we have described.[4] It is not to be expected that they will remain static. There is a well-established intention on the part of the oil-producing countries to squeeze as much revenue as they are able without "killing the goose;" and much thought is being given to how this may best be done. The directions which these efforts may take, and the limitations upon them, will be discussed at later points.

As a consequence of basing taxes on fixed posted prices, the income tax plus royalty has almost become a tax of a

Table 10.1

Middle Eastern Government Revenues per Barrel of Oil

(Cents Per Barrel of Exports)

Year	Kuwait[a]	Saudi Arabia[a]	Iran	Iraq	Total Middle East[b]	Libya
1957	79.6	88.2	86.8	93.1	85.7	
1958	81.7	81.7	89.0	88.9	84.8	
1959	77.8	75.8	83.6	82.4	79.8	
1960	76.4	75.0	80.1	78.6	77.7	
1961	74.4	75.5	75.8	76.5	75.8	62.7
1962	74.8	76.5	74.5	76.7	75.9	64.7
1963	74.3	78.7	79.7	80.7	77.9	65.0
1964	76.9	82.0	81.8	80.1	78.6	62.9
1965	78.9	83.2	82.9	81.7	79.9	83.8
1966	78.4	83.4	83.2	81.3	81.5	87.0
1967	78.5	84.2	83.5	85.4	82.1	99.7

[a] Including half Neutral Zone.
[b] Including Qatar, Abu Dhabi, and Bahrain.

Source: Petroleum Press Service, August 1968, p. 302.

[2] As noted below, certain discounts are allowed for a period of years. To that extent, the illustration overstates the effective rate of taxation.

[3] In Iraq a royalty expensing agreement has not been reached, in the midst of numerous points of dispute as yet unsettled.

[4] To follow through all the consequences for the companies' profits would require looking into the tax systems of their home countries, a complicated sphere into which we cannot enter.

fixed amount on each barrel of oil produced, judged to be on the average between 80 and 90 cents per barrel (see Table 10.1). This is from the viewpoint of the oil company a fixed unit cost item. The consequence is to establish a high tax-paid cost figure for the companies. The "real" cost of oil (i.e., excluding payments to governments) is very low. According to estimates of Professor M. A. Adelman, a close student of the subject:

> . . . The current production cost–meaning current operating expense plus a 20 percent return on the necessary development investment–is about 12 cents per barrel in the most expensive of the big five producing countries of the Eastern Hemisphere.
>
> Furthermore the fields now exploited are adequate to support much higher rates of production over the next 15 years, at a cost of about 20 cents a barrel.[5]

However, when government taxes and royalties of more or less 85 cents per barrel are added as a company cost, a figure of upwards of $1.00 per barrel defines the limit to which prices might be driven by competition, unless taxes are reduced. Prices are still well above this cost base; but they have gone down, according to Adelman's estimates, by very roughly 40 percent since early 1957, and in his view, are likely to continue downward.[6] With prices six to ten times as high as real costs, according to his estimates, it seems likely that, short of a producer countries' cartel, the tax barrier is the main protection against a deep competitive erosion of prices.

4. *Relinquishment of Concession Territory*

Apart from the revision of the financial terms of the old concessions, as noted above, the principal change in the terms has been the reduction of the area covered by the concessions. The process and degree of relinquishment, past and prospective, differs greatly from country to country. The record for the various concessions has been presented in Chapter 9.

The practices differ from country to country in the selecting of areas to be relinquished; and the subject is one that has at times led to disagreements between companies and countries. The companies naturally wish to make the choices, while the countries wish to have a hand in them. In Iraq in particular, difference on this point was one of the bones of contention which finally led to the abrupt withdrawal in 1961 of most of the area covered in the concession.

The effect of the relinquishments is to leave in the hands of the countries large areas which may be reassigned for exploitation either by national oil companies or under new agreements with foreign companies.

---

[5] Statement by M. A. Adelman in *Governmental Intervention in the Market Mechanism*, Hearings before the Subcommittee on Antitrust and Monopoly of the Committee on the Judiciary, U.S. Senate, 91st Congress, 1st Sess., *The Petroleum Industry*, Part 1, March 11, 1969, p. 8.

[6] *Ibid.*

The Organization of Petroleum Exporting Countries (OPEC)

The immediate occasion for the organization of OPEC in 1960 was collective resistance to the lowering of posted prices. But from the beginning its function was conceived more broadly as that of promoting the interests of the principal oil-producing countries by parallel lines of national policy and by cooperative action. The original members included Venezuela, Saudi Arabia, Kuwait, Iran, and Iraq, to which have been added Indonesia, Libya, Qatar, Abu Dhabi, and most recently, Algeria. The objectives were stated in its Statute as follows:[7]

> A. The principal aim of the Organization shall be the coordination and unification of the petroleum policies of Member Countries and the determination of the best means for safeguarding their interests, individually and collectively.
> B. The Organization shall devise ways and means of ensuring the stabilization of prices in international crude-oil markets with a view to eliminating harmful and unnecessary fluctuations.
> C. Due regard shall be given at all times to the interests of the producing nations and to the necessity of securing a steady income to the producing countries; an efficient, economic and regular supply of petroleum to consuming nations; and a fair return on their capital to those investing in the petroleum industry.

1. *Posted Prices and Taxation*

As we have seen, the primary initial purpose of OPEC was to stabilize, and so far as possible enlarge, the base upon which the income tax was calculated. The tax system which had been generally adopted, with variations, in the early 1950's was an income tax of 50 percent upon net company income calculated on the basis of posted prices. The action of the companies in reducing posted prices was by way of bringing them into closer conformity with the prices on current transactions. Acting with a united front through OPEC, the countries challenged the right of the companies thus to arbitrarily undermine the fiscal base. They did not succeed in having the cuts rescinded; but they did create an environment within which posted prices have not since been further reduced.

As a result of further concerted pressure through OPEC, the companies later conceded the other change, referred to earlier–the "expensing of royalties." By keeping the royalty as a separate payment and deducting it from the profit before tax, instead of for credit on tax, the total payment of royalty-plus-tax was increased substantially.

Since posted prices, as presently stabilized, present anomalies as between one country and another, and between one grade of oil and another, a further purpose of OPEC is to remove these anomalies, according to the general rule of upgrading where oil is underpriced relatively to the highest posted prices in effect. This is not just a matter of short-run advantage from higher prices, but is also

---

[7] Organization of the Petroleum Exporting Countries, *The Statute*, Chap. I, Article 2.

a matter of providing a more uniform price structure with due regard to transport costs (as among countries), to facilitate their future cooperative endeavors in relation to markets and company relationships. Adjustment to differences in crude oil quality is also involved.

The projects described above—the stabilization of posted prices and the expensing of royalties to increase revenues—are the principal substantive achievements successfully promoted by OPEC to date. The potential importance of the Organization is, however, of much wider dimensions. First of all, it is a continuing agency for the consideration of all the mutual problems and possible lines of cooperative or parallel action of the oil-producing countries. Second, it has been influential in helping to build up a body of talented and well-trained personnel, capable of matching the expertise of company representatives in negotiations and of analysis of the various lines of action open to countries individually and in their relations with companies and one another. A strong intellectual element of expertise and reason is introduced into investigations of lines of evolutionary change.

2. *The Proposed Production Program*

We may now turn to other possible lines of action which have engaged the attention of OPEC. From the beginning, the possibility of operating a joint production program as a means of supporting prices was a subject of discussion. This arose from what, from the point of view of the countries, is a fundamental weakness in the market situation—the existence of great reserves capable of supporting much larger production, the interest of each country in enlarging its exports, and a competitive situation among companies which tended to weaken the prices of refined products in world markets. An early proposal, especially favored by Venezuela, was the formation of what would have amounted to an international producers' cartel which would set the level of production and assign shares to individual countries. An OPEC Commission set up in December 1964 conducted studies bearing upon the central problems presented by the world supply situation; and after two years it came to the conclusion that the only effective solution would be a production program to which all OPEC countries would unreservedly conform.

Meanwhile, in July 1965 what was called a "transitory" Production Program was initiated under which OPEC proposed annual rates of growth applicable to supplies from each country, paying regard to the special circumstances of each. The countries were not bound by the proposed rates, but it was hoped that they would substantially conform their production rates to those proposed. The resolution establishing the Program was as follows:[8]

> The Conference,
>
> having heard the report of the Economic Commission; with a view to counteracting the continuing erosion of crude and product prices;
>
> considering that one of the contributing factors to the deterioration of crude and product prices is the unrestricted competitive use of the excess producing capacity; recognizing the need for a steady flow of oil to international markets on the basis of equitable and stable prices;
>
> considering that there is an immediate need to formulate a programme to achieve the objectives as set out in the Resolutions of the Organization,
>
> resolves
>
> 1. to adopt as a transitory measure a production plan calling for rational increases in production from the OPEC area to meet estimated increases in world demand; and
> 2. to submit a production programme to the Governments of Member Countries for approval.

The reception of the program by member countries was mixed; and after two years it was permitted to lapse in June 1967. But at the OPEC Conference of November 1967, the Conference "reaffirmed its conviction that a Program is an effective instrument for the pursuit of the Organization's objectives, and instructed the Economic Commission to undertake a comprehensive study with a view to perfecting an economically practicable system."[9] Reporting in November 1968, the Commission stated that, while some prerequisites of such a program were not yet capable of fulfillment, continued efforts should be made to overcome the obstacles to implementing an effective program.

Clearly, the effort to establish a joint production plan binding upon all OPEC member countries faces very great and possibly insuperable obstacles. The various countries find themselves in very different situations. Venezuela is more interested in higher prices than in expanded production; Libya has been primarily interested in expansion; Iran is strongly committed to securing a larger share of the market; and so on. No consensus yet exists to support a definitive production program. In addition, each country has complications of its own in dealing with the companies as to the quantities they are willing to lift.

3. *General Principles of Policy*

Beyond the subjects already mentioned—prices, revenues, and production programming—the matters in the present forefront of OPEC attention may be conveniently summarized by quoting in full a Resolution adopted at the Sixteenth Conference in June 1968, as follows:

RESOLUTION XVI. 90

DECLARATORY STATEMENT OF PETROLEUM POLICY IN MEMBER COUNTRIES

> The Conference,
>
> recalling Paragraph 4 of its Resolution 1.2; recognizing that hydrocarbon resources in Member Countries are

[8] Organization of the Petroleum Exporting Countries, Resolutions of the Ninth Conference held in Tripoli from 7th to 13th July 1965, Resolution IX.61.

[9] OPEC, *1967 Review and Record*, p. 4.

one of the principal sources of their revenues and foreign exchange earnings and therefore constitute the main basis for their economic development;

bearing in mind that hydrocarbon resources are limited and exhaustible, and that their proper exploitation determines the conditions of the economic development of Member Countries, both at present and in the future;

bearing in mind also that the inalienable right of all countries to exercise permanent sovereignty over their natural resources in the interest of their national development is a universally recognized principle of public law and has been repeatedly reaffirmed by the General Assembly of the United Nations, most notably in its Resolution 2158 of November 25, 1966;

considering also that in order to ensure the exercise of permanent sovereignty over hydrocarbon resources, it is essential that their exploitation should be aimed at securing the greatest possible benefit for Member Countries;

considering further that this aim can better be achieved if Member Countries are in a position to undertake themselves directly the exploitation of their hydrocarbon resources, so that they may exercise their freedom of choice in the utilization of hydrocarbon resources under the most favorable conditions;

taking into account the fact that foreign capital, whether public or private, forthcoming at the request of the Member Countries, can play an important role, inasmuch as it supplements the efforts undertaken by them in the exploitation of their hydrocarbon resources, provided that there is government supervision of the activity of foreign capital to ensure that it is used in the interest of national development and that returns earned by it do not exceed reasonable levels;

bearing in mind that the principal aim of the Organization, as set out in Article 2 of its Statute, "is the coordination and unification of the petroleum policies of Member Countries and the determination of the best means for safeguarding their interests, individually and collectively";

recommends that the following principles shall serve as basis for petroleum policy in Member Countries.

Mode of development

1. Member Governments shall endeavour, as far as feasible to explore for and develop their hydrocarbon resources directly. The capital, specialists and the promotion of marketing outlets required for such direct development may be complemented when necessary from alternate sources on a commercial basis.
2. However, when a Member Government is not capable of developing its hydrocarbon resources directly, it may enter into contracts of various types, to be defined in its legislation but subject to the present principles, with outside operators for a reasonable remuneration, taking into account the degree of risk involved. Under such an arrangement, the Government shall seek to retain the greatest measure possible of participation in and control over all aspects of operations.
3. In any event, the terms and conditions of such contracts shall be open to revision at predetermined intervals, as justified by changing circumstances. Such changing circumstances should call for the revision of existing concession agreements.

Participation

Where provision for Governmental participation in the ownership of the concession-holding company under any of the present petroleum contracts has not been made, the Government may acquire a reasonable participation, on the grounds of the principle of changing circumstances.

If such provision has actually been made but avoided by the operators concerned, the rate provided for shall serve as a minimum basis for the participation to be acquired.

Relinquishment

A schedule of progressive and more accelerated relinquishment of acreage of present contract areas shall be introduced. In any event, the Government shall participate in choosing the acreage to be relinquished, including those cases where relinquishment is already provided for but left to the discretion of the operator.

Posted prices or tax reference prices

All contracts shall require that the assessment of the operator's income, and its taxes or any other payments to the State, be based on a posted or tax reference price for the hydrocarbons produced under the contract. Such price shall be determined by the Government and shall move in such a manner as to prevent any deterioration in its relationship to the prices of manufactured goods traded internationally. However, such price shall be consistent, subject to differences in gravity, quality and geographic location, with the levels of posted or tax reference prices generally prevailing for hydrocarbons in other OPEC Countries and accepted by them as a basis for tax payments.

Limited guarantee of fiscal stability

The Government may, at its discretion, give a guarantee of fiscal stability to operators for a reasonable period of time.

Renegotiation clause

1. Notwithstanding any guarantee of fiscal stability that may have been granted to the operator, the operator shall not have the right to obtain excessively high net earning after taxes. The financial provisions of contracts which actually result in such excessively high net earnings shall be open to renegotiation.
2. In deciding whether to initiate such renegotiation, the Government shall take due account of the degree of financial risk undertaken by the operator

and the general level of net earnings elsewhere in industry where similar circumstances prevail.

3. In the event the operator declines to negotiate, or that the negotiations do not result in any agreement within a reasonable period of time, the Government shall make its own estimate of the amount by which the operator's net earnings after taxes are excessive, and such amount shall then be paid by the operator to the Government.
4. In the present context, "excessively high net earnings" means net profits after taxes which are significantly in excess, during any twelve-month period, of the level of net earnings the reasonable expectation of which would have been sufficient to induce the operator to take the entrepreneurial risks necessary.
5. In evaluating the "excessively high net earnings" of the new operators, consideration should be given to their overall competitive position vis-à-vis the established operators.

Accounts and information

The operator shall be required to keep within the country clear and accurate accounts and records of his operations, which shall at all times be available to Government auditors, upon request.

Such accounts shall be kept in accordance with the Government's written instructions, which shall conform to commonly accepted principles of accounting, and which shall be applicable generally to all operators within its territory.

The operator shall promptly make available, in a meaningful form, such information related to its operations as the Government may reasonably require for the discharge of its functions.

Conservation

Operators shall be required to conduct their operations in accordance with the best conservation practices, bearing in mind the long-term interests of the country. To this end, the Government shall draw up written instructions detailing the conservation rules to be followed generally by all contractors within its territory.

Settlement of disputes

Except as otherwise provided for in the legal system of a Member Country, all disputes arising between the Government and operators shall fall exclusively within the jurisdiction of the competent national courts or the specialized regional courts, as and when established.

Other matters

In addition to the foregoing principles, Member Governments shall adopt on all other matters essential to a comprehensive and rational hydrocarbons policy, rules including no less than the best of current practices with respect to the registration and incorporation of operators; assignment and transfer of rights; work obligations; the employment of nationals; training programs; royalty rates; the imposition of taxes generally in force in the country; property of the operator upon expiry of the contract; and other such matters.

Definition

For the purposes of the present Resolution, the term "operator" shall mean any person entering into a contract of any kind with a Member Government or its designated agency including the concessions and contracts currently in effect, providing for the exploration for and/or development of any part of the hydrocarbon resources of the country concerned.

The principles set forth in this Resolution may be expected to dominate the thinking of member countries in their long-range strategic maneuvers, vis-à-vis the companies, as they continue their effort to gain greater control over the exploitation of their oil resources. The effort they are making is not simply, or not even mainly, to squeeze greater revenues from present operators. That is of course an objective. But the other purpose, that of ultimately achieving effective control, is likely to provide the dominant driving force, conforming as it does to the deep-seated emotional drives in the current political situation.

The principles as laid down in the Declaratory Statement express, it will be seen, a set of aspirations and a long-run strategy for individual countries in the future development of their oil resources. Emphasis is placed upon a number of concepts and procedures:[10]

(1) the principle expressed in a United Nations resolution of "the inalienable right of all countries to exercise permanent sovereignty over their natural resources";
(2) the desirability of direct exploitation of resources by agencies of the producing countries, rather than by outside agencies;
(3) the doctrine of "changing circumstances" which justifies countries in demanding changes in the terms of existing concession agreements;
(4) under the principle of changing circumstances, the right to acquire a participating ownership share for governments under the existing concession agreements;
(5) the right of governments to alter the financial terms of agreements where companies are receiving "excessively high net earnings";
(6) the right of governments to determine posted or tax reference prices;
(7) the accelerated relinquishment of concession territory.

One thing that these "principles" almost say is that the countries have the right unilaterally to change the terms of

[10] A useful analysis of the terms of the Declaratory Statement is given by Francisco R. Parra, formerly Secretary General of OPEC, in "OPEC: Present and Future Role," *Middle East Economic Survey, Supplement*, May 16, 1969.

the concessions as they see fit in the light of "changing circumstances." This implication raises a fundamental question concerning the nature of national sovereignty in relation so the contractual character of the agreements with companies. The stated principles of policy run in direct conflict to the interpretation of the binding character of the agreements except as altered by mutual consent. It is not possible, by simple reference to the principles, to forecast what may be attempted by one country or another in the effort to put them into effect. What can be said is that they will affect the substance of demands on companies by countries in many controversial contexts. So long, however, as countries recognize a high degree of dependence on the companies for the marketing of their oil, dictates of prudence will no doubt limit the severity of their demands.

For straws in the wind, to test possible lines of country action, we need to examine especially three lines of present and impending policy:

(1) the establishment of national oil companies and the extent of their present and potential responsibilities;
(2) the character of the more recent agreements under which countries achieve a participating ownership status or procure services on a contractual basis; and
(3) the demands of countries for participation in the ownership and management of the old major concessions.

The National Oil Companies

Beginning with the National Iranian Oil Company, organized in 1951, each of the four largest producing countries in the Middle East has set up a state-owned company to perform certain functions in relation to oil production operations and collateral activities: the Iraq National Oil Company in 1961, the Kuwait National Petroleum Company in 1960, and the Saudi Arabian General Petroleum and Mineral Organization in 1962. A Libyan national company was organized in 1968. The present responsibilities of these companies differ greatly from country to country; but potentially they all constitute the agencies through which the countries may exercise an increasing degree of control over oil activities. At present, they represent countries in their partnership or contractual relations under new agreements with foreign companies. They have also undertaken an expanding range of functions as the status of the old major concessionaires undergoes progressive curtailment. Conceptually, they embody the aspiration of the countries to achieve the greatest degree of control over the exploitation of their oil resources consistent with limiting circumstances and economic interest.

1. *The National Iranian Oil Company (NIOC)*

The NIOC, organized under authorization of the Nationalization Law of 1951, became operative more or less simultaneously with the Consortium which took over operational responsibility for the former concessionary properties of the Anglo-Iranian Oil Company. NIOC absorbed, and used as an operating subsidiary, the Iranian Oil Company, which had interim responsibilities prior to the Consortium. As we have seen, NIOC became titular owner of former properties of the Anglo-Iranian Oil Company, supervised operations under the Consortium, and took over gradually the "nonbasic" services in Consortium territory—housing, health, social services, technical training, and the like. But these latter activities were secondary to its main responsibilities.

NIOC was in fact charged with complete responsibility for all phases of the oil industry other than those performed under the Consortium—exploration and production outside the restricted Consortium territory, distribution of oil products throughout the country, development of refining activities, marketing its crude oil and/or products outside the country, internal pipeline transport, external tanker transport, and the development of collateral industries such as petrochemicals. It has been more or less active in all these fields.

In promoting the discovery and development of producing properties, it has proceeded in three ways. First, it has conducted some exploratory operations on its own account. Second, it has entered into a number of partnership arrangements under which foreign companies undertake exploration at their own risk while NIOC acquires a 50 percent ownership interest in the companies exploiting successful fields. Third, it has entered into contractual arrangements with foreign companies for exploration and development under NIOC ownership. Thus NIOC is directly involved in producing operations, though the scale is still only a very minor fraction of that of the Consortium.

NIOC appears to be anxious to enter into the field of international marketing in ways which will raise the level of Iranian exports without unsettling the markets in which Consortium oil is marketed. So far, it has made barter deals on a small scale with Rumania, Bulgaria, Yugoslavia, and Hungary, and most recently at the end of 1968 a more ambitious deal with Czechoslovakia to exchange crude oil for industrial plant and equipment. A projected pipeline from a Yugoslav base through Hungary and Czechoslovakia, and possibly to Poland, in which NIOC might participate has also been under discussion. It has also interested itself in other barter deals and in possible participation in the building of refineries in developing countries of Africa and Asia. It has recently negotiated an agreement with Japanese interests for the sale of liquefied petroleum gas. It also has participation in two overseas refineries, one in India and one in South Africa.

Another direction in which NIOC initiative has expressed itself is the building up of a tanker fleet through its subsidiary Iranian Tankers, Ltd. By 1969 it had developed a fleet of several hundred thousand tons and was reported to be negotiating for two 200,000-ton tankers. It would appear that Iran has the intention of progressively taking over the transport of its oil production in its own bottoms.

NIOC is also developing plans for the utilization of natural gas as the energy base for the development of new industries; and a joint NIOC-Soviet gas pipeline to the U.S.S.R. is nearing completion. NIOC has also organized a subsidiary, the National Petrochemical Company, to operate in the petrochemical field; and plans are under consideration for joint partnership arrangements with foreign petrochemical companies.

Altogether, NIOC is engaged in a wide range of activities. Within the oil industry proper, it is (1) directly engaged in oil exploration and production, refining, pipelining, domestic marketing, and to a limited degree foreign marketing; (2) joined in a partnership relation with foreign producing companies; and (3) looking after Iran's interest in the activities of the operating and trading companies under the Consortium agreement. All these activities serve as a training ground for the much larger role that it probably aspires to in the range of activities now conducted by the Consortium, which still accounts for 95 percent of the producing operations. The present Consortium agreement provides for an initial period of twenty-five years to 1979, but also provides for three extensions of five years each at the option of the Consortium if certain specified conditions are satisfied. For the long-run future it is to be inferred that NIOC expects to achieve a much expanded participation in the operations of the industry in Iran.

2. *The Iraq National Oil Company (INOC)*

The INOC was established in 1964 by Law No. 11 to serve as the foundation for the future exploitation of the areas of which the IPC companies had been deprived in 1961. Its activities were to include exploration, production, and pipeline operations; and it was also authorized to engage in such activities outside Iraq. In 1967 it was reorganized to enable it, among other activities, to refine and distribute products within Iraq. Under the draft agreement of 1965, which was not ratified (see Chapter 9, p. 116), it would have entered into joint-venture agreements with member companies of the IPC group for exploiting a portion of the areas that had been reclaimed from the concessionary companies.

Pending some resolution of the issues between the government and the IPC companies, the powers of INOC to proceed independently were in a state of considerable uncertainty under the law. However, in 1967, by Law No. 97, the Iraq government allocated to INOC exclusive rights over all those areas of which the IPC companies had been deprived in 1961, over territorial waters relinquished earlier, and over the continental shelf. INOC was entitled to exploit these areas in conjunction with others, but was forbidden to enter into a concession-type arrangement.

Meantime, a number of companies had held discussions with INOC regarding arrangements for developing the oil resources. However, as the IPC companies maintained the validity of their contractual rights over the forcibly "relinquished" areas, it was difficult for possible new entrants to be given satisfactory assurances of INOC's title; and they were faced with possible litigation with the IPC companies. To date, INOC has succeeded in concluding only one new agreement, with the French state-owned company ERAP, in 1967. This agreement is of the contract type described below (see pp. 134 and 136).

In brief, the whole process of discovery and development of new oil resources has been at a virtual standstill since 1961. The means to ending this stalemate remain undefined. The new government established under the *coup* of July 1968 is strongly committed to direct development of oil resources by INOC. INOC does not itself have the capital to proceed on any extensive scale. It remains to be seen what power it has to attract foreign contractors on the terms permitted under the law of 1967.

Apart from its primary responsibility for exploiting oil resources, the other principal duties of INOC are the operation of a group of small refineries, which serve the domestic market, and the administration of the state-owned system of distribution of oil products—duties until recently in the hands of another agency, the Government Oil Refineries Administration.

3. *The Kuwait National Petroleum Company (KNPC)*

The KNPC was organized in 1960 with a 60 percent state share interest, the other 40 percent being subscribed by private Kuwaiti interests. Its first task was to take over the local distribution of oil products after purchase of the facilities from the Kuwait Oil Company. It also formed the Kuwait Aviation Fueling Company, jointly with British Petroleum, for aircraft refueling services formerly carried on by a British Petroleum subsidiary.

KNPC is charged with responsibility for exploring and developing concession territory relinquished by Kuwait Oil Company, either directly or in association with foreign companies. In this connection it has become a partner with the Spanish company, Hispanoil, to exploit relinquished lands, holding a 51 percent interest in the joint company.

Its most ambitious project has been to contract in 1964 for the construction of a 95,000 barrels/day refinery completed in 1968 at a cost of $150 million. The economics of this enterprise are not clear, but may be aided by royalty rebates or other government support. The building of this refinery supports other evidence that KNPC plans over time to develop into an integrated international oil company. One of its early activities was participation in a Rhodesian refinery. Its marketing department has begun to establish marketing subsidiaries abroad so far in two countries. Activities in this direction were further formalized by the setting up in 1968 of two subsidiary companies, KNPC (Far East) for trading and marketing operations and Kuwait National Petroleum Company for Trading and Transport to engage in all aspects of trade in crude oil and products, including wholesale and retail distribution and marketing, transport, storage, and refining. For several years the Kuwait Oil Tanker Company has been building up a fleet which has reached the order of 500,000 deadweight tons.

Kuwait is engaged in a dynamic program of industrial development. Separately from KNPC, but oil related, it has established chemical fertilizer and petrochemical companies, and is setting up a giant complex with harbor facilities, power and desalination plant, and other infrastructure facilities to support extensive industrial development, and is equally engaged in an intensive program of technical training to provide personnel. Because of its large oil revenues from the Kuwait Oil Company concession and the smallness of the country, Kuwait is in the strongest position of any of the countries to provide capital financing for industrial projects.

4. *The General Petroleum and Mineral Organization (Saudi Arabia)*

Petromin, established in 1962, was designed to be the mother organization for a wide range of developments not only in oil but in other minerals. It has the primary responsibility for discovery and development of new oil outside the concession territory of ARAMCO. As ARAMCO has relinquished more or less two-thirds of its original territory, and is due over time to relinquish most of the remainder, Petromin has an enormous area within which to exercise its responsibility. So far it has followed the example of Iran in entering into joint partnership arrangements with foreign companies.

Over the past few years, Petromin has taken over from ARAMCO, by stages, responsibility for domestic marketing of oil products, now covering the whole country. It has built a refinery on the Red Sea coast to supply the western region and plans one or more in the interior, but is otherwise dependent upon ARAMCO's Ras Tanura refinery for its supplies.

Though less active than the Iran and Kuwait companies, Petromin has begun to exhibit an interest in conducting foreign marketing operations on its own account and in 1968 set up a new Department of Foreign Trade.

Petromin has engaged itself zealously in promoting oil-related industries, examining possible projects with the aid of foreign experts and companies. It has a 52 percent share in the Saudi Arabia Fertilizer Company, and is engaged in planning for other industrial development, especially in the field of petrochemicals.

5. *The Libyan General Petroleum Corporation (LIPETCO)*

The Libyan national company, LIPETCO, organized in 1968, is expected to undertake a wide range of responsibilities similar to those in other countries described above.

6. *Comments on National Oil Companies*

It is too early in the history of the national oil companies to foretell what part they may come to play in the future production, refining, and marketing of crude oil and products. So far the role is quite limited. Mainly, the national companies serve as suppliers to their local markets and as the government partner in the new type of joint enterprise agreement. However, they have responsibilities for the future exploration and development of territories outside the limits of the old concessions; though only in one case, Iran, is the company, to a slight extent, exercising this responsibility directly, rather than jointly or under contract with foreign companies. They are also cautiously entering the field of international marketing, though the obstacles to moving far or fast in this direction are very great. Large capital outlays would be required; and they would be in competition with the marketing apparatus so largely in the hands of the international major companies. They are free to enter the field, but only by a rough road.

It is perhaps a sign of their aspirations that the national companies of the OPEC countries have on two occasions undertaken discussions of the possibility of joint action to build up international operations and to cooperate in other ways. At the first meeting in October 1966, the representatives presented for approval by their respective companies the following recommendations:[11]

1. That the National Oil Companies should determine the fields of training, and the technical and administrative assistance required, in order to insure the optimum utilization of their manpower and exchange personnel along the lines of their most pressing needs, through OPEC Secretariat, by means of, among others:
   a) cross-posting assignments;
   b) collective use of regional academic institutions and research centres;
   c) staging of seminars of short duration on specific fields, and
   d) assistance from OPEC experts.
2. That the National Oil Companies shall share their own experiences with problems relating to their penetration into their respective domestic markets.
3. That the National Oil Companies shall endeavour to avoid setting precedents which might be harmful to the interests of other National Oil Companies and, at the same time, follow practices which might be helpful to each other in their arrangements with foreign oil companies in their respective countries.
4. That the National Oil Companies shall endeavour to exchange on a regular basis non-confidential technical information which may be useful to them. Reports of OPEC of a similar nature should also be made available for exchange.
5. That annual meetings be held in Member Countries and the OPEC Secretariat, on a rotating basis. The Meeting also decided that a senior staff member will be appointed by each of the National Oil Companies, to act as Liaison Officer for his Company and that a senior staff member of the OPEC Secretariat should be nominated to act as Coordinator.
6. That a committee from Member Companies be set up, within the OPEC Secretariat, to study the ways and means by which National Oil Companies might, in a coordinated way, contribute effectively to the marketing of oil and its derivatives in the inter-

[11] *OPEC Bulletin*, November 1966.

> national market, as a supplement to their present and future individual activities, with a view to avoiding the effects of wasteful competition and furthering the aims of OPEC. For the purposes of meeting the costs of these studies, adequate funds should be placed by the National Oil Companies at the disposal of the Secretary General of OPEC.

All these recommendations reflect a felt need for co-operative action; but subsequent discussion has centered on the last one for the study of a coordinated approach to international marketing activities. Within the OPEC Secretariat, and in individual countries, a considerable apprehension has been felt that the national companies, in their efforts to establish a footing in international markets, might engage in competitive practices which would weaken the price structure. No plan has, however, emerged for effecting coordination.

One important role that the national companies are being called upon to fill is that of participation in the ownership, and to a degree the management, of concessionary companies. As we shall see in the next section, most of the new concessions assign to the national company a participating share in the enterprise. Potentially far more important, some of the countries are claiming the right to such participation in the old major concessions. In what ways such a relationship may develop cannot now be foreseen; but a few of the possibilities can be usefully introduced at this point.

When a national company becomes a partner in a producing company, it acquires title to a share of the oil. It can, if it wishes, permit this to be marketed by the foreign partner, merely sharing in the money proceeds. On the other hand, it can undertake to market some or all of its own oil. Up to the present the national companies have not had much oil at their disposal under the new agreements; and they have mostly left the marketing to the other partner because of the lack of outlets of their own. Some efforts at selling have been made by special deals in markets not now served by Western companies—mainly in the Communist sphere and through barter deals. But they are much handicapped in the effort to sell elsewhere. They could, of course, sell easily at cut-rate prices, and to some extent have done so, but only at the risk of eroding the price structure.

So far, the matter has been of relatively minor significance because of the small amount of oil involved. But it would enter an entirely different order of magnitude if the national companies were to become substantial partners in the old major concessions. With large amounts of oil available to them, they could hardly hope to market much of it in competition with the private companies without seriously undermining world prices—unless the market shares were underpinned by a workable producer countries' cartel.

A basic difference must be noted between the position of a national company as seller of its country's oil and the position of a private concessionaire. The concessionaire cannot profitably offtake oil at a price below tax-paid cost, say upwards of $1.00 per barrel. A national company, if free of tax and royalty, has no base to which competition among countries trying to sell their oil might drive prices, higher than the *ex tax* cost, say 20 cents. The tax-paid cost of concessionaire oil gives an assurance of country revenues which would not be assured under competitive sales by national companies.

These matters will receive later attention. Here we have wanted merely to point out certain obstacles that are likely to limit the scope of the activities that can be assigned to national companies, consistent with the economic advantage of the producing country. Just what shape these activities may take, in one country or another, will arise out of experience and cannot now be reasonably predicted.

Partnership or Participating Agreements

Starting in 1957, the considerable number of new agreements have almost universally departed from the older type of concession arrangement. Generally, except in Libya, they have followed a pattern of joint enterprises in which the foreign company entered into some form of partnership or participating arrangement with a national oil company. In other cases, the agreements take the form of a service contract between a foreign company and a national oil company which is the titular owner. The new agreements have been made for three types of territories: (1) land areas not covered by old concession agreements, (2) areas relinquished by old concessionaires, (3) offshore areas in the continental shelf outside conventional territorial waters and thus outside the coverage of old concessions. Mostly, they were confined to relatively restricted geographical limits. The participating type of agreement has been introduced by Iran, Saudi Arabia, Kuwait, Abu Dhabi, and Libya. The contract type has appeared in Iran and Iraq.

The terms of the participating agreements differ substantially from country to country. We shall summarize some of them at sufficient length to bring out the common and some of the differentiating features. Since Iran was the country which pioneered the new type of agreement and has made the most use of them, it is well to review the record in that country at some length.

1. *Iran*

Rules and procedures governing agreements of various types were provided in the Petroleum Law of 1957. The device put into effect over the following years, except in one instance, was the company jointly owned by the National Iranian Oil Company and a private company, with equal ownership shares and equal membership on the board of directors—the chairman being an Iranian; the management director, a foreigner. In most cases the private company was itself a joint venture of a number of foreign parent companies. Except in one case, Shell, none of the foreign participants in these ventures were the major companies which operate under the old concessions or in the Iranian Consortium. A considerable number of new foreign interests were thus introduced into the Iranian picture, some of them private companies, some state companies of consuming countries.

As of 1968, the following eight companies were operating under the participating type of agreement:[12]

(1) *Societe Iran-Italienne des Petroles (SIRIP)*. The first company, organized in 1957, had for its outside partner the Italian government-owned AGIP Mineria. Three areas were assigned totaling 23,000 square kilometers, of which one covering 5,600 square kilometers was located offshore in the continental shelf of the Persian Gulf, another inland, and another in the southeast.

(2) *Iran Pan American Oil Company (IPAC)*. The outside partner in this joint company, organized in 1958, was Pan American Petroleum Corporation (after 1962, American International Oil Company), a subsidiary of the Standard Oil Company (Indiana). It was assigned an offshore area of 16,000 square kilometers at the head of the Persian Gulf.

(3) *Iranian Marine International Oil Company (IMINOCO)*. The outside participants were a group consisting of the Italian AGIP, Phillips Petroleum Company, and the Indian Oil and Natural Gas Commission. Area: 7,960 square kilometers offshore.

(4) *Lavan Petroleum Company (LAPCO)*. The participants were four American companies: Atlantic Richfield, Sun, Union Oil of California, and Murphy. Area: 8,000 square kilometers offshore.

(5) *Iranian Offshore Petroleum Company (IROPCO)*. The participants were another group of American companies, seven in number: Tidewater, Superior, Skelly, Sunray DX, Kerr-McGee, Cities Service, and Atlantic. Area: 2,250 square kilometers offshore. A Spanish group was reported as a probable purchaser of a share in this group.

(6) *Dashtistan Offshore Petroleum Company (DOPCO)*. The outside participant was Royal Dutch/Shell, the only major international company among the new joint companies. Area: 6,075 square kilometers offshore.

(7) *Farsi Petroleum Company (FPC)*. The outside participant, known as Sofiran, was a combination of three French state-owned agencies, but not including Compagnie Francaise des Petroles (CFP), which as a member of the Iraq Petroleum Company had been the only French interest in Middle East oil. Area: 5,800 square kilometers offshore.

(8) *Persian Gulf Petroleum Company (PEGUPCO)*. The participant group was made up of six German companies. Area: 5,150 square kilometers offshore.

It will be seen that the territory assigned under seven of the agreements is in the offshore continental shelf, and the eighth has onshore and offshore areas. Of the eight, three include solely American interests; one, French state interests; one, Italian state interests; one, German interests; one, mixed national interests; and one, the Royal Dutch/Shell group. By 1968 only two, IPAC and LAPCO, were producing in substantial volume. SIRIP was also producing; and IMINOCO was expected to enter production in substantial volume in 1969. Exploratory results in several cases have been disappointing and development costs are reported to have been phenomenally high. The Shell operation is reported to have given up.

The financial and other terms, substantially similar except for initial bonuses (ranging from \$25 to \$59 million) and area, are worth special attention because they open up a number of questions concerning country revenue benefits and possible lines of market behavior by companies having NIOC participation.

Duration: The term of the agreements is twenty-five years from the date of first commercial production, with three possible extensions of five years each.

Exploration: Exploratory expense is undertaken by the foreign company, with minimum amounts prescribed by periods. After commercial oil is discovered, NIOC will repay half by annual assignment of a portion of its share of the oil. The agreement lapses if commercial oil is not discovered within twelve years.

Relinquishment: Territory is to be relinquished 25 percent after the first five years, another 25 percent after ten years. After twelve years only areas in actual production are to be retained.

Royalty: No royalty is prescribed.

Taxation and Pricing: Income tax is at the rate of 50 percent of net profit, payment to the government being made, respectively, by NIOC and the foreign company out of their half shares. Profits are nominally based on posted prices but, by agreement between the NIOC and the foreign partner, may be based upon realized prices or upon some other discounted price basis.

Crude Offtake: Each partner may lift 50 percent of production, and either may lift the unwanted portion of the other's share at a price halfway between posted price and unit production cost. In effect this presumably means that the foreign partner engages to market all the oil except what NIOC wishes to market through its own channels.

The most striking feature of these agreements is the tax provisions, which are flexible in that the profits on which they are based can be calculated from prices reflecting discounts from posted prices, contrary to the OPEC standard. It is our understanding that this practice has been followed.

2. *Saudi Arabia*

While Saudi Arabia has so far made only slight use of joint venture or partnership arrangements, it is headed in that direction. Three recent agreements have been of that character. The national company, General Petroleum and Mineral Organization (Petromin), will act as the agency for participation. It appears to be the intention that in one form or another Petromin will be a participant in all ventures in which foreign companies are involved, not only

[12] Another agreement with a Canadian company, Sapphire, was cancelled.

new ones but eventually in ARAMCO under the old concession.

(1) *The AUXIRAP-Petromin Agreement.* In 1965, an agreement was made with the state-owned French company, AUXIRAP, for exploration in an offshore area in the Red Sea. A one-third interest in AUXIRAP's rights was later acquired by an American company, Tenneco. If oil was found, a company would be formed in which Petromin would take a 40 percent share interest (with 50 percent voting power). The agreement contained stiff financial provisions, and provided for marketing arrangements, eventually to include Petromin participation in downstream operations. After early drilling experience which proved discouraging, this agreement was in abeyance.

(2) *The Sinclair/Natomas-Petromin Agreement.* This 1967 agreement with two American companies, also in the Red Sea coastal area, provides for a 50 percent Petromin participation in a company to be formed if oil is discovered. The detailed terms are similar to those described in the next agreement.

(3) *The AGIP-Petromin Agreement.* This 1967 agreement between the Italian state-owned company AGIP (later joined by Phillips) and Petromin is by far the most important participating agreement yet entered into by Saudi Arabia and the most elaborate in any country. It covers an area of 77,000 square kilometers in the southeast in territory relinquished by ARAMCO. The terms, which are extremely detailed, are described by *Petroleum Press Service* as follows:[13]

> The form of the ENI deal is threefold: an agreement between the Government of Saudi Arabia and Petromin; a contract under which Petromin assigns its rights to ENI's subsidiary AGIP (Saudi Arabia), while retaining legal title; and a petrochemical agreement between Petromin and the ENI affiliate ANIC.
>
> The area concerned covers 77,382 sq.kms. (nearly 30,000 sq. miles) in the Rub' al-Khali desert in the south-east of the country. (It had been relinquished by Aramco which is now drilling at Kidan near part of the area.) The exploration period is three years, with the possibility of a three-year extension, and the exploitation period 30 years, which may be extended by 10 years at the government's option. Within three years of the granting (to Petromin) of an exploitation concession following commercial discovery, 20 per cent of the unexploited acreage must be relinquished, and a further 20 per cent every five years. Surface rentals range from $10 per sq.km. for the first five years up to $625 per sq.km. from the 26th year onwards. A bonus of $2 million is payable on signature, one of $3 million on the granting of the exploitation concession, and one of $8 million when production reaches 300,000 b/d. Annual contributions are payable to the Saudi College of Petroleum and Minerals.
>
> Royalty is stiff: 14 per cent, reduced to 12.5 per cent if production does not exceed 60,000 b/d in any year or 13 per cent if it is between 60,000 b/d and 80,000 b/d. It may be taken in cash or kind, is calculated on posted prices and is fully expensed. For the fourth year of the exploitation concession and thereafter, royalty payments are subject to an annual minimum of $5 million.
>
> Tax is 50 percent—or any higher rate which may be imposed. Taxable income is calculated on the basis of posted prices; a half-cent per barrel marketing allowance is deductible but rentals, bonuses and pre-production exploration expenditure are not. Posted prices must be approved by the government.
>
> Petromin may take a 30 per cent participation in the Saudi Arabia Oil Company (SAOC) which will be set up to operate the exploitation concession, and will then pay 30 per cent of exploration expenditure. The state company may raise its participation to 40 per cent when production reaches 300,000 b/d and to 50 per cent when production reaches 600,000 b/d, in each case making the proportionate investment. If no oil is discovered, AGIP bears all the exploration outlay.
>
> From the beginning of exploitation of a long-range programme of integrated operations—production, refining, marketing and distribution—shall be worked out. When production reaches 200,000 b/d for 90 days, Petromin is committed to constructing a 30,000 b/d refinery in Saudi Arabia, or embarking on another project requiring similar investment, to be completed within three years. In addition at least 30 per cent of both partners' net profits from crude oil production must be made available for investment in integrated operations, after feasibility studies which take into account their discounted cash flow profitability compared with third-party sales of crude oil.
>
> The government is entitled to purchase a certain proportion of crude at discounted prices. AGIP must market Petromin's share of crude if required, paying Petromin on the basis of realized prices. The ENI group has numerous refining and marketing outlets in Europe and Africa and is short of crude. It has a joint-venture in Iran which has yielded only small production, and one in Egypt, where the fields are occupied by the Israelis. But in any case, AGIP is obliged to sell to independent third parties at the highest prices obtainable in free competitive markets between 10,000 and 15,000 b/d during the first three years of commercial production and thereafter 10 per cent of annual crude output or 30,000 b/d, whichever is the smaller, with a minimum of 20,000 b/d. Detailed procedures for appraising third party prices are laid down and failure by AGIP to obtain an adequate "highest market price" may entail a reduction in output in the following year, but not below the economic level which would give AGIP a 10 per cent return on assets.
>
> Disputes between the government and Petromin shall be submitted to the Saudi Board of Concession Appeals. Disputes between Petromin and AGIP shall be referred to arbitration within Saudi Arabia.
>
> The petrochemical agreement provides for a company to be set up, in which Petromin and the Saudi public shall have 50 per cent participation and ANIC 50 per cent, to construct petrochemical plants. Feasibility studies shall be completed within six months and if no feasible petrochemical project is forthcoming AGIP will, on certain conditions, forfeit $5 million or $10 million.

[13] *Petroleum Press Service*, January 1968, pp. 4–5.

It is worth examining this agreement a little further because it exhibits a number of features pointing the direction in which country policy is attempting to move.

(1) It does not freeze the tax rate at 50 percent based on posted price, but specifies any rate which may later be introduced.
(2) Posted prices must be approved by the government.
(3) It specifies a rising rate of country participation according to volume of production.
(4) Petromin may take its share in oil, but if AGIP does the marketing, it may pay Petromin on the basis of realized prices in sales to third parties.
(5) A portion of output must be sold to third parties to establish a "highest market price," presumably as a check on the transfer price to AGIP's own refining affiliates.
(6) Plants for integrated downstream operations are to be made and, under certain conditions, 30 percent of profits are to be reserved for such investment.
(7) At a certain level of production a portion of output is reserved for processing in a Saudi Arabian refinery.
(8) The agreement is linked to a petrochemical project in which an AGIP affiliate will provide half the capital.

Perhaps the most notable features are (1) the attempt to direct a part of the profits into channels of investment of direct importance to the Saudi Arabian economy, and (2) to lay a base for an extended range of activities for Petromin.

3. *Kuwait*

(1) *The Shell Concession*. In 1961, Kuwait granted to an affiliate of the Royal Dutch/Shell group (organized as the Kuwait Shell Petroleum Development Company) a concession of more or less conventional type for forty-five years, covering 1,500 square miles offshore on the continental shelf. It had, however, one feature which brought it within the new category partnership, or participating, agreements. After oil had been discovered, the Ruler of Kuwait was entitled to acquire a 20 percent interest in the company and either to take 20 percent of the oil produced, or to have it marketed through Shell's commercial channels. This was the first occasion upon which one of the great international oil companies entered into an agreement which gave the country implicit partnership status. Because part of the territory was under dispute with Iran, activity under this agreement is suspended.

(2) *The Hispanoil-KNPC Agreement*. In 1967 the Kuwait National Petroleum Company (KNPC) entered into an agreement with Hispanoil (a Spanish company formed by three state-owned or state-controlled refining companies) covering 9,262 square kilometers, onshore and offshore. The duration is thirty-five years plus a five-year extension. An unusual feature is that the national company holds a majority interest of 51 percent. Hispanoil undertakes all exploration expense, of which 51 percent is reimbursable from one-half of KNPC's profit share if oil is discovered. Royalty is 12.5 percent in cash or kind, based on posted prices, and is expensed. Income tax is 50 percent, based on posted prices. Posted prices in line with comparable postings in the Persian Gulf must be approved by the government. Hispanoil undertakes to market KNPC's share of the output, if requested, at agreed realized prices.

Another unusual feature of the agreement is that the Spanish government undertakes to reserve 25 percent of the Spanish market for fifteen years for Hispanoil's share of the output.

4. *Abu Dhabi*

After the four major producing countries, by far the most prolific area to be developed in the Middle East in the postwar period is the small sheikhdom of Abu Dhabi on the Trucial Coast. The original concession of conventional type was granted in 1939 to Petroleum Concessions, Ltd. (now the Abu Dhabi Petroleum Company), an associate of the Iraq Petroleum Company; but commercial production only began in 1962. In 1953 a similar concession was granted to the Abu Dhabi Marine Areas (ADMA), in which British Petroleum (formerly Anglo-Iranian) had a two-thirds interest and the French CFP a one-third interest, covering offshore territory on the continental shelf. The policy of participating agreements has recently been introduced.

Abu Dhabi Oil Development Company. In 1967 a portion of the offshore Abu Dhabi Marine Areas concession was relinquished and awarded to the above company, a consortium of three Japanese companies. The agreement was of the new participating type, providing that after oil is discovered the Ruler can acquire a 50 percent share interest by capital investment payable out of production proceeds. On other points, the agreement follows the general lines of recent agreements influenced by OPEC principles: a stated rate of relinquishment of territory, reaching 75 percent in ten years; stated outlays on exploration within a defined period; bonus payments; 12.5 percent royalty in cash or kind based on posted price; 50 percent income tax calculated on posted prices less discounts, with royalties expensed. The company is to market the Ruler's share of output on request at realized prices.

A considerable portion of the original land concession of Abu Dhabi Petroleum Company (ADPC) has also been relinquished and is being offered for new agreements, which will presumably include the participating feature. One such agreement has been concluded with a group of three interests: one, the Italian state-owned AGIP, the other, two American interests—Phillips Petroleum and Aminoil (made up of companies with a concession in the Neutral Zone).

5. *Libya*

The most recent agreements in Libya have all been of the participation type. The first to be concluded, in April 1968, was with the French state company, ERAP, on terms substantially more favorable to Libya than in earlier agreements. At that time it was understood that in such partnerships state-owned companies would be favored as the outside partner. In May 1968, the Libyan premier, as reported in *Petroleum Press Service*:[14]

> . . . implied that partnerships with "other states" similar to the Libyan-French agreement . . . would be particularly welcome. This decision will not affect existing contracts. . . . It appears to mean, however, that the old system of competitive bidding for acreage between individual companies or ad hoc consortia will lapse in favour of direct negotiations between LIPETCO and other state-supported groups—assuming that suitable applicants are forthcoming.

In early agreements, such preference did not work in practice. Of three agreements concluded in May 1969, only one was with a state-owned company, the Italian ENI-AGIP group. And four others approaching final form were all with private companies.

(1) *The AGIP Agreement.* This agreement, covering 20,000 square kilometers, is reported to assign to LIPETCO a participating share on a sliding scale according to the level of production.

(2) *The Ashland Agreement.* This agreement with Ashland Oil and Refining Company, covering 7,400 square kilometers, also includes the sliding-scale participation feature. LIPETCO's share will be 30 percent at 200,000 barrels/day production, rising by stages to 50 percent at 400,000 barrels/day.

(3) *The Chappaqua Agreement.* Chappaqua Oil Company is an ad hoc company put together for the purpose of acquiring Libyan acreage. The agreement covers 5,300 square miles in various locations. It contains the unusual feature that LIPETCO has the option to acquire a participation not only in producing operations, but also in any downstream operations undertaken.

Other agreements reported to be near final form as of early summer 1969 were with ESSO (New Jersey Standard), Shell, Phillips, and Occidental.

Contract Agreements

While most of the new agreements since 1957 have been of the participating type described in the preceding section, by the summer of 1969 a different type had been introduced in four cases, three in Iran and one in Iraq. In this type the foreign company comes in, not as a concessionaire, but as a contractor to the national oil company. Two of the contracts were with the same French state-owned company, Entreprise des Recherches et d'Activities Petrolieres (ERAP). As was true also of the participating, or joint-venture, type of agreement, Iran was the pioneer in introducing the contract type.

1. *The NIOC-ERAP Contract Agreement*

Under the 1966 agreement between the National Iranian Oil Company and ERAP, the foreign company explores for oil at its own risk. If oil is found, the company will lend NIOC the funds for development against repaying with interest. The oil discovered is the property of NIOC. The company has a first charge on the oil to amortize its exploration costs. After this, NIOC assigns one-half the reserves discovered to a "national reserve," either by setting aside separate fields or by reserving a part of the oil in particular reservoirs. Of oil produced from the other half, the company may purchase 35 to 45 percent (depending on location) at a price halfway between cost and realized price (or in effect 50 percent of profit calculated on realized price). The duration is twenty-five years from commercial production in each production field.

This remarkable agreement is described in more detail by *Petroleum Press Service*, as follows:[15]

> The Iranians pioneered in the Middle East the principle of partnerships between national and foreign companies. Now they have gone a step further and introduced into the region the arrangement whereby a foreign company acts as contractor to the national entity. This role has been accepted by ERAP (Entreprise des Recherches et d'Activities Petrolieres), a company wholly-owned by the French state. The agreement, it is said, gives about 90 per cent of profits to the Iranian Government, and 10 per cent to ERAP. But this is not in essence a profit-sharing type of agreement, it is a new departure: a contractor looks for oil and is paid with a share of what is discovered.
>
> ERAP is not paying any bonus—usually required for Persian Gulf acreage. It will search for oil at its own expense, losing the investment if no oil is found. On discovery, it will lend National Iranian Oil Company the money for development, and this loan together with the exploration outlay will be reimbursed over a period, possibly partly in oil. The French company's only reward consists in the right to acquire on advantageous terms 17½–22½ per cent of oil produced.
>
> The contract agreement covers offshore and onshore. Offshore, ERAP will work in 20,000 sq.kms. in Petroleum District 2, in the Lingeh-Bandar Abbas area. After a year's geological and seismic work, 50 per cent of the area will be surrendered. Three test wells will be drilled during the second and third year, after which the area will be reduced to two-thirds. During the fourth to the sixth year ERAP will spend at least 15 million French francs annually.
>
> Onshore it will survey a much larger territory, reported to be of the order of 200,000 sq.kms., in Districts 8, 11, 12, 14 and 18—inland districts east of the Consortium area. At the end of a year NIOC will study the data and itself choose 20,000 sq.kms. for further work by ERAP. At the end of the fourth year this area will be reduced to half and at the end of the sixth year to one-quarter, i.e. to 5,000 sq.kms. ERAP undertakes to carry out at least 2,000 kms. of seismic work, to drill three test wells, and to spend at least 10 million francs a year in the fourth, fifth and sixth years.
>
> If no oil is discovered, ERAP bears all this expenditure, which it is estimated will come to between 150 and 200 million francs say, £11–£14½ million—without

[14] *Petroleum Press Service*, July 1968, p. 247.

[15] *Petroleum Press Service*, September 1966, pp. 324–326.

return. If oil is discovered, the exploration outlay is regarded as an interest-free loan to NIOC to be amortized within a maximum period of 15 years at the rate of 10 cents a barrel.

Finance for development advanced by ERAP to NIOC will bear interest at the Bank of France bank rate (at present 3½ per cent) plus 2½ per cent and be repayable in cash or through the sale of crude oil within five years from the beginning of commercial production. The sale would be effected, if NIOC so desires, by ERAP exporting to world markets one million tons of oil for a year for five years for NIOC's account and crediting the proceeds, less two per cent commission and marketing charges, against the loan.

Once sufficient reserves have been discovered to amortize the exploration investment, one-half of subsequently discovered reserves shall be appropriated to NIOC as national reserves. This is to be done either by appropriating separate fields to NIOC or by reserving a share of crude oil within a reservoir.

ERAP will be entitled to purchase 35–45 per cent of the other half of the oil produced for a period of 25 years at cost price plus two per cent plus the taxes "which shall be calculated on basis of the real price." (This is obscure but presumably means that realized not posted prices shall be used as basis.) The percentage of crude oil to which the contractor is entitled depends on the remoteness of the field: if 500 kms. or more from the coast his share is 45 per cent; if 100 kms. or less then 35 per cent. There is here some recognition of the difficulty of exporting oil from the ultramontane inland deserts.

NIOC may call on ERAP to sell in the world market, out of NIOC's share of crude, three million tons a year during the first five years of commercial production and four million tons a year during the second five years, paying NIOC the actual price less two per cent commission. NIOC or the Government of Iran undertake to import French-made goods or to utilize the services of French concerns to the value of the above exported oil.

In announcing the agreement, NIOC laid great stress on the "contractor" status of ERAP–although a contractor is not normally expected to put up risk capital. Various provisions throughout seek to underline that ERAP is working directly for NIOC. No bonus is payable by the French company. For the onshore areas the choice of retained acreage is NIOC's. The stringent relinquishment provisions ensure that the "employer" has the fullest disposal of land not being worked. Operations programmes and budgets, development plans, production and export programmes, are to be drawn up in consultation between ERAP and NIOC, who must approve at all stages. All crude is the property of NIOC at the well-head and ERAP has merely the right to purchase a certain proportion of it.

. . . . . .

What do the French get out of it? As a move to obtain for the French state company additional sources of crude outside North Africa the agreement follows logically on that concluded last year with Saudi Arabia and exploration in the Red Sea area. But that was on the basis that the Saudi national company might take a 40 per cent participation at the producing stage. The French right in the present case to acquire less than a quarter of oil found does not look very attractive. The French state is already a partner with NIOC in Farsi Petroleum Company, in District I offshore (where no oil has so far been found). However, in the case of ERAP as well as NIOC wider considerations may have been taken into account. Iran's undertaking to take French goods and services to offset ERAP's marketing of NIOC oil would not have much relevance for a commercial oil company but softens the terms for a wholly state-owned entity. Again, although the French have widespread oil interests in the Middle East, through IPC, the Consortium, and in Abu Dhabi, they may wish to undertake more direct operations in the region, offering scope for the use of French equipment and techniques. For both sides, the "contract" is evidently a deal.

It is difficult to assess the relative advantages of such a contract for country or company, in comparison with a joint venture. They will depend heavily on the capital cost and amount of oil discovered in particular cases. If results are propitious in both respects, the company will receive a large amount of low-cost oil. Otherwise, it could be loaded with heavy capital costs. If oil is discovered in paying quantities, there is an advantage to the company in having no permanent capital investment and quick recovery of advances for development. ERAP secures 35–45 percent of current production (not including oil eventually produced from the "national reserve"), plus oil or cash to amortize its loan advances. ERAP pays no royalty, but in effect pays a tax of 50 percent of realized price less cost plus 2 percent.

2. *The NIOC-AREPI Contract*

In March 1969, NIOC entered into a similar contract with a group of European companies organized as Association de Recherche et d'Exploitation du Petrole d'Iran (AREPI). The participating members of the group are three national companies: ERAP (France), ENI (Italy), and Hispanoil (Spain), and two other companies, Petrofina (Belgium) and OMV (Austria). The contract covers an area of 27,260 square kilometers on the shore of the Persian Gulf and consists of part of the lands relinquished by the Consortium in 1967. Forty-five percent of the area is to be relinquished after five years, 50 percent of the remainder after eight years, and all but commercially productive fields after ten years. The exploitation period is twenty-five years from the beginning of commercial production. No bonuses are called for. Other terms of the contract are reported to be similar to the ERAP contract except as to details.

3. *The NIOC-Continental Contract*

In April 1969, a new contract agreement was reported between NIOC and the Continental Oil Company, the first American company to enter such an arrangement. The area covered is 4,965 square miles (12,860 square kilometers in southeast Iran, on lands earlier relinquished by the Consortium). After $1 million initially, periodic bonuses add up to a possible $10 million. Terms in general are similar to the two earlier contracts. The company is entitled to purchase 45 percent of production up to an output level of 275,000 barrels/day, and 30 percent above that level. The price to be paid is unit cost of production (plus 2 percent), plus one-half the difference between unit cost and realized price. Relinquishment of territory is to be 50 percent after five years and 50 percent of the remainder after seven years. The initial exploration period is five years and expenditure of $8 million is called for; two optional extensions of two years each are provided, with $2 million expenditure in each. The exploitation period after commercial oil is discovered is twenty-five years.

4. *The INOC-ERAP Contract*

The 1967 ERAP contract with the Iraq National Oil Company is in general pattern similar to the Iran contracts. The area is 10,800 square kilometers—some 8,000 in central and southern Iraq, the remainder offshore. The duration is twenty years from the start of exports. After three years, 50 percent is to be relinquished, 75 percent after five years, and all but producing areas after six years. Five years after the start of exports, INOC takes over joint management responsibility with ERAP. Exploration and development expenses are to be repaid to ERAP after commercial production begins, but at intervals for a period of ten years after discovery ERAP must make bonus payments totaling $15 million. Half the reserves discovered are assigned to a national reserve. From the other half ERAP is entitled to buy from INOC 30 percent of production, and undertakes to market INOC crude for a small marketing fee. For 12 percent of production, the price to be paid is cost plus royalty of 13.5 percent of posted price. For the other 18 percent the price is, if we understand the formula correctly, 50 percent of realized price less cost and royalty.

The four contracts described above are found in the two Middle East countries that have moved furthest in modifying the old concessionary rights. They represent the only striking departures to date from what has come to be the typical feature of most recent agreements, that of some form of partnership between the foreign companies and the national oil companies.

The Principle of Participation Universalized

As we have seen in an earlier section, almost all the postwar agreements with "newcomers" have provided for a degree of ownership participation by national oil companies. The idea has been brought up from time to time that the same principle should apply to the companies operating under the old major concessions; but it was only in 1968 that the idea began to force its way to the center of attention in the councils of OPEC. Early in June Mr. Yamani, Saudi Minister of Petroleum and Mineral Resources, was quoted as saying: "The principal change which the major companies are required to make at the present time. . . is to revise their whole system of operation in the producing countries and enter into partnership with their host governments."[16]

[16] *Middle East Economic Survey, Supplement*, June 7, 1968, p. 3.

Later in June, in its "Declaratory Statement of Petroleum Policy," the OPEC Conference included the following statement:

> Where provision for Governmental participation in the ownership of the concession-holding company under any of the present petroleum contracts has not been made, the Government may acquire a reasonable participation, on the grounds of the principle of changing circumstances.

Since such a provision is not included in any of the major concession agreements this is a very sweeping declaration which forebodes a long stage of controversy and negotiation between countries and companies.

Though not on the official agenda of the November 1968 OPEC Conference, "participation" was reported by the *Middle East Economic Survey* as the dominating subject of discussion.[17] In his opening speech, Mr 'Atiqi, Kuwait Minister of Finance and Oil, singled out the principle of host government participation as the issue most worthy of the attention of the OPEC delegates. He said: ". . . one important component of these principles [in the June 'Declaration'] —namely national participation in the operations of the companies operating in our countries—has already met with a favorable response in the requisite practical fashion, to the extent that more than one of the members of our Organization have started giving serious thought to the task of its implementation."[18]

The *Middle East Economic Survey* further stated:[19]

> But what do the Middle East countries really mean by participation in the major producing ventures with their huge reserves of low-cost oil: This is a question in respect of which international oil circles—oil companies, consumers, economic consultants and academics—have been straining at the leash to obtain a definite answer: What is the main motive? Is it government control over operational and management decisions? Is it a higher average income per barrel, in other words an additional "tax" imposed by means of guaranteed buy-back arrangements for the offtake of the government's share of production by the other partners? Is it the desire to gain access to large quantities of really cheap "cost" crude to facilitate the downstream expansion of the national oil companies? Or a mixture of all three factors—and, if so, in what proportions? And how does the expressed desire for participation in the downstream operations of the international majors fit into the picture?
>
> None of these questions are likely to receive satisfactory answers yet awhile—even assuming for argument's sake that the provision of satisfactory answers to such questions, which at the present stage of the game might well be categorized as lying on the borderline between economics and psychology, might be theoretically possible. The projected OPEC studies will no doubt help to clarify matters in due course. The most that can definitely be said at the moment—and this is of

[17] *Middle East Economic Survey*, November 15, 1968, p. 2.
[18] *Ibid.*, p. iii.
[19] *Ibid.*, pp. 3–4.

some considerable importance—is that quite a few leading oil officials from producing countries in widely separated parts of the world have a remarkably similar vision of the broad shape of things to come in the oil industry as they would like to see it in, say, 10 to 20 years from now. Such a vision may, in very general terms, be divided into two complementary facets: on the one side a reasonable participation through the national oil companies of the producing countries in the actual *mechanics* of the international oil business, commensurate with the amount of funds that can be spared for investment downstream and the level of technical expertise and know-how which may have been reached by the nationals of the various producing states at any given time; on the other side an appropriate voice in determining the overall *terms* governing the fabric of the world oil trade. By the latter is meant not some fantasy of dictatorial powers, but a practical design to create an orderly but adequately flexible framework for the administration of the oil trade through agreements or treaties between the major groups of oil exporting and consuming countries, guaranteeing security of oil supplies for the latter and markets for the former while protecting both sides against excessive fluctuations in prices, up or down. However, what remains blurred and confused in this sort of vision is, as it were, the middle distance—the forthcoming period of transition when the avenues of change will have to be worked out in hard, down-to-earth detail, almost certainly in a piecemeal and often ad hoc fashion, with all the parties involved.

It should also be noted that participation is regarded by its advocates as a political as well as an economic necessity. In the present turbulent political conditions in the Arab world, it could well be that equity participation by host governments would provide the only sure means of guaranteeing the continued smooth operation of the major oil producing ventures in times of crises. And it has in fact been demonstrated in practice that a solidly-based joint venture can survive a political storm much more easily than a 100-percent foreign concessionaire—particularly when that concessionaire belongs to a country which inspires political hostility. Moreover, this new drive for participation can hardly have come as a surprise to the majors. For years now a whole phalanx of joint ventures and other new-style deals has been building up around the old-established ventures. Inevitably, the day has now come when the citadel of the traditional concessions is called upon to surrender its exclusivity.

Subsequently, the Eighteenth Conference of OPEC in July 1969, was reported to have dispelled "any lingering doubts that there may have been about the determination of the oil producing countries to press ahead with their demand for participation in the ownership of their concession-holding companies. . . ." It was further reported:[20]

> . . . The real accomplishment of the OPEC delegates . . . was to reach agreement on a . . . framework of reference within which a detailed blueprint for a future concerted strategy can be drawn up. On this basis, the Conference decided to set up a high-powered technical committee . . . to study all aspects of the matter in depth in line with the terms of reference set by the Conference. It is understood that these terms of reference stipulate that the committee's studies should encompass participation both upstream in the producing end and downstream in the market.
>
> . . . The majority view—strongly spearheaded by Saudi Arabia—was that participation downstream was an essential prerequisite for the maintenance of oil prices . . . in the absence of which upstream participation would be likely to lead to an uncontrolled spate of f.o.b. crude sales, thereby endangering the whole price and market structure. . . .
>
> . . . And since downstream participation—unlike participation in the producing end alone—is not amenable to unilateral imposition by the producing governments, even if they wanted to do so, the obvious conclusion is that OPEC has now embarked on the first stage in the preparation of a concerted strategy for what promises to be a highly complex and delicate series of negotiations with the major international oil companies.

This issue then has the appearance of a major long-run effort at fundamental revision of the relations between countries and major concessionaires. In his original interview, quoted above, Mr. Yamani had said:[21]

> . . . the combination of changed circumstances in oil agreements and current political developments in the area is sufficient to impose the partnership principle on the major companies if they want to stay in business in this part of the world. . . .
>
> Control over all oil operations is our objective. . . . I think the major part of our long-term plan [for Saudi Arabia] will be achieved within 10 years from now. However, it may take up to 25 years to attain full control.

The purely formal aspect of the participation in view seems to be part ownership of the share capital of major concessionaires. But the substance, as it may materialize, cannot be foreseen. It might be very far-reaching, as Mr. Yamani suggests, or much more limited in affecting company operations; and it might differ greatly from one country to another. As noted, steps have been taken in OPEC to conduct a study of the problems presented, and the methods by which the desired restructuring of country-company relations can best be pursued. Presumably the effort will be made to keep the policies of different countries as closely in line as possible, as was done earlier in the matter of taxation.

The first step in achieving participation could be expected to be the acquisition by the national oil company of an ownership share in the concessionaire company, perhaps small at first and increasing. This would entail the fixing of the capital value of the shares to be acquired and the terms of payment for them. If the sole purpose of the government were to share in the profits of the company, over and above

[20] *Middle East Economic Survey, Supplement,* July 11, 1969, pp. 1–2.

[21] *Middle East Economic Survey, Supplement,* June 7, 1968, pp. 4–5.

taxes, this would be the gist of the arrangement, the companies carrying on as usual. Even so, there would be technical problems of cutting the national company into the actual profit streams, since some of the nominal concession companies are mostly nonprofit service companies, profits showing up elsewhere in the company structure of the participant parent companies in the various concessions.

But such profit-sharing is clearly only a secondary feature of what proponents have in mind. Discussing the matter, Dr. Nadim al-Pachachi, formerly Minister of National Economy in Iraq and currently oil adviser to the Ruler of Abu Dhabi, has said:[22]

> The participation we propose is not intended to be a mere device to secure additional revenues to the state concerned. It goes much further than that. The first result would be to give the governments, through their national oil companies, a voice in the policy decisions of the company and to exercise more control over their oil industry. Governments should no longer be satisfied with the role of a "sleeping partner" or "absentee landlord." The second result is the long-term implication. Producing countries should develop technical and managerial skills in the oil industry in order to enable the governments to be in a strong position when the concession agreements begin to expire. The Consortium agreement in Iran expires in 1979. It can be extended by a five-year period but I doubt very much whether Iran would agree to extend the agreement on the same terms. Any new arrangement reached with Iran will have to be extended to the other producing countries. By mid-1970 new arrangements and terms will evolve to replace the present terms of the major concessions.
>
> . . . . . .
>
> . . . I would like to see a straight participation in the present concessions. The long term objective would be 50 percent participation. I realize that this objective cannot be achieved immediately. But for the immediate future, governments should insist on the establishment of the principle of participation. Some arrangement has to be made, as in the joint ventures concluded in the Middle East, for the government's share to be marketed by the oil companies. I am not suggesting that the national oil companies should remain without outlets. Some companies are trying to acquire some outlets, but we must realize that we have to travel a long road before reaching our objectives. Major companies were not built overnight. Penetration in international markets should begin as soon as possible. Some active measures are being pursued by countries like Kuwait and Iran.

The problem presented by such a statement is the content to back up such phrases as "a voice in the policy decisions" and "to exercise more control." Dr. Pachachi's concrete suggestions go no further than educating national oil companies for larger responsibilities. Meantime, and for an undefined period, the companies are to be relied upon as the responsible agency in international markets, except as national companies may be able to secure some foothold there.

Discussing the principle of participation, Mr. Francisco Parra, former Secretary General of OPEC, has written:[23]

> Governmental participation in companies to which new areas are entrusted has, since the NIOC/ENI initiative in 1957, become such a standard matter in all but the smallest and newest producing countries that it is, inevitably, a "natural" in any consideration of terms of existing concessions. But it is still far from clear, if this principle is to be implemented shortly, what direction implementation would take. A fundamental choice would first have to be made between two alternatives, both of them consistent with the principle itself: equity shareholding, or joint-venture, whether extended to downstream operations or not. Both imply "participation in the ownership of the concession-holding company;" but, clearly, more is intended than passive ownership of shares, and an equity shareholding would no doubt have to include effective participation in certain key management decisions, particularly those concerning investment and output levels. It may be noted, en passant, that this type of equity ownership in the established concessions would be slightly more complicated than appears at first sight, since the operating companies (who are in some cases the concession-holding company, and in some not) generally realize only an operating fee as a profit. The real profit is accumulated in a two-tier system of trading companies, the whole structure essentially responding to the desire of the American companies to avoid US taxes on dividends remitted by affiliates owned less than 50 percent. Hence, equity participation means participation in the ownership of the trading companies as well as of the concession-holding companies.
>
> Once such an interest had been acquired in the real profits of the companies, agreement would need to be reached on the question of dividend policy towards "outside" shareholders—and this in itself could be presented as a strong argument in favor of an effective voice in management by the "outside" (viz., government) shareholder. A further implication of such equity shareholding is that it would once again give governments a more direct interest than they at present have in market prices—such an interest having been partially obscured by the presently existing posted price or tax-reference price system.
>
> The second alternative, that of participation through some form of joint-venture operation, possibly extending to downstream operations, no doubt implies a far closer degree of control and has wider implications for the market, inasmuch as the government would have a direct hand in operations in the market. Regardless of the degree of participation in such a joint venture, this inevitably would involve a far more basic adjustment in the present shape of the industry than would the first alternative. It is also interesting to note that the first alternative involves primarily the regulatory function of government, and would presumably devolve on the

[22] "Disparity of Concession Terms in Middle East Oil Producing Countries," *Middle East Economic Survey, Supplement*, November 1, 1968, p. 9.

[23] "OPEC: Present and Future Role," *Middle East Economic Survey, Supplement*, May 16, 1969, p. 4.

responsible ministry; the second alternative more closely concerns industry operations, including especially international marketing, and would consequently be handled more directly by the national oil companies involved. This fundamental choice between the two alternatives will have to be the first step; only afterwards can any questions be answered about how much participation, or how it is to be acquired, or when.

The countries no doubt desire an intimate involvement of their national companies in the management decisions and operational control of the concessionaire companies, short of nationalization. But just how much can be achieved in this direction remains to be demonstrated. No one has suggested any effective way in which the central role of the companies in international marketing could be greatly changed by government participation in concession ownership. The crux of the problem of participation appears to be how greater country influence on domestic producing operations and investment can be accommodated to the worldwide supply logistics and financial operations of the international parent companies which are the present participants in the concessions. It is difficult at present even to speculate upon the outcome, so many are the possible and unpredictable forms of adaptation and compromise.

The more sober minds in oil-producing countries tend to regard nationalization within the calculable future as a false goal, entailing disastrous consequences. The goals they envisage are, then, not the elimination of foreign company activities, but a state of affairs in which, as fully as possible, the role of government is progressively enlarged, in the pursuit of the economic and political interests of the countries. There is no consensus as between countries or even within countries, as to how the goals should be defined or by what steps and stages they could be most nearly attained. A fluid environment of thought and attempted action is thus provided. It could hardly be otherwise, since whatever is to be accomplished will grow out of bitterly contested negotiations, or, in the last resort, unilateral action by governments.

Even with a moderate meaning of participation, one has to differentiate the case of Libya. In the Persian Gulf countries, the problems are confined to dealings with a single major concessionaire. In Libya, revised agreements would have to be forced upon a considerable number of companies, involving difficult negotiations and certainly loading the national oil company with wide and onerous responsibilities—a prospect apparently not to the taste of the government in power prior to the revolution of late 1969.

A key issue for a country contemplating acquisition of a participating share in a major concession will be the capital required. Assuming agreement upon the value of the share to be acquired (itself no easy matter), a weighty question for the government must be the relative advantages of using its funds for this purpose as against other uses. It cannot acquire participation (except by partial expropriation) without diminishing the funds available for promoting national economic development. It has to consider whether this sacrifice will be offset by some overriding advantage from part ownership in the oil industry. As we interpret statements that have been made by responsible officials, the advantage in view is not one that can be measured in purely economic terms, but arises from the desire of governments to be less dependent on the oil companies and in the fullness of time to be masters in their own house. Even so, the temporary economic sacrifice might be considerable, and for this reason it could be conjectured that, if and when the process of gaining participation is put in train, the process would be streched out over a considerable period of time, if economic considerations prevail.

The Organization of Arab Petroleum Exporting Countries (OAPEC)

A new step toward collective action among oil-producing countries was taken early in 1968 with the organizing of OAPEC by three founding members, Saudi Arabia, Kuwait, and Libya. Little has so far been announced concerning its intended scope of action; and indeed it appears not to have any agreed program, being in process of studying what might usefully be done. Its statements, however, provide ground for interesting speculations.

In the first place, its membership is intended to be limited to Arab countries whose economies are predominantly dependent on oil. Iraq was invited but did not join.* Its exclusively Arab membership implies a political orientation which has never been made explicit in any published statement. It may in a degree have been originally intended as a device to protect the member countries against other Arab countries with different and more radical ideas concerning the use of oil as a means of political pressure and retaliation. The Organization has dissociated itself from the idea of its members joining a proposed Arab Petroleum Organization under the aegis of the Arab League with strong political overtones. An apparently abortive Organization of Revolutionary Arab Petroleum Exporting Countries is reported to have been proposed by Iraq and Algeria after the organization of OAPEC.

In an interview in September 1968,[24] Mr. Yamani of Saudi Arabia, later appointed Acting Secretary General, gave the following three purposes of the organization:

(1) To keep oil activity within the organization with a view to protecting the member states from precipitate decisions and making oil a genuine weapon to serve the interests of the producing countries and the Arab countries in general;
(2) to create opportunities for joint investment in the oil resources of the member states;
(3) to build an economic bridge between the organization and the consuming countries so as to create an outstanding economic structure which we can utilize to expand our markets and thereby

---

* Algeria, Abu Dhabi, Bahrain, Dubai, and Qatar were admitted to membership in May 1970.

[24] As reported by *Middle East Economic Survey*, September 13, 1968, pp. 6–7.

> establish a stronger political center of gravity for the states of the region.

Unlike OPEC, which has only planning and advisory functions, the Organization is set up as a legal financial entity capable of establishing "big companies capable of undertaking all types of financial and technical operations." At a meeting of the Secretariat with the heads of national companies in December 1968, the first specific proposals for joint action were reported to have been:[25] (1) a project to build a pipeline from the Persian Gulf to the Red Sea, (2) establishment of a unified tanker fleet, and (3) coordination of production and marketing of chemical fertilizers.

In opening the Conference, Mr. Yamani said, in part:[26]

> In time, the oil producing countries began to consider means of effective participation in the oil industry which would augment their profits, and finally decided to establish their own national oil companies. Having done this, these countries realized that cooperation among themselves was the best guarantee of success, and for this and various other reasons Saudi Arabia, Kuwait and Libya established OAPEC. This organization occupies an important position in the world of oil, accounting for 300 billion barrels of crude oil reserves and for a daily combined output of 8 million barrels—an output which could be doubled in a fairly short time. Thus when the Organization speaks, it does so confidently for it has the resources; and what is even more important in my view, the requisite wisdom and intelligence. We see our way clearly and are aware of the pitfalls and obstacles that lie ahead. Likewise we are well informed about the intricacies of international oil policy and strategy.

In an earlier interview,[27] Mr. Yamani made it clear that OAPEC was not starting with a "program," but was initiating a broad inquiry into all the means by which cooperative action among the members might advance their mutual interests, contribute to the well-being of the oil-producing countries generally, and aid in establishing satisfactory relations between producing and consuming countries and between both and the international oil companies. He said, "We don't have too much to talk about, but we have a lot to work for. We are people who are planning on a long-term basis, and we intend to work hard to realize our imaginative vision. We believe in cooperation with all the parties involved in the international oil industry."

Behind these hopeful general statements lies a series of troublesome problems which endanger the economic position of the oil-producing countries. The central threat is that competitive rivalry among the oil-producing countries may undermine the structure of oil prices. All other problems cluster around this central one. The new national oil companies may, by their separate efforts to break into the market, weaken prices. The major companies, squeezed between high taxes and declining prices, might be forced into abdicating their position as concessionaires. The consuming countries, heretofore in some degree amenable to price-supporting policies, might use their power to reduce their energy costs by inducing competitive rivalry among producing countries. In the presence of an easily expandable supply of low-cost oil, the economic prospects of the producing countries, as a group, all depend upon their capacity to maintain, or enhance, the margin represented by the taxes paid by the major oil companies. To do so through the growth of national companies would require the ability to limit price competition.

One may reasonably conjecture that the member countries of OAPEC have banded together in an attempt to cope with these fundamental problems. While the possibility of organizing a tight producers' cartel of all OPEC countries appears remote, the three largest Arab producing countries may feel that by acting together they may salvage the situation by action in several directions: (1) by a collective production policy; (2) by establishing a large stake in the oil transport network; (3) by curbing the competitive behavior of their national oil companies through joint marketing arrangements; (4) by working out viable relationships with concessionaire companies which will sustain a mutual interest between companies and countries; (5) by working out supply agreements with consumer countries; (6) by bringing other Arab countries into the fold; and (7) by using whatever stick-and-carrot methods are available to prevent other producing countries from upsetting the market applecart. These objectives, we must emphasize, are conjectural and have not been spelled out by OAPEC.

As we shall see (Chapter 11), some Western observers are skeptical of the ability of the producing countries to contain the competitive forces inherent in the potential supply situation, especially while pursuing national aspirations which tend to undermine the existing institutional relations between countries and companies. This is not an opinion that can be accepted by the producing countries; and it is certain that their most intelligent and experienced experts will be working to prove it mistaken. The organization of OAPEC could be taken as evidence of this effort. Collective action by producing countries appears to be a *sine qua non* of success in any such effort. The association of three producing countries may represent a beginning of such an effort at collective market control in order to dampen the competitive forces.

Mr. Yamani says:[28] "We believe in cooperation with all the parties involved in the international oil industry. We don't believe in fighting, and we will do our utmost to make the oil business a healthy industry with a minimum of conflicts." Presumably its first great problem will be to achieve a meeting of minds of its members concerning what their mutual interests are. It will be important to see what lines of action OAPEC will initiate, in relation to the policies of the more broadly based OPEC.

Appendix: The Arithmetic of Joint Ventures

The terms of agreements under which a national company has a participating ownership share in the company

[25] *Middle East Economic Survey*, December 27, 1968, p. 5.
[26] *Ibid.*, pp. 4–5.
[27] *Middle East Economic Survey*, November 22, 1968, pp. 4–7.
[28] *Ibid.*, p. 7.

which holds producing rights present certain complexities in the calculation of the income which the country receives. Since the financial results for the country are frequently misunderstood, it is necessary to examine the character of the calculations and the concepts underlying them. We shall therefore present an illustration of the profit-sharing features under a particular type of joint venture. The model can be adjusted to the special terms of particular agreements.

When Iran signed its joint-venture agreements with IPAC and ENI in 1957 and 1958, respectively, great attention was given in the trade journals and especially in the local press to the purported fact that these agreements, by giving the national company a 50 percent ownership share, gave to the host country 75 percent of the profit from the enterprise, compared with the figure of 50 percent which had by then become the standard "profit split" of the putative income from a concession. This was commonly called the 75-25 formula. Thus, it was implied—indeed, it was at times even claimed—that the host country derived half again as much revenue from such an operation as it would have received if it had granted a standard concession.

This view contains a fallacy that arises from not recognizing that two different kinds of *income* are involved, and that they must be identified explicitly. In the case of the standard concession, the "income" upon which the so-called 50-50 split is calculated is a purely "notional" profit which is based upon the posted price for the crude oil in question, a nominal price which has been running substantially above that realized in actual market transactions.[29] On the other hand, under a joint venture, two different kinds of "profit," separately calculated, are combined to make up the government share. The part contributed by the private partner is calculated upon the posted price, or upon some discount from that price. The part contributed by the government's ownership share is calculated upon the basis of the price actually realized or upon some agreed measure of realized price, and no royalty is added.

In Table A10.1 (shown graphically in Chart A), we compare hypothetical calculations of government revenue from a standard concession and from a joint venture. In panel A of the table it is assumed that posted price is $1.80 per barrel, cost (including return on capital) $0.20, royalty $0.225, and income tax rate 50 percent. In panel B, the government has a 50 percent ownership share. The revenue from the foreign concessionaire's half-interest is calculated on the same basis as in panel A. The revenue from the government's ownership is calculated upon an assumed realized price of $1.35 per barrel and the same cost of $0.20.

For the concessionary case, without government participation, the cost of production as measured in accounting terms and the royalty are both deducted from the posted price, the difference being the putative income which is subject to income taxation by the host government. At a tax rate of 50 percent, one-half of the putative income—i.e.,

Table A10.1

Hypothetical Computations of Government Oil Revenues: Concession and Joint Venture Compared

(Per Barrel)

A. *Concessionary case*		
Posted price	$1.800	
less Cost	0.200	
less Royalty (12.5 percent)	0.225	
Putative company income before tax	1.375	
Tax (50 percent)	0.688	
plus Royalty	0.225	
Government take		$0.913
B. *Joint Venture case (50 percent participation)*		
1. From foreign partner ($0.913 × ½)		$0.457
2. From country's equity share		
Realized price	$1.35	
Cost	0.20	
Accounting net income	1.15	
Country share ($1.15 × ½)		0.575[a]
Total country revenue		$1.032

[a] The "effective" revenue would be somewhat less if allowance is made for the opportunity cost to the government of the capital invested in its half share of the enterprise, which is assumed to be greater than an adequate rate of return on capital to the concessionaire company.

Chart A

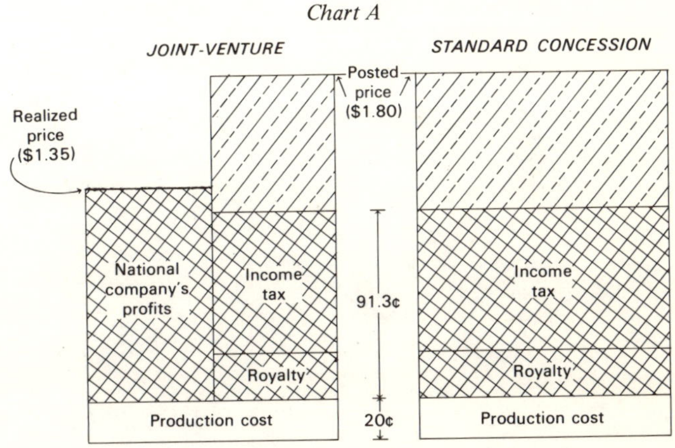

68.8 cents[30]—is added to the royalty payment, and the government's gross take is 91.3 cents per barrel, as displayed in panel A of Table A10.1.

In the case of the joint-venture agreement, each barrel is divided into two halves. On the half which "belongs" to the foreign partner, the foreigner pays royalty and income tax as is ordinarily the case for a concessionaire.[31] Thus the government receives one-half of 91.3 cents or 45.7 cents on each half-barrel which is lifted for the account of the foreign partner.

[29] In addition, since the expensing of royalties was introduced, the government take became larger than 50 percent of the notional "profit" (see pp. 121 and 123, above).

[30] The tax revenue will be somewhat less where a small discount from posted price is allowed in computing "profit," as is now generally the case, but to be phased out within a few years.

[31] It should be noted, in addition, that some joint agreements stipulate no royalty, and some provide for taxation based upon realized prices, or some other discount from posted price.

On its own half-barrel, the country's gross receipts are the actual, cash realized price in the market place (or some agreed equivalent when the company offtakes the oil), estimated here to be equal to $1.35 per barrel for a crude oil which is sold f.o.b. at terminal in the Persian Gulf at a posted price of $1.80 per barrel. The government's accounting revenue is equal to the cash sale price less the accounting cost of production (20 cents) or $1.15 per barrel for half the total production.

The apparent difference in the revenues to the government in the two cases is 11.9 cents–i.e., $0.913 in the concessionary case and $1.032 in the joint-venture case. This difference could be reduced even further by allowing for the opportunity cost of capital to the producing country; by such differences in joint-venture terms as are alluded to in footnote 31, and by a lag in the timing of sales receipts as compared to tax payments by concessionaires.

A Note on Company "Profit"

The calculations we have made show the revenue benefits to *government* under particular circumstances. They say nothing about the benefits to the foreign companies. Basically, any attribution of company profit per barrel from crude oil production is highly artificial, since the company's interest is in net integrated profit as a rate of return on investment. However, assuming that the "realized" price has some substantial meaning for the company in profit accounting under individual concessions, the presumptive take of the company can be calculated from the data in Table A10.1.

Under the standard concession, the calculation would be:

Realized price	$1.35
less cost	0.20
Accounting income	$1.15
Country take	0.913
Company take	$0.237

Under the joint venture in Table A10.1, the calculation would be:

Accounting income	$1.15
Government share	0.575
Company share	0.575
less tax and royalty	0.457
Company take	$0.118

Since the company has only half the capital investment, it enjoys the same rate of return on capital as in the other case. This assumes that the respective investment shares of company and country are in fact equal. If the country acquired its share at a low value, or by expropriation, the result would of course be different. It is also assumed that the country's oil is sold for its own account. If it were lifted by the concessionaire at some payment to the national company less than realized price, the company take would be larger. The capital transactions, by which the government acquired its share from the company through payments from the proceeds of oil as it was produced, are assumed to have been completed. The company comes out of this process financially undamaged only if it has alternative opportunities for investment of its recovered capital equally remunerative to the uses given up.

CHAPTER 11

Views on the Trend of Country-Company Relations

IN this present chapter we shall present some views on the possible directions of change in the relations of countries and the major concessionaire companies: first, from "outside" Western observers and, second, from "inside" representatives of countries and companies.

Some "Outside" Views

In recent years a number of persons in Western countries with expert knowledge have engaged in analysis of trends in country-company relationships and of some of the consequences to which they might lead. We proceed to an account of some of the views that have been expressed. The summaries are intended solely to express the views of the original authors. Any comments attributable to the authors of the present study will be identified in the text.

1. *Professor Adelman's Analysis*

Over a period of years Professor M. A. Adelman of the Massachusetts Institute of Technology has studied costs in the main producing areas, the trends of crude oil and product prices, and the causes of changing market conditions. Upon the evidence presented, he has also advanced views concerning some of the possible future consequences. The results of his studies have been brought together in *The World Petroleum Market.*[1] We shall summarize the parts of his analysis relevant to our present purpose.

Certain fundamental features stand out in Adelman's analysis: (1) Since 1957 oil product prices in world markets have been drifting downwards, and this trend may be expected to continue, though subject to resistance and slowing down from the policies of both producing and some large consuming countries. This development was due to the vast reserves easily and cheaply convertible into production. The abundance will remain or even increase at least through the early 1980's. The entry of independent newcomers into production and refining was one facet of a larger process: the pressure of competition upon the old-line integrated majors slowly moving price toward cost.

> . . . The post-Suez history may be summed up: The modest growth of oil sales outside the vertically integrated companies, by producers with a small share of total sales (integrated plus non-integrated) reacted back upon those companies. Some independence in buying and selling mobilized, *though it did not cause,* the huge and growing excess of potential capacity in the international oil trade and made it bear down upon the price.[2]
>
> . . . the current price level is an incentive for nearly every operator to produce and sell far more than he now does. Hence he must restrain his offerings in order not to spoil the market. But the incentive remains: repressed but felt. The urge for additional sales and profits puts the market under a severe continuing strain. In order to reconcile the objectives of more sales volume and no price cutting, producers give away some of the excess of price over cost by special deals for some customers; they may operate marketing-refining facilities at little or no additional profit; try to gain business at the final customer level by non-price rivalry. They tie up long-term outlets by loans or other inducements that equate to lower net prices. . . .
>
> As more and more companies appear, the circle of hands becomes wider and looser, and breaks more easily. Small producers can do increasing damage by shading prices because others may follow. The companies' strength always lay in their being slow to react. But they cannot let markets be taken away from them, and in the round of chiselling, retaliation, and further chiselling, the whole price structure threatens to crumple.[3]

(2) The taxes per barrel paid the producing countries by the companies increased sharply in the 1950's and continue to rise; and further demands in the same direction may be expected. Hence, the tax-paid costs per barrel to the companies may rise further. The essence of the various agreements, based on downwardly inflexible posted prices, is "that the income tax has been transformed into an almost pure price per-barrel tax."[4] The per-barrel tax is now the basic support of the current price level.[5]

(3) The squeeze on profits caused by declining prices and rising tax-paid costs could reduce profits to a point where some companies would have to decide whether to cease lifting oil from a particular concession and increase offtake in another, or to buy from another company. While most companies are not yet at that point, some are probably approaching it. And, since countries may be expected to make fresh demands for higher payments, others are likely to be driven into that situation in the not distant future.

(4) The countries may expect the companies to recoup by higher prices. "We always come back to the same basic question: Can the supply of crude oil, and therefore the price, be controlled or not?"[6] If country quotas could be fixed and adhered to, this would be possible. But this outcome is wholly improbable. The companies cannot operate a successful cartel; the countries have even less prospect of doing so. Thus, the competitive forces inherent in a great excess of oil potentially available at a basic cost which is

---

[1] Tentative title of study now under preparation for publication as a Resources for the Future book; all references are to his draft manuscript of 1969.

[2] *Ibid.,* Chap. VII, p. 23.

[3] *Ibid.,* Chap VII, pp. 28–29.

[4] *Ibid.,* Chap VII, p. 37.

[5] *Ibid.,* Chap. VII, p. 38.

[6] *Ibid.,* Chap. VII, p. 41.

only a small fraction of current price will keep a downward pressure on prices.

(5) Higher taxes will differentially favor low-cost sources. Elsewhere, companies will be trapped between the upper and nether millstones, and curtail output. The disadvantaged countries, attempting to maintain volume, may be forced to grant differential tax relief to companies or to undertake trading on their own account. In either case they can only maintain volume by price concessions. If tax increases plus participation on OPEC principles is attempted, a series of crises in particular company-government relations may be expected.

> . . . For governments and companies, the real issues have become tolerably clear. The companies' share cannot be pushed lower than a barely adequate return on development investment, i.e., to cost as reckoned in this study. In practice, it cannot be pushed even that low. For the more demands the government makes, the greater the political risk, and the minimum acceptable rate of return. The government must then ask: how much can we take before they decide to pack up and leave? And if they leave, can we earn as much from producing the oil as they formerly paid us. . . .[7]
>
> There has been too much talk of expropriation, nationalization, confiscation, just compensation, and the like. . . .
>
> What we must expect in the producing countries is continued *appropriation* of all the surplus between the bare minimum profit necessary to keep the company operating, and the real price of crude oil. . . .[8]

By miscalculation, governments may drive taxes above what companies can tolerate.

(6) The attempt of governments to drive company profits to the minimum will tend to remove the price-supporting quality of the posted price.

> . . . For a government which aims to take everything over and above the minimum necessary return on investment, posted prices are irrelevant. For if a posted price, less allowed cost, yields a tax which leaves the operator more than the necessary minimum, the government is committed to tax it away. On the other hand, if the tax is such that the operator cannot pay it, the government must decide how much to give up in order to keep the rest. Posted prices are no help in making either adjustment.
>
> . . . . . .
>
> After coming full circle, the wheel has made another half turn. A per-barrel royalty became a percentage royalty, then an income tax, then an excise tax . . . , and the governments are now moving toward an income tax again. . . . But as the governments now move to put themselves back to income tax, *the stabilizing effect of per-barrel payments will be lost*. . . . The closer the governments get to their objective, the more they take on the job of price-making decisions. . . .[9]

[7] *Ibid.*, Chap. VII, pp. 45–46.
[8] *Ibid.*, Chap. VII, p. 46.
[9] *Ibid.*, Chap. VII, pp. 55–56.

Companies may be forced into the position of hired contractors to produce or sell or both; if they are expelled, others will have to be hired in their place, unless governments undertake to do their own trading. "A private company, running its own 'concession,' but taxed to the economic limit, is nothing but a hired contractor-operator for the government-owner."[10]

Once governments have the responsibility of making output and price decisions, in order to maintain or increase their market shares, they will be drawn into special deals which are profitable in themselves but have the effect of undermining prices. Evidence of this is already developing in the case of the new partnership agreements. It can only be prevented from spreading by a producers' cartel, which cannot be expected to be effective.

(7) The tendency toward price cutting will be strengthened by the desire of governments to expand the market role of their national companies. So far, these companies have had little oil at their disposal; but as the amount increases, they will have to force their way in by shading prices on incremental sales. Since they are not inhibited by tax payments they have more room for price cutting while still showing highly profitable results. Whether their tactics are ultimately disadvantageous to the revenues of their countries will depend upon whether they undermine prices so severely that the revenues obtainable from concessionaire companies have to be reduced in order to retain their services. The chances of this will be increased in the degree that individual countries utilize their national companies in the effort to expand their shares in the market.

The governments have not all the same urgency for immediate revenues as compared with deferred. Some invest in the international money markets, and discount the future at a low rate. Others have high growth rates and a high social or private rate of return on new investment; they would benefit more by revenues today from a cut-price deal than they would lose a few years hence when lower prices finally affect tax per barrel. Others will have as great a sense of urgency, but from failure, not success. In any event, the diverse rates of discount reinforce diverse costs and predictions of future prices and costs, and make a cartel less likely.

(8) Professor Adelman's general conclusion is that the policies of producing countries are set upon a course of weakening prices. We quote his summary conclusion:

> The shift from income taxation to per-barrel taxation has put a floor to prices, which has been effective even at a distance. But the attempt of the governments to get an even larger share of earnings than they now receive leads inevitably back to income tax on realized prices, and the disappearance of the floor. National oil companies will by their nature be price cutters. The governments are now demanding large ready reserves for their national companies, to have them enter the business soon and on a large scale. With no automatic price-cost mechanism to regulate market shares, every country can produce much more than any quota

[10] *Ibid.*, Chap. VII, p. 55.

agreed from the others. The governments are less able to operate a successful cartel than the companies. Not only do they lack the companies' experience, but they also lack the inter-company contacts at two levels: crude production and sales . . . and in the refined product markets. These contacts are necessary for sound decisions on when to meet competition, when to beat it, when to disregard it. There is nothing to prevent the governments from going into the refining-marketing business, but the large investments needed, and the mediocre returns will slow down their entry and minimize their contact. Furthermore, they will increase the number of competitors at both levels.

Thus analysis of the actual role of the governments. . . indicates that their increasing role in the market will tend to increase competition and reduce prices. This is the contrary of what they want. It should surprise nobody with any sense either of economics or history. Adam Smith emphasized how sellers produce market results they never intended. History is sprinkled with persons and groups who exerted great power to bring about their discomfiture or even ruin.[11]

(9) The conclusions quoted in (8) above need to be qualified since the present policies of consuming countries, in Adelman's view, have the effect of supporting prices, and may be expected to continue to do so. Various energy policies in the Western European countries have been justified by the belief in the doctrine that Adelman has dubbed TALAM (that is, that oil prices are "temporarily abnormally low and anyway marginal"). In this view, rising consumption is expected to dry up the present surplus of oil, strengthening prices. The demands of producing nations for higher revenues will also force prices upward. While contrary to the evidence, as Adelman believes, this doctrine has been almost universally accepted as official truth and used as the basis for policies restricting competition in the energy markets, hence slowing the erosion.

For example, in Great Britain, Adelman argues, the case for developing an atomic energy industry for electrical generation incorporates an implicit oil reference price much higher than recent actual prices. Moreover, the implied high oil price justifies the continued use of coal for electrical generation. Oil companies can obtain all the business available to them by merely meeting the price of coal, and have no incentive to bid lower or compete with one another.

The general energy policies of a few countries have coal protectionism as perhaps the most important element. Fuel oil is heavily taxed; and they have forced the use of coal, especially in the government-owned or government-supported electrical power industry in Britain, France, Belgium, Germany, and Japan. The principal reason for government protection of coal is the social cost of permitting its rapid disappearance. The argument of national security from indigenous fuel supplies is also invoked. The use of an unrealistically high oil reference price conceals the great cost to the country of using coal. "Probably no European or Japanese coal would be produced now if it were exposed even to the limited competition now at work in oil."[12]

As coal declines in use, much of it is scheduled to be replaced by nuclear energy. The high oil reference price equally conceals the probably uneconomic character of the actual or projected installations.

In Adelman's view, particular countries have their own reasons for other policies that support the price of oil products, especially the protection of a local refining industry designed in part to supplant the control of markets by the major companies. In France the objective of state-supported integrated companies, controlling markets at home and engaging in production abroad, is a dominating influence, overriding economic considerations. "Self-sufficiency is an end in itself, to be secured by owning concessions abroad."[13] In Britain, in addition to the commitment to coal and nuclear energy, the selling price of the new natural gas from the North Sea fields is supported by a high oil price, enhancing the economic success of the state-owned Gas Council. The British government, moreover, through its majority ownership of British Petroleum has some stake in supporting the price of crude oil and products.

In his summary for Europe, Adelman says:

> Governmental policy, by which we mean government actions, not statements, seems to be moving more and more toward maintenance of high prices. Much and perhaps most of the "truly European" oil and gas production will be done by State companies. *The companies must not lose money, lest the governments lose face.* The products they sell must be valued at a high enough price to make them look successful.[14]

Adelman sees the same factors at work outside Europe. India seems committed to self-sufficiency and high prices. The United States for its own reasons supports high prices. Japan is facing both ways; but if it continues to seek self-sufficiency through its own producing companies it may be committed to high prices. The less developed countries, as committed supporters of commodity agreements to raise the relative prices of primary commodities, may for consistency be committed to the efforts of the oil-producing countries in that direction. Also, some of them will want high product prices to support domestic refining. And many of them hope to become oil producers.

Consuming countries, acting alone, cannot halt the erosion of prices, though they will slow it down. The only effective barrier to price decline would be a full-fledged commodity agreement entered into by the large producing and consuming countries, a possible but not very likely arrangement. Even if established, the temptations to "evade, chisel, and cheat" would be so great as to eventually undermine it. Short of this, the producing countries appear to be courting agreements with individual consumer countries for what they may be worth and with a view to more extensive agreements in the future.

(10) Against his picture of the forces in consuming countries tending to support prices, Adelman presents the conclusion that consuming countries could, if they wished, secure much lower prices.

[11] *Ibid.*, Chap. VII, pp. 72–73.

[12] *Ibid.*, Chap. VII, p. 78.

[13] *Ibid.*, Chap. VII, p. 99.

[14] *Ibid.*, Chap. VII, p. 108.

> . . . In general, the consuming countries will have the price they want.[15]
>
> . . . . . .
>
> If some consuming nations wished they could induce lower prices. The international oil companies would of course be strongly opposed, but as their producing profits will in any case shrink toward the minimum operation fee, so will their reason to oppose. In any case, the consuming countries can have the price they wish, down to a limit around 20 cents per barrel. But for better or worse few of these governments know what they do, or what is happening in the oil market. . . .[16]

Producing countries, aware of the threat, are already issuing calls for collective punitive action against countries attempting to undermine prices, at the moment directed mainly at Japan.

(11) Looking well ahead, Adelman reaches the following conclusion:

> Only the rate of decline is in doubt. Prices will slide and lurch downward, in the face of confident predictions that *this* time they will last or get back to "normal." Host government revenues per barrel will at first be stable but then will decline almost cent for cent with prices, as these governments in fact and later in form emerge as the owners. Some concessions will be expropriated for a time or for good, but seizure will only be the punctuation or background music for the gradual appropriation by host governments of anything above what they recognize as the adequate return on development investment. Some companies will find it cheaper to buy from other companies or from government companies than to produce from their own wells. . . . Profit margins will for some companies be dangerously squeezed, but in the long run they will all or nearly all survive and prosper for their services are essential and the margin allowed them cannot be narrowed past a certain return on investment.
>
> By 1985 the world energy outlook will have changed in ways we cannot foresee, and the current problems may be moot—or more urgent than ever. In the meantime, world oil looks to be more and more a mingled company-government affair, increasingly a chronic and politically dangerous struggle among companies and governments over markets, and between companies and governments over profits. The more they invest and make oil cheap, the more they will hold it dear. There will be much to-and-fro by eminent gentlemen to meetings wherein will be discourse of justice, oppression; equity, unfairness; threats and promises; dramatic exits and quiet re-entries; etc. Patched-up deals will last awhile; painfully drafted "final" agreements will be forgotten before the ink is dry. Somehow the oil will continue to be produced at ever-expanding rates. It will be an interesting story. No mere economist can hope to do it justice, but some will try.[17]

Professor Adelman's analysis is directed primarily toward the future of crude oil prices. But involved is the whole complex area of the policies of consuming and producing countries and of company-producing country relations. His conclusions run counter to the beliefs on which he thinks many consuming country policies have been predicated and to the interests which the producing countries are attempting to protect. But he is convinced that these policies and interests must give way before the fundamental economic situation in which great potential supplies overhang a market, in which prices are many times costs, and in which competitive forces are increasingly active. The only effective barrier would be monopolistic control over supply by a degree of collective action among producing countries which he thinks most unlikely to be achieved or, if temporarily achieved, to survive.

2. *Professor Penrose's Analysis*

The relations of the oil companies and the governments of oil-producing countries have in recent years been closely studied by Professor Edith Penrose of the University of London; and her writings on the subject are among the most cogent and knowledgeable to be found.[18] Her approach is different from that of Professor Adelman, being primarily institutional, while he is primarily concerned with trends of costs and prices. In the end, however, they occupy much common ground, since in looking toward the future the institutional approach requires attention to the economic fundamentals while the economic approach requires attention to the conditioning institutional relationships. For this reason we may deal with Professor Penrose's analysis somewhat briefly using Adelman as a point of reference. To a considerable degree, they arrive at similar conclusions, though hers are more tentatively stated than Adelman's.

With respect to the erosion of prices, Professor Penrose summarizes as follows:

> The underlying situation that has brought about the present state of affairs. . . is not fundamentally the pressure of the producing countries for greater revenues, but the erosion of the control exercised by the major companies. This control has been weakened by an increase in transactions outside the vertically integrated channels of these companies due to the rise of independent private producers, refiners and marketers, the creation of national companies in the producing countries. . . the growth of national refining and marketing companies in importing countries, the re-appearance of Russian crude exports, and competition among the majors themselves for crude oil sales to outside buyers. As a result, a genuine market price for crude oil has for the first time appeared in the industry. . . . the still strong position of the major com-

---

[15] *Ibid.*, Chap. VII, p. 116.

[16] *Ibid.*, Chap. VII, p. 117.

[17] *Ibid.*, Chap. VII, pp. 118–119.

[18] The analysis on which we have drawn is contained in her book, *The Large International Firm in Developing Countries: The International Petroleum Industry* (London: Allen and Unwin, 1968); in two lectures reproduced in *Middle East Economic Survey, Supplement,* August 2, 1968; in an article, "Government Partnership in the Major Concessions of the Middle East," *Middle East Economic Survey, Supplement,* August 30, 1968; and in "OPEC and the Changing Structure of the International Petroleum Industry," *Middle East Economic Survey, Supplement,* March 7, 1969.

> panies in the crude oil industry has been largely responsible for the comparative gentleness of the fall in prices that has occurred over the past six to eight years outside the United States, but . . . the process is only retarded and is likely to continue.[19]

Efforts by the countries to support prices through establishment of a producers' cartel, if attempted, have little prospect of success due to the dissatisfaction of individual countries with their share of the market. The level of taxation, if uniformly applied, can provide a floor to prices, although low-cost producers will still be in a better position to compete. But countries at times show a disposition to violate this floor by special terms to their concessionaires or national companies to give a high level of offtake.

Given the continuing erosion of prices, the demands of the countries for larger revenues apply a squeeze on profits which, in Professor Penrose's view, the companies may come to find unsupportable. The companies, as they may have foreseen when the so-called 50–50 profit-sharing arrangements were introduced, can no longer protect themselves by reducing the posted prices on which profits are calculated. Nor can they recoup by raising prices in markets over which they exercise diminished control.

The pressure by the countries for larger revenues may be expected to continue, further encroaching upon the integrated profits of the companies up to some breaking point. "To the extent that excess profits still exist in the industry it seems to me highly probable that the producing countries will get them."[20] The OPEC principle of leaving the companies a "reasonable profit" is likely to be interpreted differently by countries and companies.

Heavier taxation is not the only route by which the companies may be pressed beyond their powers of endurance. Professor Penrose indicates that the other, and possibly the most likely, route arises from the demands of countries for participation in the ownership of the concessionaire companies. This might, of course, be merely a way of securing larger revenues, alternative to higher taxes. But "participation" or "partnership" appears to mean much more than this to those who have taken up the cry. They want for their countries an active and influential part in the management decision-making processes of the companies. Perhaps the most direct interest the country partners might have would be to increase the offtake, and to force it into their market channels. But their pressures could take a number of directions, especially with respect to investment of funds. They would want the earnings to be used more largely for the development of their own economies, rather than being available to parent companies for investment where and for what purposes they chose. Pressure by individual countries on their principal concessionaires could severely impair the flexibility of operations by which the companies carry out the international logistics of supplying worldwide product markets from numerous sources of crude oil supply.

Driven too hard, the companies will be forced from their existing position of concessionaires to some other relationship with countries under which they will be essentially contractors. Professor Penrose is of the opinion that this is the wave of the future, though the pathway of change and the exact destination cannot be clearly foreseen—probably differing from country to country. She says:

> . . . there is no reason to assume that the steady encroachment by the governments on the freedom of action of the Companies will cease, and it seems highly probably that in the forseeable future a contractual form of relationship between Companies and governments for the production of crude oil will emerge.[21]

One possible outcome could be the disappearance of the producing operations of the concessionaire companies by transfer to governments.

> If the trends described above—increased taxes, government intervention, and falling market prices—continue, the oil Companies may well find it desirable to explore means of selling their major producing affiliates to the governments of the oil-producing countries and substituting some form of contractual relationship.[22]

If the inroads on profits and on managerial freedom of companies continue,". . . the major oil companies may well, in the not-too-distant future, look for ways of divesting themselves of their ownership rights in their chief crude oil producing affiliates."[23] Under such an altered relationship, the major companies would buy their oil supplies, but continue their role as the international marketers of oil. The fact that the countries realize the great advantage to them of the integrated distribution channels of the major companies would serve to strengthen the position of the companies in negotiations looking to such a transfer of functions.

Short of any such radical change of relationship, and no doubt preceding it, there would continue to exist the present pressures toward greater participation of countries in the operations of the major concessionaires. Presumably, the change would take place only when the parent companies of the producing affiliates found the situation intolerable. All the parent companies of joint-ventures, such as ARAMCO, would have to work in concert.

It may be assumed that countries could successfully cope with their producing problems, presumably with the aid of outside technical assistance and the newer forms of exploratory contracts and participation. But perhaps the great threat that might deter countries from pressing matters to the point of driving the major concessionaries out is the danger that it would lead to a heavy fall in the price of crude oil, and of their revenues, through the character of the contracts for the sale of this oil. Competitive forces

[19] "The International Oil Industry in the Middle East," *Middle East Economic Survey, Supplement,* August 2, 1968, p. 17.

[20] *Ibid.*, p. 14.

[21] *The Large International Firm in Developing Countries, op. cit.* (n. 18), p. 216.

[22] *Ibid.*, p. 255.

[23] *Middle East Economic Survey, Supplement,* August 2, 1968, p. 16.

might be unleashed which would further undermine the role that the great integrated companies have historically had in maintaining, or retarding the fall, of crude oil and product prices. The companies would also probably entertain the same fear.

Except in the unlikely establishment of an effective producers' cartel, the chances lie strongly on the side of weakening prices. This prospect, deplored by countries and major companies alike, could well be the force preventing so drastic a change as that outlined above. ". . . Many responsible Arab oil experts are themselves skeptical of the effectiveness of a producers' cartel, and this partly explains the unwillingness of many of them in all countries to support proposals that would undermine the position of their present concessionaires in this way."[24]

Assuming the absence of an effective producers' cartel, it would appear that the countries might diminish, rather than increase, their revenues by driving too hard bargains on taxation and ownership participation. But revenue considerations do not wholly govern the behavior of countries. As a matter of principle, they desire effective control over the development of their resources, and this desire comes through most sharply in the drive toward participation, and eventual control, in management decisions. Professor Penrose says:

> Among the issues in which principles become important for each side, the most difficult relate to local participation in the ownership of the oil-producing companies, and to the degree of control over the conduct of the affairs of the companies by the local governments, including control over prices.
>
> . . . . . .
>
> For the producing countries the issues involved in "participation and control" raise principles related to their own concepts of independence and national sovereignty, while for the Companies they impinge on their "commercial freedom of action" and have far-reaching implications for the organization of their integrated operations as a whole.[25]

In a subsequent article, Professor Penrose reiterates her view that:

> . . . it is highly unlikely that [the major international companies] would accept any arrangement which contained the reality of partnership, that is, one which gave any effective influence to the government partner over operations, and especially over offtake, prices and even investment.
>
> . . . . . .
>
> . . . The special difficulties that arise with respect to the price of crude oil, and to equity participation for governments in the crude oil producing companies are both traceable to the same source: the fact that the producing companies are not simple crude oil producers and sellers but are an integral part of, and are owned by, larger groups which are engaged in refining and the marketing of products all over the world. Most of the revenues of the integrated groups come from the sale of products, not from the sale of crude oil. . . .
>
> . . . . . .
>
> . . . no individual government is in a position to appraise the overall circumstances of its major concessionaires and thus to take an active part in the determination of the international distribution of their supplies of crude oil. . . .
>
> . . . . . .
>
> . . . it is highly probable that if the major companies accept government equity participation, the acceptance will mark the advent of a policy of withdrawal from participation in the ownership of the companies providing their major sources of crude oil . . . becoming mere offtakers or perhaps also contractors to governments.[26]

Professor Penrose subscribes to the view that there is no necessity that matters be pressed to the same conclusion in all countries. But, in all, the political pressures for a much greater participation in control of their resources will be strong, and may in some unforeseeable exigencies push the responsible authorities into positions that go beyond their own dictates of economic prudence. In the case of Libya, the existence of multiple concessionaires adds a new dimension to the problem.

Meantime, as matters now stand, while pressing their case for participation in the major concessionary companies, the countries are beginning to assert their independence of action: through market penetration by their national oil companies, through their new nonmajor concessionaires, and through partnership arrangements with national companies of some consuming countries. These activities are at present relatively minor in comparison with the market activities of the major companies. Nevertheless, they introduce new elements of competition which seem well-designed to further depress the price of oil. At the same time they are providing an education to the nationals of countries in the problems of management and in the intricacies and pitfalls of the international oil business, a matter of great importance if the countries do in fact attempt to assume a much greater degree of control.

3. *Dr. Frankel's Views*

Dr. Paul H. Frankel, the head of a consulting firm in London, has long been one of the best known and widely read analysts of international oil. A recent contribution[27] was inspired by the paper of Mr. Yamani (see Chapter 10, p. 137, dealing with "participation," and by the "Declaratory Statement" of OPEC (see Chapter 10, pp. 124ff.). Dr. Frankel's analysis is primarily concerned with the conflict

[24] *Ibid.*, p. 18.

[25] *The Large International Firm in Developing Countries, op. cit.* (n. 18), pp. 212–213.

[26] "Government Partnership in the Major Concessions of the Middle East: The Nature of the Problem," *Middle East Economic Survey, Supplement,* August 30, 1968.

[27] Paul H. Frankel, "The Current State of World Oil," *Middle East Economic Survey, Supplement,* September 6, 1968. In an acknowledgement to Professor Penrose, he states that "some of our conclusions are virtually identical," referring to her paper in *Middle East Economic Survey, Supplement,* August 2, 1968.

of interest between producing and consuming countries, and the ways in which these groups might arrange, or rearrange, their relations with the international companies in the effort to serve their own respective interests.

Historically, companies and producing countries had a common interest in the company policies which kept prices high through what amounted to an informal system of proration. Given this situation, it was natural, and justifiable, for the countries to take for themselves as much of the large margin above cost as they could, consistent with the maintenance of price-sustaining policies. Through concerted action on OPEC principles, they have been highly successful in this.

At the same time, new elements of competition have weakened the international price structures—on the reasons Dr. Frankel does not differ from Adelman or Penrose. By their tax policies, the countries have protected their own interests; but taxes and declining prices have reduced company profits to more or less one-third of what they had been. [*Sic.* Presumably per barrel, not rate of return on investment.] If this erosion continues, the interests of both companies and countries will suffer. It is therefore to their mutual advantage to work closely together to check it.

In spite of this fact, producing countries are moving along lines in the opposite direction—in particular by unilateral efforts to force a larger offtake of their oil by national companies or concessionaires. Their interest, on the contrary, is to act in alliance with the companies to support price stability. Anything approximating a tight cartel is impractical; but Dr. Frankel appears to think that the forces of competition could be considerably damped by concerted action of countries and major concessionaires. Without predicting, he appears to hold that outcome well within the range of possibilities (in contrast to the suggestion by Adelman and Penrose of relentless pressure on the companies to the point of their departure from the fold). In speaking of the newly active principle of "participation," for example, he says:

> . . . the word may mean anything or possibly nothing: it is not even clear what Mr. Yamani had in mind when he asked for it . . .
>
> It may well be that nothing much is to come of such a campaign, but it might be considered as one more turn of the screw, a tightening of the vise in which the companies have found themselves for some time past. It could also be a step towards that "alliance," participation providing the operational opportunities for an alignment of interests.[28]

He says further:

> Price maintenance by way of posted prices, though endangered by volume-minded governments, may become a more likely bet if the mutual sympathy of producer-country governments and oil companies leads to a de facto setup in which the several interests are "harmonized" in such a way as to maintain, possibly to enhance, tax take *and* profit margins.
>
> Although such an understanding would still be subject to problems bound up with any attempt to maintain prices and may not work perfectly, its comprehensive yet diversified and flexible nature would give it a more than reasonable chance of achievement in the course of a fairly long run.[29]

So much for the producing countries and companies. What of the consuming countries? Consumer country interests, Dr. Frankel finds, have never been articulate. For reasons of their own, the more important (Britain, France, Italy) have tended to regard themselves as de facto producer countries; and no serious attention has been given to how they might acquire their oil more cheaply. Given the fact that, in his view, every practicable device has been, and is being, used, to make them pay a monopoly price, he regards this as a remarkable oversight. Oddly enough, also, some consumer countries are using their state-owned companies to bid on extravagant terms for concessions likely to produce high-cost oil.

Though he makes the point less emphatically than Professor Adelman, Dr. Frankel credits the consuming countries with power to secure oil on better terms. He says:

> It is evident that in the circumstances only governmental policies in the consuming countries can provide a modicum of countervailing power. I believe I have just shown that one such line of approach—the acquisition of "cost oil"—does not, or at least not any longer, fulfill the task of reducing substantially the cost of imported energy. Only by remembering what a smokescreen has almost obscured, that in a market where potential supply greatly exceeds current demand it is the buyer who is in a strong position, can one devise an adequate consumer-country policy.
>
> Japan . . . has also had the right idea . . . that it should take a hand in securing acceptable terms. . . . Once the objective is understood the means to reach it are not difficult to envisage: the obvious idea of a counter-OPEC of consumer countries is not necessarily relevant; it would be sufficient for any one of the biggest oil importing countries to make it known that oil imports would not be acceptable if their cost were to exceed a certain level; if the tax levied by producer countries were to rise, making it impracticable to keep within stated limits, short- and medium-term measures would be taken to concentrate on imports from such countries which would prefer to maximize their revenue by increasing the volume of these exports. As long as such policies were pursued judiciously and with moderation they would fit almost painlessly into the existing setup of international oil, without dispossessing the oil companies whose role is, and for some considerable time will be, indispensable.[30]

Dr. Frankel does not pursue the subject to the point of speculating upon how the relationships of companies with producing and consuming countries, respectively, might be affected under various contingencies. He does, however, quote an anonymous company executive[31] to the effect

[28] "The Current State of World Oil," *Middle East Economic Survey, Supplement*, September 6, 1968, p. 9.

[29] *Ibid.*, p. 10.

[30] *Ibid.*, p. 12.

[31] *Ibid.*, pp. 12–13.

that, if crude oil prices continue to erode and producing country revenues are undermined, perhaps the international majors should reexamine their policies towards producing countries' negotiations, with the view that their ultimate long-term success might be more closely associated with their markets in the consuming countries. ". . . the oil companies with markets may find themselves in a position with increased profits resulting from the ever-decreasing crude prices, and, as a matter of fact, eventually come out much better off than they are at present."

This line of thought suggests that companies are considering their strategy pending the time when the exactions of countries may make it preferable for them to abandon their role as concessionaires and to become mere offtakers or contractors. It is precisely this possibility that the producing countries have to face where their interest is no longer protected by taxes based on posted prices and where the companies' interest would lie in low offtake prices enforced by competitive selling by the countries.

Dr. Frankel's line of thought in this recent article differs considerably from that in his book written in 1962.[32] The oil industry, in his view, opposed by others, was "inherently unstable" and "could not operate under competitive conditions." The large integrated companies, some of them already ensconced in the high-price United States market where controls protected the high-cost sources, were able to secure control of the reserves in the new low-cost regions. Following pricing policies for which the guidelines were set by high U. S. prices, they were able to exact enormous economic rents from their low-cost sources. In Frankel's view, the companies were compelled, in due course, by the host countries to share these gains through heavy taxation, while the consuming countries for reasons of their own failed to take advantage of the opportunity they possessed to secure oil from low-cost sources at lower prices.

In his 1962 book, Dr. Frankel proposed a "rational" approach to sharing the benefits—to companies, producing countries, and consuming countries—through some sort of regulatory body representing the producing and consuming countries.

He appears now to have abandoned that conception, leaving the producing and consuming countries to protect, or impair, their own interests in their own way. The producing countries and the companies in alliance can attempt to shore up the competitive situation to their advantage; or the countries, by too great exactions on the companies, and forced to face the market directly, can cut their own economic throats, since they can probably not support an effective cartel. The consuming countries can attempt to protect their own interests by shopping around. And in certain contingencies, the companies, if too hard pressed by the producing countries, might come over to their side. One cannot forecast with any assurance the outcome of these trials of strength.

[32] Paul H. Frankel, *Oil: The Facts of Life* (London: George Weidenfeld and Nicolson, 1962). E. T. Penrose, *The Large International Firm, op. cit.* (n. 18), pp. 165ff., gives an excellent summary of his and opposing views.

4. *The Views of Mr. Hartshorn*

An excellent general account of the international oil industry is given by Mr. J. E. Hartshorn,[33] formerly of *The Economist* and now a member of an oil consulting firm. In addition to his descriptive account, he presents in depth the many issues upon which companies and host countries have had to reach an accommodation and the outstanding issues under present changing circumstances. In looking to the future, he asks many questions to which he does not attempt assured answers; he simply explores alternative possibilities.

Unlike Adelman, who feels sure that prices are headed slowly but strongly downward unless supported by the energy policies of consuming countries, Hartshorn rather hedges his position. He recognizes the strength of the competitive influences pressing downward on prices: surplus potential supplies, newcomers, national companies seeking a place in the market, the efforts of countries to expand their offtake, and increased price competition in the worldwide product markets of the major companies. At the same time, he also parades the set of influences tending to support prices: high taxes acting as a cost barrier to price competition, the strategies of major companies to retain their oligopoly position in product markets, energy policies of consuming countries acting to hold prices up, and attempted measures by producing countries to impose some cartel-like production control. One judges that, within this parallelogram of forces, he thinks the trend of prices will be downward; but he reserves judgment concerning the relative strength of the forces at work in opposite directions.

Within the general situation, he ventures a judgment concerning the market influence of the policies of host governments. He says: ". . . I believe it a shade more likely that the net influence on prices of their incursion into the market will be downward through competition rather than upward—through a cartel."[34]

Hartshorn at a number of points returns to the theme of a producers' cartel. He notes the tentative efforts of OPEC and says: ". . . the logic of OPEC's situation will drive it towards perseverance with the idea of production control—unless prices come to be stabilized in some other way."[35] Without being too explicit, his view appears to be that, while attempts will continue to be made, the obstacles to successful operation are probably insuperable. The intrinsic difficulties are those of reconciling the different interests of different producing countries. In addition, collective action by producing countries, if it showed signs of being effective, could very well lead to countervailing collective action by consuming countries.

Exploring the possibilities for collective action at both producer and consumer ends, Hartshorn says:

> . . . The governments from now on, more than ever before, will have the choice between taking individual

[33] J. E. Hartshorn, *Politics and World Oil Economics*, rev. ed. (New York: Frederick A. Praeger, 1967); English title: *Oil Companies and Governments* (London: Faber and Faber).

[34] *Ibid.*, p. 165.

[35] *Ibid.*, p. 345.

> advantage of that competition, or sticking together to try to stabilize and then push up prices for every producer. Which they choose . . . will depend . . . mainly on the chances of making a producers' cartel, or maybe a commodity agreement bringing in consumers, really stick. Not all the consuming countries in the world, certainly, are very anxious to see oil prices come down further. But whether that will make them more amenable to price-fixing by producers alone is highly debatable.[36]

Mr. Hartshorn discusses seriously the possibilities (1) that European countries might join in some supranational policy toward oil imports and (2) that organizations of producing and consuming countries might enter into some sort of joint commodity agreement. In the end he discounts the prospects for either, as he does also for a producers' cartel, partly on the grounds of the diversity of interests, but also because they would impair the logistic functions of the international majors which for the calculable future will continue to be needed. His sense of the importance of these functions is perhaps his most strongly held conclusion and supports his view that, though governments will be increasingly "breathing down their necks," the companies are unlikely soon to be displaced from their primary role. He says on his last page:

> . . . there still seems to me no prospect of any alternative structure formed out of overlapping nationalism that offers a hope of handling this international job of logistics and of deploying investment with anything like the same, reasonably impartial degree of efficiency.[37]

Mr. Hartshorn does not raise as the subject for serious discussion the question—that lies so near the center of Professor Penrose's analysis—whether there is a likelihood that the exactions of the producing countries will ultimately force the major companies to give up their concessionaire status. We get the impression, however, that, in his view, in spite of nationalistic political pressures, in the end common economic prudence will prevent the countries from squeezing out the companies that market their oil, assure them of large revenues, and contribute to the stabilization of prices.

He does nevertheless point out the possibility of some such outcome as that which Professor Penrose examined when he says:

> Responsible government partnership in local producing companies in which the other partners are internationally integrated companies may put considerable strain on the links of integration inside companies producing oil in a large number of different places. . . . it could be argued that responsible government participation in producing companies could only become effective if the internationally integrated companies were to "divorce," in effect if not in form, their logistic function of matching demand from their central choice of supply, and in effect to put their requirements out to tender from their own (and other?) producing companies. . . .
>
> . . . The integrated middlemen (ignoring for a moment all the problems of disentanglement) would retain the countervailing power of going elsewhere for their oil. . . .
>
> . . . These governments have to consider whether they would do better, as at present, out of revenues tied to artificially high posted prices for crude oil than by taking greater commercial responsibility—and commercial risks. . . .[38]

He does not suggest a serious possibility that countries will adopt the alternative route, though elsewhere he notes that nationalistic sentiments at times override economic interests. ". . . the deep desire of these countries to take a fuller responsibility for the development of their national oil resources . . . remains stronger than any strictly economic calculation of the easiest way to get the highest revenue with the least effort."[39]

In the end, Mr. Hartshorn projects a larger role for national oil companies, a larger element of government partnership with established companies and newcomers, and a large place for "national preference" companies of consuming countries. He expects greater diversity in the form of company-country relationships from country to country. And all bargaining patterns will for the calculable future be in a state of flux. Only a "rash man" would attempt to say more.[40] But, he judges, the oil companies will still be there, because the countries cannot dispense with their services without severe damage to their own economic interest, assuming countries recognize their own long-run economic interests as passive tax collectors—but will they?

Some "Inside" Views

In the preceding section we reviewed some ideas of a few Western observers concerning trends of future relationships between host countries and major concessionaires. In the present section we turn to the thoughts of some persons representing the interests of the producing countries and, in one case, of concessionaires.

Whoever is thinking about them, certain fundamental factors in the situation are the same for all: the easily expandable low-cost supplies, the new elements of competition, the efforts of the countries to enlarge their revenues, the desire of the countries to achieve a larger degree of effective control over the exploitation of the resources, the role of the companies in international markets, and the efforts of companies to protect their own interests. But these factors are seen from different viewpoints and their significance for the future trend of events may be interpreted differently. The representatives of countries are engaged in attempting to formulate programs of action, both programs of collective action among countries and programs for individual countries. Moreover, each country

---

[36] *Ibid.*, p. 364.
[37] *Ibid.*, p. 388.
[38] *Ibid.*, pp. 354–356.
[39] *Ibid.*, p. 361.
[40] *Ibid.*, p. 363.

has its own history of relations with companies and its own problems of adjustment of these relations. There is no uniformity of thought concerning the proper direction of collective action and no uniformity of thought within individual countries concerning how best to resolve their separate problems. The aspirations of countries can be stated in fairly general terms; the views on the means of bringing them closer to fruition cannot. The most that we can do is to expose to view some lines of thought that are being currently expressed.

1. *The Views of Mr. Lutfi*

In a recent book,[41] Mr. Ashraf Lutfi, an official of the Kuwait government and a former Secretary-General of OPEC, has applied his mind to one central question: How the oil-producing countries can effectively control the international market for oil in their own interest. His conclusion is that the only answer is Joint Production Programming by the OPEC countries under which total production and the production of each country would be centrally determined—in other words, a tightly organized producers' cartel. In this conclusion he follows the recommendation of the OPEC Economic Commission in 1966.

> . . . its main conclusion regarding the vital problem of establishing some sort of order in the world oil market centered on the one and only course left open to the OPEC member countries—namely a joint production program requiring the unreserved and unqualified support of all members. According to the Commission, there was no other remedy. . .; if conceived, planned and launched effectively the program would place the initiative in OPEC's hands, would create market conditions such as to enable OPEC to phase the achievement of its objectives in the preferred order of priorities, and would at the same time help the oil companies themselves to operate and grow in healthy conditions.[42]

By piecing together points made at various places in his book, one can build up the following list of objectives that Mr. Lutfi has in mind:

(1) a restoration of posted prices to the level originally sought by OPEC;
(2) the elimination of discounts from posted prices;
(3) the gradual phasing of national oil companies into participation as shareholders in concession companies up to 50 percent;
(4) satisfying the aspiration that the national oil companies should attain a large place in the international market;
(5) promoting the expansion of refining in the source countries;
(6) providing funds to support the general economic development of the countries;
(7) protecting the interests of both newcomer and established companies, and tying their interests more firmly to the host countries;
(8) countering the efforts of consuming countries to reduce the price of oil.

Failing the cartel solution, Mr. Lutfi entertains a considerable pessimism concerning the prospects for all concerned—countries, national oil companies, and new and established concessionaires. His emphasis is upon forces tending to erode prices.

Under present conditions, the national oil companies have no ability to establish a firm and substantial position in the market. To force their way in, they would have to sell at cut-rate prices, thus adding further deterioration. This could be profitable to them on the terms at which they acquired royalty or partnership oil, but would run counter to the interest of their governments in firmer markets. "Above all, they may prove a complete nuisance to OPEC, i.e., their own governments, when they find that in order to enter any market they have to undercut the price of the old-established marketers there."[43] To Mr. Lutfi this is a grave situation because these companies were organized, not simply for commercial reasons, but to fulfill an aspiration to enter the world market—an objective so strongly felt as to make "all other considerations, including possible additional financial benefits, appear to have been of secondary importance."[44] But to pursue the objective strongly under present circumstances would have damaging consequence. ". . . there is only one possible development that could bring about a reconciliation between the objectives of OPEC and its national oil companies, namely, the transformation of the oil market into a seller's market . . . brought about by *action* on the part of OPEC. . ." through joint production programming.[45]

The entrance of newcomers into the market has had an adverse effect upon prices. But their situation is precarious for another reason. Their agreements specify heavy financial obligations; their margin of profits will generally be lower than for the major companies. The majors do not favor "coexistence" with them and may attempt to undermine their market position.

There are, then, at work a set of forces which bode no good for the price structure.

> . . . the next decade will see the interplay of quite a few sets of different groups vying for a big slice of the cake on the international scene: the old, well-entrenched majors, the comparatively new group of independents who may have just succeeded in holding their own, the OPEC national oil companies and, of course, the local independents in the consuming countries.[46]

Reenforcing the competitive interplay of these groups, Mr. Lutfi finds, are the policies of consuming countries, individually and collectively, designed to lower crude oil prices. He says, for example:

[41] Ashraf Lutfi, *OPEC Oil* (Beirut: Middle East Research and Publishing Center, 1968).
[42] *Ibid.*, p. 54.
[43] *Ibid.*, p. 5.
[44] *Ibid.*, p. 2.
[45] *Ibid.*, p. 36.
[46] *Ibid.*, p. 14.

> ... The EEC [European Economic Community] has, of course, been active in creating such conditions as would ensure the continued weakening of oil prices. . . .
>
> One must admit that the efforts of such consumer groups, together with the hitherto passive attitude of OPEC in not having taken any measures to eliminate the causes of price weakness in the market, have played a major part in steering oil prices on a downward trend.[47]

(To insert a comment, one notes how this view contrasts with the view of Professor Adelman and Mr. Hartshorn that the policies of Western governments have served to bolster the price structure. Yet on the basic analysis, that either there will be prorationing or price erosion, there seems to be agreement.) Mr. Lutfi thinks that "the apparent conflict of interest as regards prices" between consumer and producer countries "invites discussion"; but, since "present market conditions place the exporters at a disadvantage," the dialogue should be postponed until this situation has been rectified—that is to say, until cartelized programming has been put into effect.[48]

In the body of his book Mr. Lutfi places the whole weight of his argument against any early attempt at nationalization of the industry. Nationalization by an individual country would be "suicidal." Even if all OPEC countries were to nationalize simultaneously, they would damage their situation.

> ... each country would instantly be forced [faced?] with the problem of disposing of all its production in an open market where all facilities, from shipping to refining and marketing outlets, were owned by foreigners and where potential supply would be well in excess of demand. . . . the whole price structure would collapse. . . .[49]

The disastrous consequences might be avoided, or mitigated, if a joint production program had been fully worked out in advance. But given such a program, better results could be achieved through partnership arrangements with the companies, at the same time building up the role of national companies.

In a postscript to the book written after the Arab-Israeli war of 1967, Mr. Lutfi takes another look at nationalization, on the consideration that some future upheaval might force the issue. He states three prerequisites: (1) that the operation should include all the principal Arab producing countries, (2) that a joint marketing organization be set up to avoid competitive selling, and (3) that the countries should have built up a tanker fleet capable of moving a large fraction of the oil. In a later article,[50] he restates his argument with some elaboration; and, since "the key to the success of any act of nationalization lies in the solution of the delivery problem," he proposes that steps be taken to build up a tanker fleet.

Mr. Lutfi confines his general analysis to a statement of the benefits to be achieved through a tightly organized producers' cartel. He does not undertake any analysis of the possible future relations between countries and companies if such organization should turn out not to be feasible, nor of the means by which the countries could best safeguard their position in the absence of collective control of the market.

2. *The Views of Mr. Yamani*

During recent years Mr. Ahmad Zaki Yamani, the Oil Minister of Saudi Arabia, has been prominent in two connections: (1) as spokesman for the idea of participation of governments in the major concessionary companies (see Chapter 10, pp. 136ff. above); and (2) as the Acting Secretary General, in the role of spokesman for the newly formed Organization of Arab Petroleum Exporting Countries (see Chapter 10, pp. 139ff. above). In these connections he has made a number of statements bearing upon the lines of thought of the "outside" observers reported in an earlier section.

While Mr. Yamani is firm in his anticipation of partnership for countries in the major concessions and in general support of the OPEC Principles of June 1968, he is notably moderate in his statements concerning the gradual approach of countries to a larger role. He is a forthright opponent of nationalization which, he says, is nothing but "an empty slogan." ". . . nationalization would be an impossible move for the oil producing countries; sooner or later it would result in suicide. . . ."[51] As reported by the *Middle East Economic Survey,* "Mr. Yamani categorically rejected current suggestions that the oil companies should disengage from their producing concessions by means of a sort of voluntary nationalization and be transformed into mere offtakers of oil."[52] Further to this point, he says:

> An idea has emanated from the West to the effect that the host governments should nationalize the oil companies operating in their territories and conclude agreements with them ... with a view to giving the companies the legal status of offtakers of oil rather than concession holders. Some major companies are covertly working to bring about such a development which would entail fundamental changes in the oil industry affecting the nature of relations between the companies and the producers on the one hand, and between the companies and the consumers on the other hand, as well as between the producers and the consumers.
>
> The Government of the Kingdom of Saudi Arabia is opposed to this trend on several grounds, the most important of which are its aversion to the principle of nationalization per se and its awareness of the great harm that would befall the producing countries as a consequence of such change in the existing framework of relations. . . .[53]

---

[47] *Ibid.,* p. 21.

[48] *Ibid.,* p. 22.

[49] *Ibid.,* pp. 43–44.

[50] "Nationalization of Oil: The Problems Involved," *Middle East Economic Survey, Supplement,* November 29, 1968.

[51] *Middle East Economic Survey,* November 15, 1968, p. 12.

[52] *Middle East Economic Survey,* November 29, 1968, p. 5.

[53] *Ibid.,* p. 6.

In another connection, commenting upon the OPEC Resolution that governments should, "as far as feasible," develop their oil resources directly, Mr. Yamani specifies the nature of the harm:

> I do not believe that it would be possible for the oil producing countries to develop this resource directly without the assistance of the international oil companies for a long time to come and when they have developed markets for their production. If the oil producing countries develop their own oil resources and then attempt to sell their crude without first securing the necessary marketing facilities, this would result in a drastic drop in oil prices and constitute a calamity for the producing countries which presently depend to a very large extent on income derived from the oil companies. This notwithstanding, direct exploitation is an aim towards which all of us should strive. In Saudi Arabia we have a well-studied plan which will certainly lead towards the attainment of this objective. I should point out, however, that the fulfillment of this aim would not be at odds with the existing oil agreements.[54]

The danger that he defines, and is at pains to warn against, is exactly the one that Professor Adelman and Professor Penrose foresee and appear to regard as a not unlikely outcome. Again there is agreement between the producer government and the academic view of what is happening, though not on what should be done about it. They suggest that the pressure for participation and for enlarged revenues will apply a squeeze which will drive the major companies to withdraw from their concessionary status. Mr. Yamani's approach prescribes, but leaves unadumbrated, a country policy specifically designed to prevent this result. Acknowledging the strength of company resistance to the participation idea, he thinks "the day will come when the companies will share our view that participation between them and ourselves is the only way to establish a stable oil industry and to protect both parties against the dangers of the new trends that have manifested themselves in the consumer countries."[55] The "alliance" of company and country is a necessary condition to stave off the threatening instability of the market, damaging to both.

In the matter of market stability, Mr. Yamani also assigns great importance to the newly formed OAPEC. Of it, he says, ". . . there is no doubt that OAPEC has become the mainspring of stability in the Middle East oil industry."[56] As we have seen, an operational program for this agency is in the formative stages; but apart from committing the three member countries to certain joint ventures, little has been disclosed concerning its modus operandi. However, in an interview Mr. Yamani expressed the desire for OAPEC to enter into arrangements with consumer countries to assure security of supplies. He said:

> . . . there is an area of mutual interest between ourselves and the consumers, and . . . the apparent conflicts of interest could be overcome if only we could arrive at some arrangement between the two sides on the basis of a package deal. The main objectives of energy policy in the consuming countries are twofold: firstly security of supplies, and secondly cheap sources of energy. For our part we, as producers, are mainly concerned to secure two things: firstly assured access to growing markets, and secondly fair and stable prices for our crude oil exports.
>
> . . . .
>
> . . . If in one way or another the consuming countries could assure themselves, by means of cooperation with OAPEC, of a secure and permanent supply of oil, they would not then need to put up all that unnecessary investment for new facilities and diversification. . . . At the same time such an arrangement could also enable us, the oil producing countries, to attain our objectives. . . .
>
> . . . we would prefer to deal with blocs or groups of consuming countries if possible. But we have no objection to starting with deals with individual countries. . . .[57]

Taken in conjunction with the idea of a close alliance of interest between companies and countries, these statements suggest some sort of quasi-cartel, at least to the extent of trying to dampen competition for markets among the large producing countries of the Arab group. However, Mr. Yamani discounts grandiose concepts of collective programming. His views are those of a gradualist who applies his mind to step-by-step stages of progress toward producer country objectives without breaking down the still essential role of the major concessionaire companies. There seems to be in his statements an undertone of warning that, unless countries are brought into a close participating relationship with major concessionaires, the explosive Arab-nationalist political forces might create a rupture to the great disadvantage of both parties.

3. *The Views of Mr. Parra*

On various occasions, Mr. Francisco R. Parra has written on the activities of OPEC with which he was associated for a number of years as Economic Adviser and Secretary General. In a 1969 paper,[58] he presented an interpretive analysis of the Declaratory Statement of Petroleum Policy adopted at the OPEC Conference in June 1968 (see Chapter 10, pp. 124ff., above). In the course of his analysis he outlined some general views bearing upon the course of future relations between host governments and the major concessionaire companies.

Having cited the first principle of the Declaratory Statement (that "Member governments shall endeavour, as far as feasible, to explore for and develop their hydrocarbon resources directly"), Mr. Parra went on to say:

[54] *Middle East Economic Survey,* August 2, 1968, p. 9.

[55] *Middle East Economic Survey,* January 17, 1969, p. 1.

[56] *Ibid.,* p. 2.

[57] *Middle East Economic Survey,* November 22, 1968, p. 6.

[58] "OPEC: Present and Future Role," presented at the American University of Beirut and reported in the *Middle East Economic Survey, Supplement,* May 16, 1969, pp. 2–3.

. . . the development, directly by the State itself, is seen as the ultimate goal, because it would ensure complete control over an industry essential to the national economy and because it would yield the State the full profit from the operation. But this is postulated only as an ideal, psychologically very important in providing the motive force behind the general drift of things towards an increasing share by the government in decisions previously reserved to the oil companies. It is not however an ideal likely to be fulfilled for many years. The companies, and more particularly the majors, are unlikely to find themselves out of a job at the producing end, not because they are a useful "buffer" between producing and consuming country, nor because they possess the technical skills and the capital necessary to maintain a going operation, nor because they can do things more efficiently, but quite simply because they have access to relatively plentiful supplies of risk capital for exploration and, more important, a profound hold on the markets. At the producing end, the companies' position may be slowly stripped down to its essentials, but it is unlikely to be eliminated. . . .

From other responsible sources in the oil-producing countries—Mr. Atiqi, Kuwait Minister of Finance and Oil, Mr. Pachachi, oil adviser to Abu Dhabi, among others—statements have come in general agreement with one of Mr. Yamani's and Mr. Parra's central points: namely, that the countries have no early intention of dispensing with the concessionaire status of the major oil companies. All agree that country participation in the companies must come, along with other revisions in concession terms; but they contemplate the continuance of working agreements with the companies.

When outside observers and analysts foresee a different outcome, it must be because they think it more likely that countries will force results which are no part of their intention: first, by making demands beyond the limits of company tolerance, and second, by compounding the intolerability through competitive behavior that will undermine prices.

4. *Another Point of View*

Not all the views expressed by well-informed persons in the producing countries are couched in such moderate terms as those of Mr. Yamani. One who adopts a more stridently nationalistic tone is Shaikh 'Abd Allah al-Tariki, a former Saudi Minister and now an oil consultant in Beirut. From reports in the *Middle East Economic Survey,* one can piece together some of his views, either as directly stated or as appearing in the publication of his firm, the *Arab Oil and Gas Journal.*

In an interview reported by the *Middle East Economic Survey* of June 28, 1968,[59] Mr. Tariki's views, referring specifically to Iraq, were given as follows:

. . . Shaikh 'Abd Allah al-Tariki expressed the opinion that, in effect, there had been a virtual nationalization of Iraq's oil resources.

The Iraqi people, he explained, had managed to wrest from the companies 99.6 percent of the area which they had controlled but left unexploited, considering it a reserve on which they could draw after they had exhausted the oil resources of neighboring countries. There is no need to fear that, with such a small portion of Iraqi territory in the company's hands, oil output from Iraq will be adversely affected. . . . IPC production can therefore be confidently expected to continue, and this will give scope to INOC to expand its operations in the south and in a few years Iraq's oil income will increase by not less than 25-30 percent.

It would be the aim of any independent people . . . to nationalize an oil industry owned by foreign companies under outmoded agreements, but nationalization must be preceded by experience in managing such an industry, and this can be acquired through the Iraq National Oil Company (INOC).

In an article appearing in *Arab Oil and Gas Journal,* the following views were expressed:[60]

[The *Journal*] dismissed the call by Saudi Arabia and Kuwait for state participation in the established concessions in their territory as an impractical scheme incapable of implementation "because the present concessions guarantee for the oil companies an independent existence from the economies of the countries in which they operate."

. . . only nationalization of the operating oil companies will free these commercial enterprises from the control of their parent companies, and translate into reality the aspirations of the oil ministers of Saudi Arabia and Kuwait. But if this is not feasible, . . . the host governments can make use of their royalty crude to enter the international market and gain all the practical experience they are seeking. The journal recommends further that the independent entry of the host governments into the market should be accompanied by an expansion of the refining and petrochemical industries in the centers of production because of the benefits this would bring to the local economies of their respective countries.

In conclusion, the journal contrasts these benefits with "the enormous economic burden which the host governments would have to shoulder if they were to acquire an equity interest in the transport, refining and marketing installations of the parent firms of their operating companies, which would not yield a return on invested capital of more than 10 percent."

In a later article in the *Arab Oil and Gas Journal,*[61] it is suggested that:

. . . the only sound way to fulfill the objectives of the principle of participation . . . would be to nationalize the existing old-style concessions and form new companies to operate these concessions and undertake the exploration of areas where no oil had been found. . . .

At the same time, the journal opposes any participa-

[59] Pp. 10–11.

[60] As reported by the *Middle East Economic Survey,* November 1, 1968, p. 7.

[61] As reported in the *Middle East Economic Survey,* December 13, 1968, pp. 10–11.

tion in downstream operations outside the producing countries. . . .

"The best advice we can give the advocates of participation," the journal continues, "is to urge them to emulate the model provided by Algeria's Sonatrach in its recent deal with Getty, under which the latter ceded 51 percent of its assets—both equipment and reserves—to Sonatrach and the two parties agreed to develop jointly both the old acreage and the new areas awarded to them by the government."

We are unable to interpret from the statements of Mr. Tariki and his *Journal* exactly how the sequential process for gaining government control of oil operations would unfold. At least in the earlier statement of June 1968, reliance appeared to be placed on revenues from the concessionaire company while the national company was gaining experience. In the later statements, Mr. Tariki appears to propose partial expropriation of the concessionary companies' interests and the organization of a new operating company in which the national company would have a controlling interest. What is far from clear is his view on how the operating company would expect to market its oil internationally on terms as rewarding to countries as the revenues they now receive. Or alternatively, if nationalization of existing concessions is not feasible, how the national company could enter the international market without the disruptive price effects which others have foreseen.

5. *A Company View*

It is not customary for companies to make public their views concerning country-company relations except in a public relations sense; but in two 1968 papers[62] Mr. George Glass, Head of Trade Relations for the Middle East in Shell International Petroleum Company, has touched upon some of the issues with which we are concerned. He pictures the role of the international companies as "the reconcilers of the objectives of producer and consumer countries" and states that "the machine set up by the international companies is so valuable that it is in the interests of both sides to ensure that it stays in business." Noting the rise of national oil companies and the entrance of new companies without well-developed "downstream" integration, he foresees that they will operate at a serious cost and marketing disadvantage. He also notes the rise of agreements of national companies of consumer countries with governments of producer countries; and he suggests that, "if such policies were to proliferate, the countries concerned would eventually find themselves competing with each other in trying to find outlets for their crude. And they might not be able to assess the point at which a situation could arise that would endanger the flexible system of the independent international companies upon which they all depend. Iraq, for example, would be unable to entertain such deals for a moment without the secure income provided by the international companies. . . ."[63]

At present, "the bilateral and government-to-government deals that have become increasingly fashionable in some countries draw only on marginal pockets of the industry's resources"; and "it is tacitly assumed by all concerned that the international companies will always be there, carrying on the bulk of the business and shouldering the major risks and costs."[64] But, in their increasing desire to be "masters in their own house," and by policies that weaken the market and place increasing financial burdens on the companies, the countries may topple the whole structure which provides their economic well-being. "The danger is that governments, by adopting policies that make perfectly good sense in isolation, could, if such policies proliferate, undermine the base provided by the international companies with unhappy consequences for the countries most immediately concerned."[65]

This point is pursued in Mr. Glass's second paper, where he says:[66]

> . . . a small number of very large and integrated competing organizations . . . constitute the basis upon which all other forms of international oil trading rest. All the new agreements now held up as examples were made in a world where this commercial fact constitutes the single most important feature of international trading in oil. If attempts are made, in pursuit of an elusive alternative, to chip away the base of the main structure on which the industry's continuing success clearly depends, the result may be the collapse of the entire edifice.

In support of his position he quotes the statement of Dr. Pachachi as follows:[67]

> . . . none of the Middle East governments, except perhaps Iraq at certain periods during the last ten years, is willing to dispense with the majors or to force them to withdraw by imposing conditions that would reduce the profitability of crude oil production to a point where it would no longer yield an adequate return on their investment. . . .

The immediate context of Mr. Glass's paper is the benefit that governments may hope to get under the new types of participating agreement or contract. What he suggests is that the possible benefits are very limited: (1) because the old types of concession offer equal opportunities of benefit in *local* producing operations, (2) because

[62] "The Changing Dimensions of Government-Company Relationships," *Middle East Economic Survey, Supplement,* May 17, 1968, and "Contract Versus Concession," *Middle East Economic Survey, Supplement,* November 15, 1968.

[63] *Middle East Economic Survey, Supplement,* May 17, 1968, p. 3.

[64] *Ibid.*, p. 6

[65] *Ibid.*

[66] *Middle East Economic Survey, Supplement,* November 15, 1968, p. 6.

[67] *Ibid.*

country participation in share ownership cannot give any control over the *international* operations of the great companies, and (3) because the new companies offering participation have a very small place in the international market.

Mr. Glass's statements are a muted and mildly stated warning to countries of the dangers of proceeding much further in the direction set by their recent policies relating to national companies, new types of agreement, and participation under old concessions. He presents a view of the role of the large international companies, thereby also defining the limitations upon effective action by the countries, as appraised by an official of one such company.

STATISTICAL APPENDIX

Projected World Energy Consumption and Trade Flows, 1975 and 1980, with Historical Background Data

STATISTICAL APPENDIX

Projected World Energy Consumption and Trade Flows, 1975 and 1980, with Historical Background Data

INTRODUCTION

Purpose

THIS statistical appendix contains historical data and rough projections of energy consumption, by major fuel category, for various regions of the world. For liquid fuels and for solid fuels—the energy sources which figure most prominently in international trade—the past and future export capacity or import dependence of the different regions have also been estimated. And world flows in liquid and solid fuels are depicted in the form of matrix tables which indicate the trade of each major world region with every other region. The twelve basic tables containing these estimates are preceded by the present text discussion dealing with substantive, methodological, and conceptual aspects of the statistical presentation.

A major focus of this study is, of course, the two-way relationship between the leading countries of the world dependent upon oil imports (those in Western Europe, treated as a group, and Japan), on the one hand, and their principal oil suppliers (countries in the Middle East and North Africa), on the other. Similarly, the primary purpose of this statistical appendix is to permit drawing out the magnitude of oil flows as they involve these particular areas. In order to quantify these prospective flows, it is desirable to construct a framework sufficiently broad to identify those factors to which European and Japanese oil imports and their geographical sources are particularly sensitive. Thus, although the estimates that are finally required pertain to the likely increase in petroleum imports, their derivation requires underlying estimates of the expansion of energy consumption generally and of the shifting role of alternative fuels in meeting these requirements. In turn, the growth of total energy consumption requires attention to economic growth as a whole in the areas of particular concern.

Nor is it sufficient to consider only trends in those consuming areas highlighted by this study. For developments elsewhere in the world have a bearing on the outlook for the areas under study. For example, the comparative ability of North America to meet its own oil consumption needs will influence the conditions under which supplies from exporting countries may be available to Western Europe, Japan, and other parts of the world; U. S. coal can be—for some range of uses—substituted for imported oil in Western Europe and Japan; or Australia's prospective shift from a position of oil import-dependence to one of self-sufficiency may release new supplies of Middle Eastern oil into world trade channels, while, conversely, the European Communist area's shrinking export balance may absorb flows that would otherwise seek markets in Western Europe.

In short, although the focus of the study remains on oil and on a limited group of world regions, the wider quantitative setting of this statistical appendix provides what we regard as a necessary underpinning for ourselves and for the critical reader. It should permit the evaluation of a projected figure in terms of underyling elements on which the projection depends. As one example, evaluation of the reasonableness of projected Middle Eastern oil flows to Western Europe could be made against the background of Western Europe's projected economic growth rate; the interrelationship between its economic growth rate and its growth of energy demand; and the degree of its prospective relative reliance on indigenous natural gas, coal, and nuclear energy, and energy supplies from other geographic areas to satisfy that demand.

Brief Summary of the Approach

Procedurally, the projections begin with an analysis of energy-GNP "elasticities" (which relate energy consumption growth rates to GNP growth rates) for various world regions. Assumed future trends in gross national product and in these elasticities yield projected growth rates and future levels of energy consumption around the world (Tables I through III). The resulting projections of total energy demand are apportioned among the major energy sources in Table IV. The extent to which regional energy requirements are likely to be fulfilled from indigenous or from foreign sources is shown for oil in Table V and for coal in Table IX. World flows in oil, by region of origin and destination, are presented in Tables VI through VIII; while similar calculations for coal appear in Tables X through XII. Preceding these twelve numbered appendix tables, a tabular exhibit, along with appended notes, clarifies different definitions of energy consumption. (See pp. 168–170 below.)

The following section takes up the procedural and substantive underpinnings of the projections in greater detail.

Making the Projections: Problems, Methods, Sources

The projections necessarily reflect judgments, as of 1969, concerning a reasonable basis upon which to extrapolate future lines of development. The presentation of only one set of figures, rather than the display of alternatives, is a matter of practicality. If one were to join alternative GNP and energy growth rates, alternative fuel shares, alternative import-consumption ratios, and alternative geographic supply sources, the proliferation of possible alternatives would become self-defeating for purposes of analysis.

Projections of energy consumption *totals* were done in an aggregative fashion so that the detailed underlying factors are only implicitly represented. It would have been preferable, were the facts available, to "build up" the projected energy consumption total in terms of specific energy-using sectors of the economy and their demands, such as electricity or automotive fuel, instead of simply relating total energy consumption to total GNP. As for the fuel shares in the total, it would have been desirable, again if the facts were available, to specify the relevant price or policy assumptions which would be consistent with the projected demand for particular fuels. In some such cases, we were aided in making the projections by consulting other studies in which some of these underlying factors are explicitly considered. In numerous cases, however, we have had to content ourselves with what are essentially trend extrapolations in which such elements are implicit.

By and large, the projections draw heavily upon the trends implied in the underlying statistical time series. However, they also allow for developments which are not yet indicated (or barely so) in the historical data presented here. Announcements of planned nuclear power plant construction, of North Sea gas developments, and Australian oil discoveries are examples of such factors that have been taken into account. One factor now emerging which, in particular, could bring about a fundamental change—oil from Northern Alaska and the Canadian Arctic—was purposely omitted from the projections. It seemed far better at this stage to give separate examination of this possibility (as we do principally in Chapter 7) in order to show explicitly to what extent a major new development could affect the trends that might otherwise be expected to unfold.

Any projection involves some degree of circular reasoning. In our projections, for example, an assumed continuation in the shift from West European consumption of coal to imported oil necessarily presupposes the corresponding availability of oil supplies from the producing countries and within some acceptable price range. On this count, we are reasonably confident, except for the problem of supply interruptions. It should be made clear that, though our projections "disclose" balance between oil import demands and oil export supplies around the world, our primary emphasis was put on the demand side. Especially in the allocation of oil exports among particular producing regions and countries (Tables VI through VIII), it was not feasible to do justice to all the factors (e.g., future trends in new discoveries or changing relative costs among regions or oil company practices regarding offtake) bearing on the evolving geographic pattern of oil exports. Again, our emphasis was on depicting *one* scheme of world flows, but one whose chance of materializing in its broad outlines is assigned a reasonable probability, within the limitations cited. As to specific details, the departure of actual events from the projections could be quite sharp.

In the paragraphs that follow, we review the methods and assumptions which went into the projections in the appended tables. Other materials and projection studies which were consulted will be referred to. However, we will not cite sources for historical data; these appear in detailed footnotes to the tables. Nor will we describe simply mechanical computations ("Table II.A equals coal equivalent figures divided by 1.5") which, likewise, are handled in footnotes to the tables. Most of the discussion centers on those aspects and areas of the world having greatest relevance to this study.

Tables I to III[1]

Projected growth in total energy requirements is related to future growth of GNP by means of energy-GNP "elasticities." GNP growth-rate projections contained in other publications were reviewed and, where appropriate, adapted—in some cases in modified form—to our study.

Of course, international GNP comparisons (even if only in terms of growth rates) are somewhat conjectural undertakings—and the present effort should be viewed as a fairly crude exercise sufficient to provide some rough basis for the 1975 and 1980 energy estimates.

For the United States, the projected GNP growth rate conforms to that in *U. S. Economic Growth to 1975: Potentials and Problems,*[2] though somewhat above that in *An Energy Model for the United States.*[3] The 4.4 percent which we have adopted appears reasonable; this rate may be slightly above a "consensus" figure, but erring on the side of overestimating rather than underestimating energy and oil demand might be defended in the context of this study.

The projected growth in Canadian output conforms closely to the figure postulated to 1975 in a 1967 report of the Economic Council of Canada.[4]

Western Europe's and Japan's GNP growth rates correspond essentially to those assumed in OECD, *Energy Policy: Problems and Objectives* (1966); the Japanese rate also averages out to the figure postulated in the 1967 *Report on Energy Policy* by the Japan Advisory Committee for Energy. (Western Europe's growth through 1968 appears to be somewhat below the implied long-term trend, while that for Japan is a bit above trend. In neither case is the departure yet sufficiently large to invalidate the OECD projection.)

The projected Soviet GNP rate falls midway within a

---

[1] A tabular exhibit, illustrating different concepts and preceding the numbered Tables I through XII, appears, along with a discussion, on pp. 168–170 below.

[2] U. S. Congress, Joint Economic Committee (Washington: Government Printing Office, 1966).

[3] The full title of this publication is *An Energy Model for the United States, Featuring Energy Balances for the Years 1947 to 1965 and Projections and Forecasts to the Years 1980 and 2000,* by W.E. Morrison and C.L. Readling, (U. S. Department of the Interior, Bureau of Mines Information Circular 8384, [Washington: Government Printing Office, July 1968]).

[4] *The Canadian Economy from the 1960's to the 1970's*, Fourth Annual Review (Ottawa: Queen's Printer, 1967).

range indicated in a recent U. S. Government-issued review, *Soviet Economic Performance: 1966–1967.*[5]

For most other regions, we have had to rely on historical trends, collateral analytical material, and simple judgment in making the GNP projections. Here we were partially guided by "medium" United Nations population projections (*World Population Prospects,* Population Studies No. 41 [New York, 1966], Tables A3.2 and A3.8) and assumed growth in per capita GNP as the basis for projecting future trends. (In the absence of historical GNP data for the two African subregions shown in Table I, trends in total population and in commerical fuel consumed per capita were used to project energy consumption.)

Energy-GNP elasticities change over time in response to improvements in the efficiency of energy use and to changes in the "mix" of the economy's final demand and of its energy inputs. We have for the most part allowed for a continuation of the moderately downward historical trends in elasticities–though at a dampened pace, especially (as in the case of the United States) where the ratio has already reached a fairly low level. This assumption signifies growth in total energy consumption near or below GNP growth for most of the highly industrialized countries; while an elasticity still substantially in excess of 1.0 for the less advanced regions reflects, among other things, accelerated mechanization, a disproportionately fast rise of energy-intensive industries such as manufacturing and transport (conversely, a relative decline of agriculture) and patterns of consumption associated with increased uses of electricity. In addition, numerous developing areas are still undergoing a transition from partial dependence on primitive energy forms (e.g., firewood) to predominant reliance upon the commercial energy sources covered in this study, which would cause relatively high (commercial) energy-GNP elasticities for such areas.

The resulting energy growth projections are compared in Table III with estimates of the OECD and European Coal and Steel Community, in studies cited in the footnote to the table. Our estimates are seen to match quite closely those of the OECD, while tending to exceed in moderate degree the projections of the ECSC.

Table IV

This table estimates total projected use as among the major energy sources. It postulates a continued slippage in the share of solid fuels (particularly in the areas where coal has continued to figure prominently) generally offset by rising shares for the other energy sources except hydro, whose worldwide proportion remains essentially unchanged. The specifics vary among different places. Although some leveling-off is likely within a decade or so, the worldwide liquid fuels share of total energy consumption is expected to continue rising. The Communist Eastern European countries, Japan, and–to a somewhat lesser extent–Western Europe are all projected as areas with marked increases in liquid fuels portions of total energy use. Natural gas shows a significantly rising share in Western Europe and Oceania, while nuclear energy, starting as it does from a negligible base in most places, is expected to increase markedly for a number of advanced regions. It should be emphasized that the realism of the nuclear power plans of most countries is hard to assess.[6]

The projection of fuel shares reflects an eclectic approach, in which the extrapolation of historical trends was modified on the basis of our own judgment as well as the analysis contained in numerous studies which were consulted. We discuss below some of the specific sources which were available, and which we reviewed–and used in varying degree–to make projections of the different fuel categories for the indicated world areas.

United States. W. E. Morrison and C. L. Readling, *An Energy Model for the United States, op. cit.* The projected fuel shares in this document were modified by the (higher) nuclear projections since developed by the U.S. Atomic Energy Commission, which suggest a 1980 installed nuclear capacity of 150,000 MW. (See, e.g., "Looking Ahead in Nuclear Power," speech by Dr. Glenn T. Seaborg, AEC Press Release, dated June 9, 1969.)

Canada. First National City Bank, *Energy Memo* (New York, April 1968). The nuclear portion was modified (upward) on the basis of International Atomic Energy Agency (IAEA), *Power and Research Reactors in Member States* (Vienna, 1968). The latter publication provides a forecast of installed nuclear electric generating capacity through the mid-1970's. The projected level of oil consumption (100 million tons in 1980) agrees with an estimate of the Alberta Oil and Gas Conservation Board (see *Oil and Gas Journal*, December 23, 1968, p. 24).

Western Europe. Substantial reliance was put on the Organisation for Economic Co-operation and Development (OECD) report, *Energy Policy: Problems and Objectives* (Paris, 1966). (The OECD projections, which only estimate imports of *total* energy supplies, had to be supplemented by judgmental estimates of imports of specific fuels.) In the projected solid fuels share, we were guided by the fact that Europe's 1968 coal consumption already fell below the lower limit of OECD's forecast for 1970. In the case of natural gas, recent developmental activity in the gas fields along with marketing plans suggest a considerably higher future level of natural gas consumption than was envisaged in the 1966 OECD report. Specifically an analysis by the OECD released in 1969 (*Impact of Natural Gas on the Consumption of Energy in the OECD European Member Countries* [Paris, 1969]) suggests ranges of natural gas consumption of 115–160 million tons oil equivalent in 1975 and 140–195 million tons oil equivalent in 1980. We have adopted the midpoint in these ranges. The nuclear share was derived from more recent estimates contained in European Nuclear Energy Agency, *Uranium–Production and Short Term Demand* (Paris, 1969), p. 21. There the 1975 and 1980 projections indicate ranges of installed

[5] U. S. Congress, Joint Economic Committee (Washington: Government Printing Office, 1968).

[6] Our indicated total of world nuclear energy consumption (92 million coal equivalent tons in 1975 and 235 million tons in 1980) compares with estimates of the International Atomic Energy Agency (Press Release dated December 18, 1968) implying coal-equivalents of 96 million and 263 million tons, respectively.

nuclear capacity of 29,000–43,000 and 88,000–118,000 MW, respectively. Partly in order not to understate the extent of possible future oil demands, we have leaned to the lower end of these ranges, assuming 30,000 MW in 1975 and 90,000 in 1980. (A plant factor of 80 percent was used to equate these figures to nuclear generation.)

Japan. OECD, *Energy Policy, op. cit.* The solid fuels share indicated by OECD was reduced in order to allow for the planned cutback in indigenous production under a more recently promulgated government coal mining financial program. (See Institute of Energy Economics, *Energy in Japan*, Quarterly Report No. 7 [Tokyo, April 1969], p. 6.) The natural gas share was raised somewhat (particularly taking into account LNG imports from Alaska) on the basis of data in *Oil and Gas Journal,* January 13, 1969, p. 65. The nuclear portion was raised somewhat on the basis of Advisory Committee for Energy, *Report on Energy Policy* (Tokyo, February 1967). The latter document set a target of 6,000 MW of installed nuclear capacity for 1975 and a range of 30,000–40,000 MW for 1985. The 1975 target was accepted for our projections. However, viewed as a probability, rather than a target, the somewhat conservative figure of 15,000 MW has been adopted for 1980. (A plant factor of 80 percent was applied to equate this figure to nuclear generation and fossil fuel equivalents.)

Other Asia. United Nations, *Economic Bulletin for Asia and the Far East,* December 1965; and IAEA, *Power and Research Reactors in Member States, op. cit.*

Oceania. Australian energy shares to the mid-1970's are estimated in Department of National Development (Canberra), "Trends in the Supply and Demand of Energy in Australia" (paper presented at World Power Conference, Moscow, August 1968). These projections were extrapolated forward to 1980. Projections for New Zealand were taken from *Petroleum Press Service,* December 1968, pp. 462–464.

U.S.S.R. The projected shares are essentially judgmental, though some support is provided by (1) Soviet planned output in 1970 and 1980 shown in Robert W. Campbell, *The Economics of Soviet Oil and Gas* (Baltimore: Johns Hopkins Press, 1968), pp. 4–5; (2) by J. G. Polach, "Outlook for Energy in the Communist World" (unpublished paper, November 2, 1966); and (3) IAEA, *Power and Research Reactors in Member States, op. cit.*

Communist Eastern Europe. Some basis for the nuclear estimate is provided in J.G. Polach, *Nuclear Power in Eastern Europe* (Resources for the Future, Reprint No. 74 [Washington, D.C., June 1968]). A recent review of the area's trends in oil and gas consumption (*Oil and Gas Journal*, February 3, 1969, pp. 54–55) suggests that in another decade the combined energy share of these two fuels is likely to range between 35 and 40 percent of total energy consumption.

Communist Asia. There is little to vouch for the reliability of historical data, let alone information from which to make estimates of future trends. Much uncertainty centers on the reliability of statistics on coal—the major energy source—while knowledge about other fuels is virtually nonexistent. (For example, moderate amounts of natural gas are believed to be used in Mainland China, although little quantitative information on this score exists.) The projections should be viewed in this light.

Table V

Given the projections of liquid fuels consumption in Table IV, it becomes a relatively straightforward matter to estimate the likely trends in imports for the principal consuming areas, most of which are either wholly dependent (or almost so) on foreign oil, on the one hand, or largely self-sufficient on the other. The most important region falling in between these extremes is the United States, where it has been assumed (admittedly with some arbitrariness in view of uncertainty over the development of reserves and U.S. import policy; see the discussion in Chapter 2) that the level of imports will hold near import-consumption ratios of recent years. The projected constancy in the share of imports in U. S. liquid fuels consumption is assumed also in the U. S. Department of the Interior, *Energy Model of the United States,* cited earlier, as well as in the quantitative framework of the study by the U.S. Department of the Interior, *United States Petroleum Through 1980* (1968), discussed in detail in Chapter 2.

Among the more problematic areas—in the sense of breaking with past trends—are Oceania and Canada. In the case of Oceania, where an historic net import balance is assumed to drop to zero by 1975, the imminence of self-sufficiency, as a result of Australian discoveries, is widely accepted.[7] As for Canada, that country's falling import-consumption share is projected as continuing its rapid decline—an assumption consistent with the fact that the estimated 1980 level of consumption is no larger than *current* levels of productive capacity.[8] And this without regard to the Canadian Arctic.

Some considerable conjecture attaches also to the prospective net trade position of the Communist countries—taken as a group. All indications point, at most, to no more than an internal balancing of supply and demand within the group (which we have assumed). A variant of the same assumption of an overall net trade balance of zero would have the Soviet Union continuing some crude oil sales to the West, with the oil-deficient Eastern European countries having to make compensatory purchases on international markets. The calculations on interregional flows (Tables VI to VIII) ignore the latter course of events.[9]

---

[7] See, e.g., *Oil and Gas Journal*, October 21, 1968, p. 57, where it was estimated that Australia would supply almost three-fourths of its oil needs early in the 1970's. In spite of the assumption of a zero net trade balance by 1975, it is believed possible that exports of *light* Australian crudes may have to be balanced off against equivalent imports of *heavy* Middle Eastern crudes, since Australian refineries are oriented toward the latter. However, this possibility is not reflected in the projections of interregional flows in Tables VI to VIII.

[8] See *Oil and Gas Journal*, December 23, 1968, pp. 24–25, report on recent forecasts of the Alberta Oil and Gas Conservation Board.

[9] We also ignore—as a simplification and because of its statistically negligible importance—the likelihood of continued Soviet oil shipments to Cuba (the latter being treated in our tables as in the Latin American rather than the Communist grouping).

Tables VI to VIII

These tables depict the origin and destination of the international flow of crude oil and oil products. It is one thing to try to specify the future volume of liquid fuels consumption in different world regions, as in Tables IV and V. Any further attempt to pinpoint the quantities of oil that are likely to originate in particular regions of the world—let alone individual countries—obviously is very much more speculative. Even though sheer size of production and reserves, and the momentum achieved, suggest that certain regions and countries are bound to remain significant oil producers and exporters, a quantitative assessment of their relative importance depends on a host of unknown factors, including, among others: future production and delivered cost and price differences among different areas; exploration, discovery, and development, which will determine the future productive capacities of the various regions; and future policy developments, both governmental and private, particularly as they affect host government-oil company relationships. When things move as dramatically as they have in recent years (witness Libya's rapidly rising share in West European imports since the early 1960's), the problem can not be handled by the "assumed continuation of historical trends."

These difficulties notwithstanding, it was necessary for the purposes of this study to make some estimates—however fragile—of the quantities of oil exports that might originate in various regions of the world.[10]

The principal assumptions in these projections include: (1) The share of North African oil in West European imports—which has been rising sharply throughout the 1960's—will over the next decade remain essentially stable. In 1962, North Africa provided little more than 3 percent of Europe's crude oil imports. By the mid-1960's, the share had risen to over one-fifth; and by 1968, to over 30 percent. (Conversely, the Middle East's share in European imports declined from nearly three-fourths in the early 1960's to under one-half in 1968.) The pace setter in the North African surge has been Libya, which accounts for the predominant share of total North African oil exports (over two-thirds in 1968) and for a still higher portion of recent incremental growth. The way in which one views Libya's long-term oil prospects is consequently crucial to future expectations regarding North Africa as a whole. Having joined the list of oil producers only in 1961, the country has already become one of the front-ranking exporters, and at the end of 1969 was close to achieving an exportable output of about 3½ million barrels/day. Libya's reserves, production, and exports have shown annual increases in the 30–40 percent range during recent years. Clearly, however, one cannot assume a continuation of anywhere near such rates of advance. The size of *present* proved reserves (30 billion barrels in 1969) implies a reserve-production ratio of about 30 to 1, compared to an average Middle Eastern ratio more than twice as high. Underlying the projected growth of North African oil exports from some 90 million tons in 1965 to 290 million in 1975 and 370 million in 1980 (Table VI), we have assumed Libyan exports growing at a rate of about 6½ percent per annum during 1969–75[11] and a bit over 5 percent thereafter—roughly similar to the projected export growth in the 1970's of the Middle East. In the absence of further knowledge one cannot prudently assume that North African exports will continue to grow relative to the Middle East, given the latter's strong reserve position. It should be kept in mind that part of the reason for the very rapid annual increases in Libyan output since 1967 stemmed from the closing of the Suez Canal. Although the country's west-of-Suez proximity to Europe will remain an attraction (as will the low-sulfur quality of its crude oil), the increase in size of tankers will probably tend to make Libyan oil somewhat less competitive—other things being equal. But even with decelerated growth of North African exports, it has been implicitly assumed in these calculations that Libya's reserve figures are conservatively estimated and that additions to reserves will serve to sustain the output increases assumed, though they may partly be achieved at the cost of some fall in the reserve-output ratio.

(2) The "Other Africa" projections assume greatly expanded production capabilities in West Africa—largely in Nigeria, but also in Gabon and Cabinda.

(3) The Middle East's total growth in oil exports, having been outpaced by North Africa in recent years, is assumed to grow commensurately to that region in the 1970's, as noted above. Part of the stimulus for rising Middle Eastern output is expected to come from rapidly rising production in Abu Dhabi and other Trucial States, as well as Qatar, Bahrain, and Oman. This latter group of countries produced nearly 68 million tons in 1969—up by around 15 percent over 1968. (*Oil and Gas Journal*, December 29, 1969, p. 94.) All told, substantial relative expansion possibilities seem indicated for this group of countries.

(4) The projections allow for significant increases from Indonesia, whose agreement to favorable joint-venture terms and to an important modification of the basis for its tax computations through a downward revision of posted prices substantially adds to the competitive attractiveness of Indonesian oil in the years ahead. The low-sulfur Indonesian crude oil has attractions for the increasingly pollution-conscious Japanese market.

Tables IX to XII

These tables deal with consumption and trade flows in solid fuels. Western Europe's sharply increased imports from the United States, Japan's rapidly rising imports from Oceania (Australia) and her continuously increasing shipments from the United States and Canada, and the Communist area's

[10] In our projected matrix of world oil flows (Table VI), we have sought to include only *significant* movements. The fact that a particular projected cell is zero does not mean that we exclude the possibility of *any* quantities between the indicated regions. For example, some North African oil may flow to Japan, though—according to present indications—not in sizeable volumes.

[11] The base year in Tables VI to VIII is 1965 since it proved difficult to construct additional years' data in a form consistent with these tables. However, developments prior and subsequent to 1965 have been taken into account in making the projections. For example, Libya's exports rose from around 60 million tons in 1965 to an annual rate of nearly 175 million by the end of 1969.

leveling off in its shipments to the rest of the world are the major elements in these estimates. Underlying assumptions are that lessened subsidization will continue to rechannel West European coal demand (falling in any case) away from high-cost indigenous supplies; that Japan's high economic and energy growth rates, accompanied by indications of substantial cutbacks in domestic capacity, will require continuation of a reasonable level of solid fuels imports—particularly of coking coal for the steel industry; and that, taking into account the growing energy stringency in the East European Communist countries, the Communist area as a whole will substantially reduce its exports.

The main effort in making these projections was directed at coming up with some reasonable judgments about the future coal production-consumption (hence, net trade) picture in various areas. (Production estimates are implicit in the consumption and net trade figures of Table IX and are spelled out in Table XII.) The projected origin and destination of regional flows (depicted in Tables X and XI) follow largely historical patterns, although we did allow for modifications where available evidence so indicated.

The level of U.S. production estimated for 1980 amounts to around 710 million tons—about 60 million tons below the level foreseen in the U.S. Department of the Interior study, *An Energy Model for the United States,* cited earlier (p. 162). Our reduced coal figure is offset by the assumed higher level of nuclear output, as explained under the notes to Table IV. On the other hand, our assumed 1980 level of U.S. exports of about 80 million tons—which was arrived at primarily by looking at prospective overseas import demand—compares to Interior's projection of 70 million.

Assumptions regarding Western European coal are of considerable importance in any determination of future world energy flows. While European coal reserves are large enough to sustain levels of output far higher than anyone foresees, production costs are high (and likely to keep rising relative to those in the United States), and the long-term outlook is decisively dependent on policy actions.

In what are basically extensions of past trends through 1968, the falling absolute level of West European coal (bituminous coal and lignite) consumption (Table IV) is expected to be accompanied by a steadily increasing reliance on imports (Table IX)—with the share of the United States growing relative to that of Communist Eastern Europe (Tables X and XI). The resultant West European solid fuels output envisaged for 1980 comes to around 280 million tons (Table XII). While this figure is exceeded by the 1980 estimated production range of 300–435 million tons in estimates of the European Coal and Steel Community, and of 340–440 million tons in OECD projections,[12] it should be noted that, already in 1968, production fell below the OECD 1966 estimate of the *minimum* level of production in 1970. Moreover, as discussed in the notes under Table IV above, we have adopted higher natural gas projections than those indicated in the 1966 OECD report, and this to some extent occurs at the expense of coal.

For Japanese coal, as in the case of Western Europe, we have allowed for a substantially reduced level of consumption and a steadily rising import share (Table IX), and for a significant scaling down of domestic output—reaching production of around 35 million tons for the period 1975–80, as assumed under a recently promulgated coal mining reconstruction program (Table XII).[13]

The other aspect of Japan's coal picture which deserves mention is the increasing relative importance of Canada and Australia as suppliers. Contractual negotiations involving large long-term commitments with these countries have been reported throughout the past several years, the outlook being particularly favorable for Australia, from which imports are expected to exceed more-distant U.S. shipments to Japan by the mid-1970's.[14]

Notes on Definitions, Conversion Procedures, and Comparability with Other Studies Cited

The historical data in the tables have, for the most part, been cast within the framework of the United Nations, *World Energy Supplies*, Series J, whose definitions and geographical classifications have likewise been generally followed. Although, as noted below, UN data have some limitations, they do have the dual virtue of (1) global coverage and of (2) extending over a reasonably long time span, both of which features were important for purposes of this study. Moreover, it was possible partially to adjust for some of the UN data deficiencies.

In the UN system, energy consumption refers to "inland" consumption—i.e., it excludes bunker fuels absorbed by maritime vessels and foreign-bound aircraft. (Western European liquid fuels bunkers in 1965 approximated 35 million tons—an amount equal to about one-eighth of inland liquid fuels consumption.) As measured by the United Nations, energy consumption also excludes that portion of crude oil supplies diverted to nonenergy uses such as petrochemical feedstocks. (Nonenergy liquid fuels consumption in Western Europe in 1965 amounted to about 27 million tons—about 7 percent of liquid fuels consumption for energy purposes.) Both omissions are a drawback from the standpoint of this study, which focuses on *total* petroleum requirements regardless of ultimate disposition. However, it was possible to overcome these limitations. In the case of bunkers, liquid fuels consumption is redefined, in line (2) of Table V, to include an allowance for oil bunkers, and oil imports are related to this broader definition of consumption. And as for petroleum for nonenergy uses, that problem is minimized by the fact that, in this study,

[12] European Coal and Steel Community, *Review of the Long-Term Energy Outlook for the European Community* (Luxembourg, 1966), pp. 46–47; and Organisation for Economic Co-operation and Development, *Energy Policy: Problems and Objectives* (Paris, 1966), p. 177.

[13] See remarks under discussion on Table IV, p. 164 above.

[14] See *Mining Journal* (London), December 20, 1968, p. 485; and *Coal Age*, October 1968, pp. 59–60.

primary interest centers on Western Europe's and Japan's oil-*import* requirements; and the crude oil component (which is the predominant portion) of imports, as shown in Tables V through VIII, *does* include crude oil going into nonenergy uses (although excluded from consumption as measured by the United Nations).

The UN definition of "Western Europe" includes Yugoslavia; it excludes Turkey, which is included under "Middle East." The fact that in OECD data, "OECD Europe" (which we have taken in the text of this study to represent "Western Europe") includes Turkey, and excludes Yugoslavia and Finland (also part of the UN "Western Europe") distorts very slightly some of the statistical series.

In the UN conversion of the different energy forms into hard coal equivalents,[15] hydro and nuclear are figured at the inherent heat value of the electricity generated—i.e., at a factor of 0.125 metric tons hard coal equivalent = 1,000 KWH. The alternative approach would be to convert primary electricity at the prevailing "heat rate" at thermal generating stations—in 1965 amounting to approximately 0.38 metric tons of hard coal per 1,000 KWH. (The OECD uses still another procedure for converting hydroelectricity, as described further on.)[16] The UN approach distorts energy comparisons among areas having markedly disparate proportions of energy drawn from hydro sources. For example, in the UN method, Britain's per capita energy consumption in 1965 was about 45 percent higher than Norway's. Under the alternative approach described above, Norway's per capita energy consumption exceeds that of Great Britain by one-third. Great Britain is, of course, virtually without hydroelectricity, while in Norway, hydro is of great importance. In our tables we conform to UN practice, although where comparison with other studies necessitates, we have made the appropriate alternative calculation.

Since energy data, although historically standardized in UN statistics in terms of coal equivalents, are needed in oil equivalents for this study, Tables II.A, IV.A, IX.A, X.A, and XII.A present data in oil-equivalent metric tons corresponding to their companion tables in coal equivalents. (In terms of relative heat content, 1 oil-equivalent metric ton = 1.5 coal-equivalent tons.)

Where oil comparisons are made between data expressed in barrels per day and in metric tons per year, the conversion factor is that of 1 barrel/day = 50 metric tons/year. Implicitly, this equates 1 metric ton to about 7.2 barrels, assuming (for convenience) an average worldwide specific gravity.

The accompanying tabular exhibit (and notes thereto) which precedes the twelve appendix tables proper summarizes a number of points which have been made in this section by illustrating the effect of different definitions of energy and oil consumption and of different hydro and nuclear conversion methods. Data in this tabular exhibit are referred to in those places in the main text of the study where alternative statistical formulations are called for.

[15] "Hard coal equivalents" refer to the energy content of bituminous coal; in the U. S., the latter is commonly termed "soft coal."

[16] See the explanation of panel (H) on p. 170, below.

Primary Energy Consumption, by Source: United States, Western Europe, and Japan—1965, 1975, 1980 under Alternative Formulations

	United States			Western Europe			Japan		
	1965	1975	1980	1965	1975	1980	1965	1975	1980
(A) MILLION METRIC TONS COAL EQUIVALENT, INLAND ENERGY CONSUMPTION, HYDRO AND NUCLEAR MEASURED BY THERMAL EQUIVALENT OF THE ELECTRICITY GENERATED									
Total energy	1,790.3	2,525.0	3,000.0	1,045.3	1,547.0	1,882.0	174.9	374.0	547.0
Solid fuels	425.2	549.0	626.0	515.4	370.5	340.0	64.5	64.6	69.6
Liquid fuels	723.2	1,010.6	1,167.2	462.7	900.0	1,151.0	98.8	284.6	440.1
Natural gas	616.8	879.8	1,041.3	28.0	202.5	255.0	2.7	6.5	10.7
Hydro	24.6	32.5	42.9	36.6	48.0	57.0	8.9	13.0	13.5
Nuclear	0.5	53.1	122.6	2.5	26.0	79.0	–	5.2	13.1
(B) MILLION METRIC TONS OIL EQUIVALENT, INLAND ENERGY CONSUMPTION, HYDRO AND NUCLEAR MEASURED BY THERMAL EQUIVALENT OF THE ELECTRICITY GENERATED									
Total energy	1,193.6	1,683.4	2,000.0	696.9	1,031.4	1,254.7	116.6	249.4	364.7
Solid fuels	283.5	366.0	417.4	343.6	247.0	226.7	43.0	43.1	46.4
Liquid fuels	482.2	673.8	778.2	308.5	600.0	767.3	65.9	189.8	293.4
Natural gas	411.2	586.6	694.2	18.6	135.0	170.0	1.8	4.3	7.1
Hydro	16.4	21.7	28.6	24.4	32.0	38.0	5.9	8.7	9.0
Nuclear	0.3	35.4	81.7	1.7	17.0	53.0	–	3.5	8.8
Percentage distribution of (A) or (B)									
Total energy	100.0	100.0	100.0	100.0	100.0	100.0	100.0	100.0	100.0
Solid fuels	23.7	21.7	20.9	49.3	23.9	18.1	36.9	17.3	12.7
Liquid fuels	40.4	40.0	38.9	44.3	58.2	61.2	56.5	76.1	80.5
Natural gas	34.5	34.8	34.7	2.7	13.1	13.5	1.5	1.7	2.0
Hydro	1.4	1.3	1.4	3.5	3.1	3.0	5.1	3.5	2.5
Nuclear	–	2.1	4.1	0.2	1.7	4.2	–	1.4	2.4
(C) MILLION METRIC TONS COAL EQUIVALENT, INLAND ENERGY CONSUMPTION, HYDRO AND NUCLEAR MEASURED BY FUEL INPUT THAT WOULD HAVE BEEN REQUIRED AT FOSSIL FUEL FIRED POWER PLANTS									
Total energy	1,841.5	2,672.2	3,271.5	1,125.0	1,674.3	2,106.6	193.0	405.2	590.6
Solid fuels	425.2	549.0	626.0	515.4	370.5	340.0	64.5	64.6	69.6
Liquid fuels	723.2	1,010.6	1,167.2	462.7	900.0	1,151.0	98.8	284.6	440.1
Natural gas	616.8	879.8	1,041.3	28.0	202.5	255.0	2.7	6.5	10.7
Hydro	74.9	88.4	113.3	111.2	130.6	150.6	27.0	35.4	35.6
Nuclear	1.4	144.4	323.7	7.7	70.7	210.0	–	14.1	34.6
(D) MILLION METRIC TONS OIL EQUIVALENT, INLAND ENERGY CONSUMPTION, HYDRO AND NUCLEAR MEASURED BY FUEL INPUT THAT WOULD HAVE BEEN REQUIRED AT FOSSIL FUEL FIRED POWER PLANTS									
Total energy	1,227.6	1,781.4	2,180.9	750.0	1,116.2	1,404.4	128.7	270.2	393.7
Solid fuels	283.5	366.0	417.3	343.6	247.0	226.7	43.0	43.1	46.4
Liquid fuels	482.1	673.7	778.1	308.5	600.0	767.3	65.9	189.8	293.4
Natural gas	411.2	586.5	694.2	18.7	135.0	170.0	1.8	4.3	7.1
Hydro	49.9	58.9	75.5	74.1	87.1	100.4	18.0	23.6	23.7
Nuclear	0.9	96.3	215.8	5.1	47.1	140.0	–	9.4	23.1
Percentage distribution of (C) or (D)									
Total energy	100.0	100.0	100.0	100.0	100.0	100.0	100.0	100.0	100.0
Solid fuels	23.1	20.5	19.1	45.8	22.1	16.1	33.4	15.9	11.8
Liquid fuels	39.3	37.8	35.7	41.1	53.8	54.6	51.2	70.2	74.5
Natural gas	33.5	32.9	31.8	2.5	12.1	12.1	1.4	1.6	1.8
Hydro	4.1	3.3	3.5	9.9	7.8	7.1	14.0	8.7	6.0
Nuclear	0.1	5.4	9.9	0.7	4.2	10.0	–	3.5	5.9

Panels (A), (B), (C), (D) explained on p. 170.

Primary Energy Consumption, by Source: Alternative Formulations–*Continued*

	United States			Western Europe			Japan		
	1965	1975	1980	1965	1975	1980	1965	1975	1980
(E) MILLION METRIC TONS COAL EQUIVALENT, TOTAL CONSUMPTION IN ENERGY USES, HYDRO AND NUCLEAR MEASURED BY FUEL INPUT THAT WOULD HAVE BEEN REQUIRED AT FOSSIL FUEL FIRED POWER PLANTS									
Total energy	1,859.9	2,700.0	3,303.6	1,176.7	1,779.6	2,211.6	209.3	447.5	669.9
Solid fuels	425.2	549.0	626.0	515.4	370.5	340.0	64.5	64.6	69.6
Liquid fuels	741.6	1,038.4	1,199.3	514.4	1,005.3	1,256.0	115.1	326.9	519.4
Natural gas	616.8	879.8	1,041.3	28.0	202.5	255.0	2.7	6.5	10.7
Hydro	74.9	88.4	113.3	111.2	130.6	150.6	27.0	35.4	35.6
Nuclear	1.4	144.4	323.7	7.7	70.7	210.0	–	14.1	34.6
(F) MILLION METRIC TONS OIL EQUIVALENT, TOTAL CONSUMPTION IN ENERGY USES, HYDRO AND NUCLEAR MEASURED BY FUEL INPUT THAT WOULD HAVE BEEN REQUIRED AT FOSSIL FUEL FIRED POWER PLANTS									
Total energy	1,239.9	1,800.0	2,202.4	784.5	1,186.4	1,474.4	139.5	298.3	446.5
Solid fuels	283.5	366.0	417.3	343.6	247.0	226.7	43.0	43.1	46.4
Liquid fuels	494.4	692.3	799.5	342.9	670.2	837.3	76.7	217.9	346.2
Natural gas	411.2	586.6	694.2	18.7	135.0	170.0	1.8	4.3	7.1
Hydro	49.9	58.9	75.5	74.1	87.1	100.4	18.0	23.6	23.7
Nuclear	0.9	96.3	215.8	5.1	47.1	140.0	–	9.4	23.1
Percentage distribution of (E) or (F)									
Total energy	100.0	100.0	100.0	100.0	100.0	100.0	100.0	100.0	100.0
Solid fuels	22.9	20.3	18.9	43.8	20.8	15.4	30.8	14.4	10.4
Liquid fuels	39.9	38.5	36.3	43.7	56.5	56.8	55.0	73.0	77.5
Natural gas	33.2	32.6	31.5	2.4	11.4	11.5	1.3	1.4	1.6
Hydro	4.0	3.3	3.4	9.4	7.3	6.8	12.9	7.9	5.3
Nuclear	0.1	5.4	9.8	0.7	4.0	9.5	–	3.2	5.2
(G) MILLION METRIC TONS COAL EQUIVALENT, OECD DEFINITION OF TOTAL CONSUMPTION AND OF HYDRO AND NUCLEAR CONVERSION									
Total energy	1,911.0	2,798.7	3,435.9	1,157.7	1,795.6	2,246.9	204.7	461.8	708.9
Solid fuels	425.2	549.0	626.0	515.4	370.5	340.0	64.5	64.6	69.6
Liquid fuels	834.0	1,184.9	1,391.3	554.6	1,091.9	1,370.7	125.5	360.4	577.1
Natural gas	616.8	879.8	1,041.3	28.0	202.5	255.0	2.7	6.5	10.7
Hydro	33.2	40.6	53.6	49.5	60.0	71.2	12.0	16.2	16.9
Nuclear	1.8	144.4	323.7	10.2	70.7	210.0	–	14.1	34.6
(H) MILLION METRIC TONS OIL EQUIVALENT, OECD DEFINITION OF TOTAL CONSUMPTION AND OF HYDRO AND NUCLEAR CONVERSION									
Total energy	1,274.0	1,865.8	2,290.5	771.8	1,197.0	1,498.0	136.5	307.9	472.6
Solid fuels	283.5	366.0	417.3	343.6	247.0	226.7	43.0	43.1	46.4
Liquid fuels	556.0	789.9	927.5	369.7	727.9	913.8	83.7	240.3	384.7
Natural gas	411.2	586.5	694.2	18.7	135.0	170.0	1.8	4.3	7.1
Hydro	22.1	27.1	35.7	33.0	40.0	47.5	8.0	10.8	11.3
Nuclear	1.2	96.3	215.8	6.8	47.1	140.0	–	9.4	23.1
Percentage distribution of (G) or (H)									
Total energy	100.0	100.0	100.0	100.0	100.0	100.0	100.0	100.0	100.0
Solid fuels	22.3	19.6	18.2	44.5	20.6	15.1	31.5	14.0	9.8
Liquid fuels	43.6	42.3	40.5	47.9	60.8	61.0	61.3	78.0	81.4
Natural gas	32.3	31.4	30.3	2.4	11.3	11.3	1.3	1.4	1.5
Hydro	1.7	1.5	1.6	4.3	3.3	3.2	5.9	3.5	2.4
Nuclear	0.1	5.2	9.4	0.9	3.9	9.3	–	3.1	4.9

Panels (E), (F), (G), (H) explained on p. 170.

Primary Energy Consumption, by Source: Alternative Formulations—*Continued*

(A) This panel conforms to the United Nations statistical practice, which is also the approach generally followed in the Appendix Tables. Consumption refers to "inland" consumption–that is, it excludes bunker fuel for maritime vessels and foreign-bound aircraft. Consumption, as measured by the United Nations, further excludes that portion of primary energy consumed in nonenergy uses, such as petrochemical feedstocks. The figures in panel (A) are taken from Appendix Table IV. Hydro and nuclear have been converted to coal equivalents on the basis of the thermal content of electricity generated (i.e., 1,000 Kwh = 0.125 metric tons of coal equivalent).

(B) Figures in panel (A) divided by 1.5 in order to approximate oil equivalents. The figures in panel (B) are taken from Appendix Table IV.A.

(C) This panel differs from panel (A) in that hydro and nuclear electricity are expressed in terms of coal equivalents, using the estimated quantity of fuel required, on the average, to generate the same amount of electricity at thermal power stations for the year in question. The assumed conversion factors and corresponding efficiencies are as follows:

	Coal (oil) equivalents in metric tons per 1,000 Kwh	Efficiency (= Btu's generated as electricity divided by Btu's of primary fuels required, all expressed in percent)
1965	0.38 (0.25)	33.0%
1975	0.34 (0.23)	36.5%
1980	0.33 (0.22)	38.0%

The 1965 conversion factor is taken from J. Darmstadter and others, *Energy in the World Economy: A Statistical Review of Trends in Output, Trade, and Consumption Since 1925* (Baltimore: Johns Hopkins Press for Resources for the Future, forthcoming). The 1975 and 1980 figures are based (partly by interpolation) on OECD, *Energy Policy: Problems and Objectives* (Paris: 1966), p. 168. For convenience, the same conversion rate is applied to each of the three areas.

(D) Figures in panel (C) divided by 1.5 in order to approximate oil equivalents.

(E) This panel differs from panel (C) in that total consumption is raised by the inclusion of liquid bunker fuels. (Coal bunkers are negligible and can be ignored.) The bunkers estimate comes from Appendix Table V, where it equals the difference between lines (1) and (2).

(F) Figures in panel (E) divided by 1.5 in order to approximate oil equivalents.

(G) This panel attempts to adapt the historical and projected figures in the present study to OECD concepts and practices (as reflected in *Energy Policy, op. cit.*). First, it further raises the liquid fuels consumption of panel (E) by adding an estimated allowance for nonenergy uses. (Historical figures on nonenergy disposition of oil appear in OECD, *Basic Statistics of Energy, 1951-1965* (Paris: 1967). These historical data were used to make rough projections of nonenergy demand.) Second, it expresses hydroelectricity at its inherent thermal value (as is done in panel (A)) but, following the OECD, it then raises this figure by assuming an efficiency of 74 percent (1965) and 80 percent (1975 and 1980) from the falling water to generating station terminals. (See *Energy Policy, op. cit.*, p. 168.) Third, it translates nuclear into coal equivalents for 1965 by assuming an efficiency of 25 percent from nuclear heat released to kilowatt-hours generated (compared to the 33 percent efficiency indicated for 1965 under panel (C)). For 1975 and 1980, it handles nuclear electricity as in panels (C) and (E) by conforming to the OECD assumption that future "efficiency of nuclear plant will probably gradually approach that of conventional equipment." (*Ibid.*)

(H) Figures in panel (G) divided by 1.5 in order to approximate oil equivalents.

APPENDIX TABLES

Table I

Growth Rates in GNP and Energy Consumption, by Selected Countries and World Regions, 1950–65 and 1965–80

	Average annual percentage growth in GNP			Elasticity: percentage growth in energy consumption divided by percentage growth in GNP			Average annual percentage growth in energy consumption		
	1960–65	1950–65	1965–80	1960–65	1950–65	1965–80	1960–65	1950–65	1965–80
United States	4.8	3.7	4.4	0.83	0.81	0.80	4.0	3.0	3.5
Canada	5.5	4.5	5.0	1.40	1.13	1.10	7.7	5.1	5.5
Western Europe	4.9	4.8	4.5	1.14	0.92	0.90	5.6	4.4	4.0
Japan	10.7	9.9	7.5	1.05	1.00	1.05	11.2	9.9	7.9
Middle East	6.5	5.8	6.3	1.23	1.76	1.50	8.0	10.2	9.4
Other Asia	4.1	4.1	5.1	1.78	1.75	1.60	7.3	7.6	8.2
Oceania	4.8	4.2	4.2	1.25	1.19	1.15	6.0	5.0	4.8
Latin America	4.6	4.9	5.0	1.17	1.55	1.48	5.4	7.6	7.4
Caribbean	NA	NA	5.8	NA	NA	1.45	5.4	8.6	8.4
Other Latin America	4.0	4.3	4.4	1.32	1.51	1.35	5.3	6.5	5.9
Africa	4.4	4.4	5.0	1.30	1.23	1.30	5.7	5.4	6.5
North Africa	NA	NA	NA	NA	NA	NA	4.4	4.6	5.6
Other Africa	NA	NA	NA	NA	NA	NA	6.0	5.6	6.7
U.S.S.R.	5.2	5.9	5.4	1.27	1.25	1.20	6.6	7.4	6.5
Communist East Europe	4.0	4.9	4.2	1.25	1.18	1.10	5.0	5.8	4.6
Communist Asia	–1.0	5.8	4.6	--	2.48	1.65	6.5	14.4	7.6

NA: Data not available.

Source: Historical data: J. Darmstadter and others, *Energy in the World Economy, op. cit.*

Projections: On projections of GNP growth rates and elasticities, see accompanying text. Energy growth rate projections obtained by multiplication of GNP rates and elasticities.

Table II

Energy Consumption, by Selected Countries and World Regions, Selected Years, 1955–80

(Million Metric Tons Coal Equivalent)

	Million metric tons coal equivalent						Percentage distribution					
	1955	1960	1965	1966	1975	1980	1955	1960	1965	1966	1975	1980
North America	1,368	1,555	1,940	2,040	2,781	3,335	42.5	36.7	37.1	37.0	32.2	29.8
United States	1,285	1,454	1,790	1,890	2,525	3,000	39.9	34.3	34.2	34.3	29.3	26.8
Canada	83	101	150	158	256	335	2.6	2.4	2.9	2.9	3.0	3.0
Western Europe	726	833	1,045	1,085	1,547	1,882	22.5	19.7	20.0	19.7	17.9	16.8
Japan	66	109	175	193	374	547	2.0	2.6	3.3	3.5	4.3	4.9
Middle East	16	30	45	49	111	173	0.5	0.7	0.9	0.9	1.3	1.5
Other Asia	73	105	148	155	325	483	2.3	2.5	2.8	2.8	3.8	4.3
Oceania	39	46	62	65	99	125	1.2	1.1	1.2	1.2	1.1	1.1
Latin America	98	138	184	193	375	538	3.0	3.3	3.5	3.5	4.3	4.8
Caribbean	51	77	104	106	233	349	1.6	1.8	2.0	1.9	2.7	3.1
Other Latin America	47	61	80	87	142	189	1.5	1.4	1.5	1.6	1.6	1.7
Africa	56	68	88	92	165	226	1.7	1.6	1.7	1.7	1.9	2.0
North Africa	10	13	16	18	27	36	0.3	0.3	0.3	0.3	0.3	0.3
Other Africa	46	55	72	74	138	190	1.4	1.3	1.4	1.3	1.6	1.7
U.S.S.R.	441	607	833	883	1,564	2,142	13.7	14.3	15.9	16.0	18.1	19.1
Communist East Europe	235	295	372	380	583	730	7.3	7.0	7.1	6.9	6.8	6.5
Communist Asia	101	446	338	366	703	1,014	3.1	10.5	6.5	6.6	8.1	9.1
World	3,220	4,233	5,231	5,509	8,627	11,195	100.0	100.0	100.0	100.0	100.0	100.0
Excluding Communist countries	2,443	2,885	3,688	3,880	5,777	7,309	75.9	68.2	70.5	70.4	67.0	65.3

Source: Historical data from United Nations, *World Energy Supplies,* Series J, various numbers. The 1975 estimate is by interpolation. The 1980 projection is the result of applying projected growth rates in Table I to 1965 absolutes shown above.

Table II. A

Energy Consumption, by Selected Countries and World Regions, Selected Years, 1955–80

(Million Metric Tons Oil Equivalent)

	Million metric tons oil equivalent						Percentage distribution					
	1955	1960	1965	1966	1975	1980	1955	1960	1965	1966	1975	1980
North America	912	1,037	1,293	1,360	1,854	2,223	42.5	36.7	37.1	37.0	32.2	29.8
United States	857	969	1,194	1,260	1,683	2,000	39.9	34.3	34.2	34.3	29.3	26.8
Canada	55	67	100	105	171	223	2.6	2.4	2.9	2.9	3.0	3.0
Western Europe	484	555	697	723	1,031	1,255	22.5	19.7	20.0	19.7	17.9	16.8
Japan	44	73	117	129	249	365	2.0	2.6	3.3	3.5	4.3	4.9
Middle East	11	20	30	33	74	115	0.5	0.7	0.9	0.9	1.3	1.5
Other Asia	49	70	99	103	217	322	2.3	2.5	2.8	2.8	3.8	4.3
Oceania	26	31	41	43	66	83	1.2	1.1	1.2	1.2	1.1	1.1
Latin America	65	92	123	129	250	359	3.0	3.3	3.5	3.5	4.3	4.8
Caribbean	34	51	69	71	155	233	1.6	1.8	2.0	1.9	2.7	3.1
Other Latin America	31	41	53	58	95	126	1.5	1.4	1.5	1.6	1.6	1.7
Africa	37	45	59	61	110	151	1.7	1.6	1.7	1.7	1.9	2.0
North Africa	7	9	11	12	18	24	0.3	0.3	0.3	0.3	0.3	0.3
Other Africa	31	37	48	49	92	127	1.4	1.3	1.4	1.3	1.6	1.7
U.S.S.R.	294	405	555	589	1,043	1,428	13.7	14.3	15.9	16.0	18.1	19.1
Communist East Europe	157	197	248	253	389	487	7.3	7.0	7.1	6.9	6.8	6.5
Communist Asia	67	297	225	244	469	676	3.1	10.5	6.5	6.6	8.1	9.1
World	2,147	2,822	3,488	3,673	5,752	7,464	100.0	100.0	100.0	100.0	100.0	100.0
Excluding Communist countries	1,629	1,923	2,459	2,587	3,852	4,873	75.9	68.2	70.5	70.4	67.0	65.3

Source: Table II coal equivalent figures divided by 1.5 in order to approximate oil equivalents.

Table III

Comparison of Projected Growth Rates in Energy Consumption, OECD, ECSC, and Present Study

(Average Annual Percentage Rates of Change)

	OECD	Present study			ECSC
	(Period shown)			1965–1980	
North America	3.7	(1964–80)	3.7	3.7	3.2
Western Europe	4.1	(1964–80)	4.0	4.0	3.8
Japan	8.0	(1964–80)	7.9	7.9	7.2
Communist countries	5.9	(1962–80)	6.4	6.3	6.3
All other	6.7	(1962–80)	7.2	7.4	6.2
World	5.1	(1962–80)	5.3	5.2	4.8
Excluding Communist countries	4.8	(1962–80)	4.8	4.6	4.1

Source: Present study based on data in Table II, except where comparison with OECD required use of base year other than 1965; in that case, base-year figure was obtained from United Nations, *World Energy Supplies.* OECD from *Energy Policy, op. cit.,* pp. 150–151 and 174–176. ECSC refers to the European Coal and Steel Community, *Review of the Long-Term Energy Outlook for the European Community, op. cit.,* p. 35. Since the ECSC document shows only a 1960 base-year figure taken from the above-cited United Nations source, we substituted the 1965 figure for purposes of this comparison.

Table IV

Energy Consumption[a] by Fuel, and by Selected Countries and World Regions, 1955–80

(Thousand Metric Tons HCE)

	Thousand metric tons HCE						Percent of total consumption				
	Total energy	Solid fuels	Liquid fuels	Natural gas	Hydro	Nuclear[b]	Solid fuels	Liquid fuels	Natural gas	Hydro	Nuclear[b]
United States											
1955 ...	1,285,090	403,680	515,450	350,900	15,060	–	31.4	40.1	27.3	1.2	–
1957 ...	1,336,000	390,820	533,110	394,890	17,180	–	29.2	39.9	29.6	1.3	–
1960 ...	1,453,960	359,190	593,270	482,170	19,265	65	24.7	40.8	33.2	1.3	–
1963 ...	1,626,791	382,988	659,290	562,947	21,163	402	23.6	40.5	34.6	1.3	–
1965 ...	1,790,337	425,203	723,245	616,792	24,640	457	23.7	40.4	34.5	1.4	–
1975 ...	2,525,000	549,000	1,010,600	879,800	32,500	53,100	21.7	40.0	34.8	1.3	2.1
1980 ...	3,000,000	626,000	1,167,200	1,041,300	42,900	122,600	20.9	38.9	34.7	1.4	4.1
Canada											
1955 ...	82,890	29,310	38,920	5,680	8,990	–	35.3	47.0	6.9	10.8	–
1957 ...	91,970	27,990	45,430	8,650	9,890	–	30.4	49.4	9.4	10.8	–
1960 ...	101,420	20,080	52,910	15,830	12,590	–	19.8	52.2	15.6	12.4	–
1963 ...	124,079	19,564	62,658	28,960	12,886	11	15.8	50.5	23.3	10.4	–
1965 ...	150,075	23,305	71,314	40,806	14,634	15	15.5	47.5	27.2	9.8	–
1975 ...	256,000	30,180	113,500	88,320	21,500	2,500	11.8	44.3	34.5	8.4	1.0
1980 ...	335,000	34,170	147,060	119,600	27,470	6,700	10.2	43.9	35.7	8.2	2.0
Western Europe											
1955 ...	726,270	568,080	131,530	7,000	19,660	–	78.2	18.1	1.0	2.7	–
1957 ...	783,720	585,750	167,170	9,290	21,510	–	74.7	21.3	1.2	2.8	–
1960 ...	833,310	540,250	248,130	15,900	28,754	276	64.8	29.8	1.9	3.5	–
1963 ...	976,295	556,780	363,725	21,946	32,930	914	57.0	37.3	2.2	3.4	0.1
1965 ...	1,045,268	515,414	462,745	27,974	36,590	2,545	49.3	44.3	2.7	3.5	0.2
1975	1,547,000	370,500	900,000	202,500	48,000	26,000	23.9	58.2	13.1	3.1	1.7
1980 ...	1,882,000	340,000	1,150,950	255,000	57,050	79,000	18.1	61.2	13.5	3.0	4.2
Japan											
1955 ...	65,870	46,110	13,540	210	6,010	–	70.0	20.6	0.3	9.1	–
1957 ...	84,160	55,330	21,780	320	6,730	–	65.7	25.9	0.4	8.0	–
1960 ...	108,640	62,240	37,990	1,020	7,390	–	57.3	35.0	0.9	6.8	–
1963 ...	146,973	64,791	71,272	2,512	8,397	–	44.1	48.5	1.7	5.7	–
1965 ...	174,937	64,534	98,819	2,688	8,896	–	36.9	56.5	1.5	5.1	–
1975 ...	374,000	64,600	284,650	6,500	13,000	5,250	17.3	76.1	1.7	3.5	1.4
1980 ...	547,000	69,600	440,105	10,670	13,500	13,125	12.7	80.5	2.0	2.5	2.4
Middle East											
1955 ...	16,050	4,270	11,750	–	30	–	26.6	73.2	–	0.2	–
1957 ...	21,740	4,810	15,920	960	70	–	22.1	73.2	4.4	0.3	–
1960 ...	29,960	4,820	21,690	3,310	140	–	16.1	72.4	11.0	0.5	–
1963 ...	37,545	5,312	27,355	4,535	343	–	14.1	72.9	12.1	0.9	–
1965 ...	44,838	5,720	33,553	5,189	375	–	12.8	74.8	11.6	0.8	–
1975 ...	111,000	10,880	83,580	15,100	1,440	–	9.8	75.3	13.6	1.3	–
1980 ...	173,000	13,840	131,310	24,900	2,940	–	8.0	75.9	14.4	1.7	–

Table IV–*Continued*

	Thousand metric tons HCE						Percent of total consumption				
	Total energy	Solid fuels	Liquid fuels	Natural gas	Hydro	Nuclear[b]	Solid fuels	Liquid fuels	Natural gas	Hydro	Nuclear[b]
Other Asia											
1955 ...	73,480	45,070	24,730	2,730	950	–	61.3	33.7	3.7	1.3	–
1957 ...	85,590	51,660	29,130	3,610	1,180	–	60.4	34.0	4.2	1.4	–
1960 ...	105,340	63,500	35,580	4,530	1,730	–	60.3	33.8	4.3	1.6	–
1963 ...	133,851	81,066	44,484	5,654	2,648	–	60.6	33.2	4.2	2.0	–
1965 ...	147,692	85,824	51,443	7,084	3,342	–	58.1	34.8	4.8	2.3	–
1975 ...	325,000	182,980	112,120	19,500	9,100	1,200	56.3	34.5	6.0	2.8	0.4
1980 ...	483,000	268,060	167,600	30,430	14,490	2,000	55.5	34.7	6.3	3.0	0.5
Oceania											
1955 ...	38,900	24,370	13,630	–	900	–	62.7	35.0	–	2.3	–
1957 ...	39,260	24,510	13,650	–	1,100	–	62.4	34.8	–	2.8	–
1960 ...	45,890	27,860	16,640	–	1,380	–	60.7	36.3	–	3.0	–
1963 ...	52,585	30,283	20,374	4	1,924	–	57.6	38.7	–	3.7	–
1965 ...	62,357	34,287	25,863	5	2,202	–	55.0	41.5	–	3.5	–
1975 ...	99,000	40,990	44,650	9,700	3,660	–	41.4	45.1	9.8	3.7	–
1980 ...	125,000	47,490	55,380	16,500	4,880	750	38.0	44.3	13.2	3.9	0.6
Caribbean America											
1955 ...	51,380	3,370	39,270	7,940	800	–	6.6	76.4	15.5	1.5	–
1957 ...	63,360	4,011	46,610	11,860	880	–	6.3	73.6	18.7	1.4	–
1960 ...	77,040	3,830	54,940	17,050	1,230	–	5.0	71.3	22.1	1.6	–
1963 ...	91,074	4,758	63,832	20,891	1,592	–	5.2	70.1	22.9	1.8	–
1965 ...	104,339	4,600	72,205	25,522	2,011	–	4.4	69.2	24.5	1.9	–
1975 ...	233,000	8,150	154,480	65,240	5,130	–	3.5	66.3	28.0	2.2	–
1980 ...	349,000	9,770	227,200	104,000	8,030	–	2.8	65.1	29.8	2.3	–
Other Latin America											
1955 ...	46,960	7,230	36,170	1,680	1,880	–	15.4	77.0	3.6	4.0	–
1957 ...	52,640	6,870	41,430	1,830	2,510	–	13.1	78.7	3.4	4.8	–
1960 ...	61,360	5,910	47,600	4,690	3,160	–	9.6	77.6	7.6	5.2	–
1963 ...	72,615	5,479	54,492	8,991	3,653	–	7.6	75.0	12.4	5.0	–
1965 ...	80,198	5,995	59,552	10,212	4,439	–	7.5	74.3	12.7	5.5	–
1975 ...	142,000	9,520	100,570	22,860	8,800	250	6.7	70.8	16.1	6.2	0.2
1980 ...	189,000	11,910	128,960	33,450	14,180	500	6.3	68.2	17.7	7.5	0.3
North Africa											
1955 ...	10,160	1,120	8,890	20	140	–	11.0	87.5	0.2	1.3	–
1957 ...	10,560	1,290	9,110	10	150	–	12.2	86.3	0.1	1.4	–
1960 ...	13,010	1,040	11,740	30	200	–	8.0	90.2	0.2	1.6	–
1963 ...	14,145	727	12,521	558	337	–	5.1	88.5	4.0	2.4	–
1965 ...	16,074	917	13,369	1,363	426	–	5.7	83.2	8.5	2.6	–
1975 ...	27,000	1,080	19,490	5,400	1,030	–	4.0	72.2	20.0	3.8	–
1980 ...	36,000	720	24,700	9,000	1,580	–	2.0	68.6	25.0	4.4	–
Other Africa											
1955 ...	46,010	37,010	8,740	–	260	–	80.4	19.0	–	0.6	–
1957 ...	50,770	39,990	10,330	–	440	–	78.8	20.3	–	0.9	–
1960 ...	55,430	42,210	12,490	10	720	–	76.1	22.5	–	1.3	–
1963 ...	61,569	45,121	15,336	55	1,059	–	73.3	24.9	0.1	1.7	–
1965 ...	72,087	52,048	18,674	114	1,249	–	72.2	25.9	0.2	1.7	–
1975 ...	138,000	86,250	42,640	5,520	3,590	–	62.5	30.9	4.0	2.6	–
1980 ...	190,000	110,200	62,890	11,400	5,510	–	58.0	33.1	6.0	2.9	–

Table IV–*Continued*

	Thousand metric tons HCE						Percent of total consumption				
	Total energy	Solid fuels	Liquid fuels	Natural gas	Hydro	Nuclear[b]	Solid fuels	Liquid fuels	Natural gas	Hydro	Nuclear[b]
U.S.S.R.											
1955 . . .	441,450	337,250	89,540	11,760	2,900	NA	76.4	20.3	2.7	0.6	NA
1957 . . .	514,400	372,140	112,840	24,490	4,930	NA	72.3	21.9	4.8	1.0	NA
1960 . . .	607,120	384,170	156,570	60,020	6,360	NA	63.3	25.8	9.9	1.0	NA
1963 . . .	726,846	391,010	207,192	119,255	9,389	NA	53.8	28.5	16.4	1.3	NA
1965 . . .	832,794	414,496	238,778	169,525	9,991	NA	49.8	28.7	20.3	1.2	NA
1975 . . .	1,564,000	582,200	482,600	472,400	25,000	1,800	37.2	30.9	30.2	1.6	0.1
1980 . . .	2,142,000	713,900	645,200	741,500	36,400	5,000	33.3	30.1	34.6	1.7	0.2
Communist Eastern Europe											
1955 . . .	234,830	212,850	12,220	9,220	550	–	90.6	5.2	4.0	0.2	–
1957 . . .	260,270	233,690	14,900	11,130	550	–	89.8	5.7	4.3	0.2	–
1960 . . .	294,530	258,210	20,360	15,160	810	–	87.7	6.9	5.1	0.3	–
1963 . . .	351,620	300,738	28,988	20,945	949	–	85.5	8.2	6.0	0.3	–
1965 . . .	372,564	308,590	36,974	25,631	1,368	–	82.8	9.9	6.9	0.4	–
1975 . . .	583,000	404,270	116,600	58,300	2,330	1,500	69.3	20.0	10.0	0.4	0.3
1980 . . .	730,000	451,980	182,500	87,600	2,920	5,000	61.9	25.0	12.0	0.4	0.7
Communist Asia											
1955 . . .	100,730	96,580	3,380	–	770	–	95.9	3.3	–	0.8	–
1957 . . .	141,360	134,690	5,380	–	1,290	–	95.3	3.8	–	0.9	–
1960 . . .	446,370	431,050	12,910	–	2,410	–	96.6	2.9	–	0.5	–
1963 . . .	299,940	285,084	12,819	–	2,037	–	95.0	4.3	–	0.7	–
1965 . . .	337,731	320,352	14,817	–	2,563	–	94.9	4.4	–	0.7	–
1975 . . .	703,000	646,760	52,730	–	3,510	300	92.0	7.5	–	0.5	–
1980 . . .	1,014,000	917,670	91,260	–	5,070	900	90.5	9.0	–	0.5	0.1
World											
1955 . . .	3,220,170	1,816,330	947,820	397,140	58,900	–	56.4	29.4	12.3	1.8	–
1957 . . .	3,535,890	1,933,600	1,066,850	467,040	68,400	–	54.7	30.2	13.2	1.9	–
1960 . . .	4,233,480	2,204,410	1,322,870	619,720	86,139	341	52.1	31.3	14.6	2.0	–
1963 . . .	4,716,084	2,173,761	1,644,435	797,253	99,307	1,327	46.1	34.9	16.9	2.1	–
1965 . . .	5,231,467	2,261,321	1,921,491	932,910	112,728	3,017	43.2	36.7	17.8	2.2	0.1
1975 . . .	8,627,000	2,987,660	3,517,910	1,851,140	178, 590	91,900	34.6	40.8	21.5	2.1	1.1
1980 . . .	11,195,000	3,615,310	4,622,315	2,485,350	236,920	235,575	32.3	41.3	22.2	2.1	2.1

NA: Data not available.

[a] Refers to inland energy consumption. See tabular exhibit, with accompanying notes (pp. 168–170, above).

[b] For numerous world regions–particularly outside the industrialized West–the nuclear power outlook to 1975-80 is virtually impossible to assess at this time. The absence of a projected nuclear share reflects this ignorance, and not necessarily the judgment that it will be zero.

Source: All historical data from United Nations, *World Energy Supplies,* Series J, various numbers. Projections of total energy from Table II. As is described more fully in the accompanying text, the projected allocation by type of fuel reflects an eclectic approach, in which the extrapolation of historical trends was modified on the basis of our own judgment as well as the analysis contained in numerous studies which were consulted.

Note: Natural gas includes imports of natural as well as manufactured gas; hydro includes geothermal; hydro and nuclear include imports of electricity (apportioned between them according to their relative importance in domestic production). Hydro and nuclear have been converted to HCE on the basis of the thermal content of electricity generated (i.e., 1,000 KWH equals 0.125 metric tons HCE).

Since the projected data on total energy are more rounded off than those on the separate fuel categories, the figures may not add precisely to the indicated totals.

Table IV.A

Energy Consumption[a] by Fuel, and by Selected Countries and World Regions, 1955-80

(Thousand Metric Tons Oil Equivalent)

	Thousand metric tons oil equivalent						Percent of total consumption				
	Total energy	Solid fuels	Liquid fuels	Natural gas	Hydro	Nuclear[b]	Solid fuels	Liquid fuels	Natural gas	Hydro	Nuclear[b]
United States											
1955 ...	856,770	269,133	343,651	233,945	10,040	–	31.4	40.1	27.3	1.2	–
1957 ...	890,711	260,560	355,424	263,273	11,454	–	29.2	39.9	29.6	1.3	–
1960 ...	969,355	239,472	395,533	321,463	12,844	43	24.7	40.8	33.2	1.3	–
1963 ...	1,084,582	255,338	439,549	375,317	14,109	268	23.6	40.5	34.6	1.3	–
1965 ...	1,193,618	283,483	482,187	411,215	16,427	305	23.7	40.4	34.5	1.4	–
1975 ...	1,683,420	366,020	673,770	586,560	21,670	35,400	21.7	40.0	34.8	1.3	2.1
1980 ...	2,000,000	417,350	778,170	694,240	28,600	81,740	20.9	38.9	34.7	1.4	4.1
Canada											
1955 ...	55,263	19,541	25,948	3,787	5,994	–	35.3	47.0	6.9	10.8	–
1957 ...	61,316	18,661	30,288	5,767	6,594	–	30.4	49.4	9.4	10.8	–
1960 ...	67,617	13,387	35,275	10,554	8,394	–	19.8	52.2	15.6	12.4	–
1963 ...	82,723	13,043	41,774	19,308	8,591	7	15.8	50.5	23.3	10.4	–
1965 ...	100,055	15,537	47,545	27,205	9,756	10	15.5	47.5	27.2	9.8	–
1975 ...	170,680	20,120	75,670	58,880	14,330	1,670	11.8	44.3	34.5	8.4	1.0
1980 ...	223,340	22,780	98,040	79,740	18,310	4,470	10.2	43.9	35.7	8.2	2.0
Western Europe											
1955 ...	484,204	378,739	87,691	4,667	13,107	–	78.2	18.1	1.0	2.7	–
1957 ...	522,506	390,520	111,452	6,194	14,341	–	74.7	21.3	1.2	2.8	–
1960 ...	555,568	360,185	165,428	10,601	19,170	184	64.8	29.8	1.9	3.5	–
1963 ...	650,896	371,205	242,495	14,631	21,954	609	57.0	37.3	2.2	3.4	0.1
1965 ...	696,880	343,627	308,512	18,650	24,395	1,697	49.3	44.3	2.7	3.5	0.2
1975 ...	1,031,380	247,000	600,000	135,000	32,000	17,000	23.9	58.2	13.1	3.1	1.7
1980 ...	1,254,730	226,670	767,340	170,000	38,030	53,000	18.1	61.2	13.5	3.0	4.2
Japan											
1955 ...	43,916	30,742	9,027	140	4,007	–	70.0	20.6	0.3	9.1	–
1957 ...	56,109	36,889	14,521	213	4,487	–	65.7	25.9	0.4	8.0	–
1960 ...	72,430	41,495	25,328	680	4,927	–	57.3	35.0	0.9	6.8	–
1963 ...	97,987	43,196	47,517	1,675	5,598	–	44.1	48.5	1.7	5.7	–
1965 ...	116,630	43,025	65,883	1,792	5,931	–	36.9	56.5	1.5	5.1	–
1975 ...	249,350	43,070	189,770	4,330	8,670	3,500	17.3	76.1	1.7	3.5	1.4
1980 ...	364,680	46,400	293,400	7,110	9,000	8,750	12.7	80.5	2.0	2.5	2.4
Middle East											
1955 ...	10,701	2,847	7,834	–	20	–	26.6	73.2	–	0.2	–
1957 ...	14,494	3,207	10,614	640	47	–	22.1	73.2	4.4	0.3	–
1960 ...	19,974	3,213	14,461	2,207	93	–	16.1	72.4	11.0	0.5	–
1963 ...	25,031	3,542	18,238	3,023	229	–	14.1	72.9	12.1	0.9	–
1965 ...	29,893	3,814	22,370	3,460	250	–	12.8	74.8	11.6	0.8	–
1975 ...	74,000	7,250	55,720	10,070	960	–	9.8	75.3	13.6	1.3	–
1980 ...	115,340	9,230	87,540	16,600	1,960	–	8.0	75.9	14.4	1.7	–

Table IV.A–*Continued*

	Thousand metric tons oil equivalent						Percent of total consumption				
	Total energy	Solid fuels	Liquid fuels	Natural gas	Hydro	Nuclear[b]	Solid fuels	Liquid fuels	Natural gas	Hydro	Nuclear[b]
Other Asia											
1955 . . .	48,989	30,048	16,487	1,820	633	–	61.3	33.7	3.7	1.3	–
1957 . . .	57,063	34,442	19,421	2,407	787	–	60.4	34.0	4.2	1.4	–
1960 . . .	70,230	42,335	23,721	3,020	1,153	–	60.3	33.8	4.3	1.6	–
1963 . . .	89,238	54,047	29,657	3,770	1,765	–	60.6	33.2	4.2	2.0	–
1965 . . .	98,466	57,219	34,297	4,723	2,228	–	58.1	34.8	4.8	2.3	–
1975 . . .	216,680	121,990	74,750	13,000	6,070	800	56.3	34.5	6.0	2.8	0.4
1980 . . .	322,020	178,720	111,740	20,290	9,660	1,330	55.5	34.7	6.3	3.0	0.5
Oceania											
1955 . . .	25,935	16,247	8,907	–	600	–	62.7	35.0	–	2.3	–
1957 . . .	26,175	16,341	9,100	–	733	–	62.4	34.8	–	2.8	–
1960 . . .	30,595	18,574	11,094	–	920	–	60.7	36.3	–	3.0	–
1963 . . .	35,058	20,190	13,583	3	1,283	–	57.6	38.7	–	3.7	–
1965 . . .	41,573	22,859	17,243	3	1,468	–	55.0	41.5	–	3.5	–
1975 . . .	66,000	27,330	29,770	6,470	2,440	–	41.4	45.1	9.8	3.7	–
1980 . . .	83,340	31,660	36,920	11,000	3,250	500	38.0	44.3	13.2	3.9	0.6
Caribbean America											
1955 . . .	34,255	2,247	26,181	5,294	533	–	6.6	76.4	15.5	1.5	–
1957 . . .	42,242	2,675	31,075	7,907	587	–	6.3	73.6	18.7	1.4	–
1960 . . .	51,363	2,553	36,628	11,367	820	–	5.0	71.3	22.1	1.6	–
1963 . . .	60,719	3,172	42,557	13,928	1,061	–	5.2	70.1	22.9	1.8	–
1965 . . .	69,563	3,067	48,139	17,016	1,341	–	4.4	69.2	24.5	.1.9	–
1975 . . .	155,340	5,430	102,990	43,500	3,420	–	3.5	66.3	28.0	2.2	–
1980 . . .	232,680	6,510	151,470	69,340	5,350	–	2.8	65.1	29.8	2.3	–
Other Latin America											
1955 . . .	31,308	4,820	24,115	1,120	1,253	–	15.4	77.0	3.6	4.0	–
1957 . . .	35,095	4,580	27,621	1,220	1,673	–	13.1	78.7	3.4	4.8	–
1960 . . .	40,909	3,940	31,735	3,127	2,107	–	9.6	77.6	7.6	5.2	–
1963 . . .	48,412	3,653	36,330	5,994	2,435	–	7.6	75.0	12.4	5.0	–
1965 . . .	53,468	3,997	39,703	6,808	2,959	–	7.5	74.3	12.7	5.5	–
1975 . . .	94,670	6,350	67,050	15,240	5,870	170	6.7	70.8	16.1	6.2	–
1980 . . .	126,010	7,940	85,970	22,300	9,450	330	6.3	68.2	17.7	7.5	–
North Africa											
1955 . . .	6,774	747	5,927	13	93	–	11.0	87.5	0.2	1.3	–
1957 . . .	7,040	860	6,074	7	100	–	12.2	86.3	0.1	1.4	–
1960 . . .	8,674	693	7,827	20	133	–	8.0	90.2	0.2	1.6	–
1963 . . .	9,430	485	8,348	372	225	–	5.1	88.5	4.0	2.4	–
1965 . . .	10,717	611	8,913	909	284	–	5.7	83.2	8.5	2.6	–
1975 . . .	18,000	720	12,990	3,600	690	–	4.0	72.2	20.0	3.8	–
1980 . . .	24,000	480	16,470	6,000	1,050	–	2.0	68.6	25.0	4.4	–

Table IV.A–*Continued*

	Thousand metric tons oil equivalent						Percent of total consumption				
	Total energy	Solid fuels	Liquid fuels	Natural gas	Hydro	Nuclear[b]	Solid fuels	Liquid fuels	Natural gas	Hydro	Nuclear[b]
Other Africa											
1955 ...	30,675	24,675	5,827	–	173	–	80.4	19.0	–	0.6	–
1957 ...	33,848	26,661	6,887	–	293	–	78.8	20.3	–	0.9	–
1960 ...	36,955	28,141	8,327	7	480	–	76.1	22.5	–	1.3	–
1963 ...	41,048	30,082	10,224	37	706	–	73.3	24.9	0.1	1.7	–
1965 ...	48,060	34,700	12,450	76	833	–	72.2	25.9	0.2	1.7	–
1975 ...	92,000	57,500	28,430	3,680	2,390	–	62.5	30.9	4.0	2.6	–
1980 ...	126,670	73,470	41,930	7,600	3,670	–	58.0	33.1	6.0	2.9	–
U.S.S.R.											
1955 ...	294,315	224,845	59,696	7,840	1,933	NA	76.4	20.3	2.7	0.6	NA
1957 ...	342,950	248,106	75,230	16,327	3,287	NA	72.3	21.9	4.8	1.0	NA
1960 ...	404,767	256,126	104,385	40,015	4,240	NA	63.3	25.8	9.9	1.0	NA
1963 ...	484,588	260,686	138,135	79,507	6,260	NA	53.8	28.5	16.4	1.3	NA
1965 ...	555,224	276,344	159,193	113,022	6,661	NA	49.8	28.7	20.3	1.2	NA
1975 ...	1,042,720	388,150	321,750	314,950	16,670	1,200	37.2	30.9	30.2	1.6	0.1
1980 ...	1,428,070	475,940	430,160	494,360	24,270	3,330	33.3	30.1	34.6	1.7	0.2
Communist Eastern Europe											
1955 ...	156,561	141,907	8,147	6,147	367	–	90.6	5.2	4.0	0.2	–
1957 ...	173,522	155,801	9,934	7,420	367	–	89.8	5.7	4.3	0.2	–
1960 ...	196,363	172,149	13,574	10,107	540	–	87.7	6.9	5.1	0.3	–
1963 ...	234,425	200,502	19,326	13,964	633	–	85.5	8.2	6.0	0.3	–
1965 ...	248,388	205,737	24,651	17,088	912	–	82.8	9.9	6.9	0.4	–
1975 ...	388,690	269,510	77,730	38,870	1,550	1,000	69.3	20.0	10.0	0.4	0.3
1980 ...	486,690	301,320	121,670	58,400	1,950	3,330	61.9	25.0	12.0	0.4	0.7
Communist Asia											
1955 ...	67,157	64,390	2,253	–	513	–	95.9	3.3	–	0.8	–
1957 ...	94,245	87,798	3,587	–	860	–	95.3	3.8	–	0.9	–
1960 ...	297,595	287,381	8,607	–	1,607	–	96.6	2.9	–	0.5	–
1963 ...	199,970	190,066	8,546	–	1,358	–	95.0	4.3	–	0.7	–
1965 ...	225,165	213,579	9,878	–	1,709	–	94.9	4.4	–	0.7	–
1975 ...	468,690	431,180	35,150	–	2,340	200	92.0	7.5	–	0.5	–
1980 ...	676,030	611,780	60,840	–	3,380	600	90.0	9.0	–	0.5	0.1
World											
1955 ...	2,146,887	1,210,947	631,912	264,773	39,269	–	56.4	29.4	12.3	1.8	–
1957 ...	2,357,378	1,289,131	711,269	311,376	45,602	–	54.7	30.2	13.2	1.9	–
1960 ...	2,822,461	1,469,680	881,957	413,167	57,429	227	52.1	31.3	14.6	2.0	–
1963 ...	3,144,213	1,449,246	1,096,345	531,529	66,208	885	46.1	34.9	16.9	2.1	–
1965 ...	3,487,819	1,507,623	1,281,058	621,971	75,156	2,011	43.2	36.7	17.8	2.2	0.1
1975 ...	5,751,620	1,991,780	2,345,290	1,234,100	119,060	61,270	34.6	40.8	21.5	2.1	1.1
1980 ...	7,463,710	2,410,220	3,081,560	1,656,910	157,950	157,050	32.3	41.3	22.2	2.1	2.1

Source: Table IV coal equivalent figures divided by 1.5 in order to approximate oil equivalents. Notes to Table IV apply here as well.

Table V

Liquid Fuels: Consumption and Net Trade, 1955–80

(1) = Inland[a] consumption (1,000 metric tons);
(2) = Total[a] consumption (1,000 metric tons);
(3) = Net imports or net exports(–) (1,000 metric tons);
(4) = Net imports or net exports(–) as percent of total consumption[b]

	1955	1957	1960	1963	1965	1975	1980
United States							
(1) ...	343,651	355,424	395,533	439,549	482,187	673,770	778,170
(2) ...	356,020	370,650	408,900	451,560	494,410	692,261	799,532
(3) ...	49,890	58,500	85,210	99,000	117,840	165,580	188,400
(4) ...	14.0%	15.8%	20.8%	21.9%	23.8%	23.9%	23.6%
Canada							
(1) ...	25,948	30,288	35,275	41,774	47,545	75,670	98,040
(2) ...	27,440	32,160	37,300	44,050	50,000	79,450	102,942
(3) ...	14,870	12,150	15,920	11,890	12,270	8,170	6,680
(4) ...	54.2%	37.8%	42.7%	27.0%	24.5%	10.3%	6.5%
Caribbean							
(1) ...	26,181	31,075	36,628	42,557	48,139	102,990	151,470
(2) ...	34,680	42,140	46,720	54,080	60,920	130,072	190,848
(3) ...	−97,650	−121,280	−126,300	−144,200	−152,580	−197,780	−220,860
(4) ...	−281.6%	−287.8%	−270.3%	−266.6%	−250.5%	−152.1%	−115.7%
Other Latin America							
(1) ...	24,115	27,621	31,735	36,330	39,703	67,050	85,970
(2) ...	25,430	28,940	33,080	37,720	41,160	69,393	88,853
(3) ...	18,790	20,740	18,090	15,740	19,540	34,940	40,000
(4) ...	73.9%	71.7%	54.7%	41.7%	47.5%	50.4%	45.0%
Western Europe[b]							
(1) ...	87,691	111,452	165,428	242,495	308,512	600,000	767,340
(2) ...	104,950	129,900	187,980	271,180	342,950	670,173	837,310
(3) ...	109,630	133,450	190,800	281,490	364,340	709,390	893,090
(4) ...	104.5%	102.7%	101.5%	103.8%	106.2%	105.9%	106.7%
North Africa							
(1) ...	5,927	6,074	7,827	8,348	8,913	12,990	16,470
(2) ...	7,680	7,170	8,920	9,830	10,330	15,202	19,019
(3) ...	6,090	3,411	−2,070	−41,500	−80,240	−279,090	−360,420
(4) ...	79.3%	47.6%	−23.3%	−422.2%	−776.8%	−1,835.9%	−1,895.1%
Other Africa							
(1) ...	5,827	6,887	8,327	10,224	12,450	28,430	41,930
(2) ...	8,680	11,360	11,910	14,180	17,150	39,229	57,544
(3) ...	8,820	12,929	−640	8,530	2,550	−45,170	−81,700
(4) ...	101.6%	113.8%	−5.4%	60.2%	14.9%	−115.1%	−142.0%
Middle East							
(1) ...	7,834	10,614	14,461	18,238	22,370	55,720	87,540
(2) ...	15,950	19,610	27,020	33,030	37,410	88,595	133,936
(3) ...	−133,760	−147,550	−222,840	−303,030	−377,300	−694,020	−894,250
(4) ...	−838.6%	−752.4%	−824.7%	−917.4%	−1,008.6%	−783.4%	−667.7%
Japan[b]							
(1) ...	9,027	14,521	25,328	47,517	65,883	189,770	293,400
(2) ...	10,150	16,170	29,490	55,730	76,730	217,893	346,219
(3) ...	10,640	17,780	31,840	61,970	83,500	237,400	375,620
(4) ...	104.8%	110.0%	108.0%	111.2%	108.8%	109.0%	108.5%

Table V–*Continued*

	1955	1957	1960	1963	1965	1975	1980
Other Asia							
(1) . . .	16,487	19,421	23,721	29,657	34,297	74,750	111,740
(2) . . .	20,450	24,950	28,280	34,850	40,250	86,893	129,052
(3) . . .	4,480	5,280	1,920	8,470	11,260	25,260	12,830
(4) . . .	21.9%	21.2%	6.8%	24.3%	28.0%	29.1%	9.9%
Oceania[b]							
(1) . . .	8,907	9,100	11,094	13,583	17,243	29,770	36,920
(2) . . .	10,620	10,640	12,650	15,410	19,580	33,488	40,981
(3) . . .	10,970	10,940	13,030	16,350	20,270	–	–
(4) . . .	103.3%	102.8%	103.0%	106.1%	103.5%	–	–
Communist countries							
(1)[c] . . .	70,090	88,740	126,560	166,000	193,710	434,580	612,640
(2)[c] . . .	70,090	88,740	126,560	166,000	193,710	434,580	612,640
(3) . . .	–4,600	–8,010	–22,980	–37,660	–47,220	–	–
(4) . . .	–6.6%	–9.0%	–18.2%	–22.7%	–24.4%	–	–
World							
(1) . . .	631,912	711,269	881,957	1,096,345	1,281,058	2,345,290	3,081,560
(2) . . .	692,210	782,510	958,890	1,187,970	1,384,740	2,557,229	3,358,879

[a] "Inland" consumption excludes, while "total" consumption includes, fuels consumed as bunkers.

[b] Since consumption, as calculated in lines (1) and (2) refers to consumption of energy fuels per se, while line (3) includes imports of crude oil partially diverted to petrochemical feedstocks and nonenergy uses, the figures of line (4) may for certain regions (importing virtually all of their liquid fuels consumption) exceed 100 percent.

[c] Bunkers figures for Communist countries–not separately available–are assumed to be included in line (1), to which line (2) has therefore been made equal.

Source: All historical data from United Nations, *World Energy Supplies,* Series J, various numbers. Projections of line (1) taken from Table IV.A. Projections of line (2) on basis of historical relationships to line (1) but allowing for some change in bunkering patterns arising from assumed changes in trade routes. Projections of imports and import-consumption shares [lines (3) and (4)] are discussed in the accompanying text (Statistical Appendix, p. 164). Note, however, that only net import consumption shares were explicitly projected. Net exports were derived following the allocation of imports by geographic sources (in Table VI). When projected net exports–so derived–appeared to require additional adjustment, this computational procedure was repeated one or more times.

Table VI

Interregional Shipments of Liquid Fuels, by Origin and Destination, 1965, 1975, 1980

(Thousand Metric Tons)

Importing area	Exporting area											
	United States	Canada	Caribbean	Western Europe	North Africa	Other Africa	Middle East	Japan	Other Asia	Communist countries	Total imports	Net imports or net exports(–)[a]
						1965						
United States	–	16,250	83,200	150	3,750	900	19,750	–	3,330	–	127,330	119,443
Canada	1,150	–	20,250	200	–	–	5,900	–	150	–	27,650	11,360
Caribbean	2,000	–	–	50	–	–	3,200	–	–	9,170	14,420	–144,805
Other Latin America	–	–	10,000	150	750	750	7,000	–	–	4,330	22,980	22,980
Western Europe	2,250	–	38,050	–	85,200	15,850	202,150	–	1,410	27,520	372,430	368,658
North Africa	50	–	500	1,250	–	250	6,250	–	–	1,500	9,800	–81,060
Other Africa	200	–	5,150	1,050	500	–	17,850	–	–	–	24,750	6,500
Middle East	50	–	200	400	–	500	–	–	600	150	1,900	363,306
Japan	2,187	40	1,875	22	60	–	67,826	–	7,551	1,762	81,323	80,543
Other Asia	–	–	–	–	–	–	20,710	780	–	140	21,630	3,109
Oceania	–	–	–	–	–	–	14,570	–	5,480	–	20,050	20,050
Communist countries	–	–	–	500	600	–	–	–	–	–	1,100	–43,472
Total exports	7,887	16,290	159,225	3,772	90,860	18,250	365,206	780	18,521	44,572	725,363	
						1975						
United States	–	29,810	122,980	–	6,440	4,680	18,290	–	4,660	–	186,860	167,250
Canada	1,980	–	25,680	–	1,170	1,980	8,920	–	–	–	39,730	7,600
Caribbean	6,800	–	–	–	2,680	5,120	13,340	–	–	–	27,940	–189,810
Other Latin America	–	–	19,130	–	1,890	2,400	15,020	–	–	–	38,440	38,440
Western Europe	7,160	–	35,810	–	275,590	60,980	338,950	–	–	–	718,490	716,560
North Africa	–	–	400	800	–	1,000	5,890	–	–	–	8,090	–281,910
Other Africa	370	–	7,470	750	1,850	–	26,390	–	–	–	36,830	–45,170
Middle East	50	–	250	380	380	780	–	–	820	–	2,660	–666,050
Japan	3,250	2,320	6,030	–	–	–	170,740	–	49,520	–	231,860	227,830
Other Asia	–	–	–	–	–	5,060	71,170	4,030	–	–	80,260	25,260
Oceania	–	–	–	–	–	–	–	–	–	–	–	–
Communist countries	–	–	–	–	–	–	–	–	–	–	–	–
Total exports	19,610	32,130	217,750	1,930	290,000	82,000	668,710	4,030	55,000	–	1,371,220	
						1980						
United States	–	36,580	140,960	–	8,790	5,630	17,940	–	5,380	–	215,280	190,300
Canada	2,330	–	29,780	–	1,660	2,190	10,490	–	–	–	46,450	6,210
Caribbean	9,970	–	–	–	6,110	8,360	15,650	–	–	–	40,090	–211,960
Other Latin America	–	–	24,280	–	2,260	2,310	15,150	–	–	–	44,000	44,000
Western Europe	8,950	–	40,290	–	347,070	98,700	407,920	–	–	–	902,930	902,110
North Africa	–	–	300	300	–	630	4,710	–	–	–	5,940	–364,060
Other Africa	–	–	8,770	–	3,580	–	30,950	–	–	–	43,300	–81,700
Middle East	70	–	350	520	530	920	–	–	1,160	–	3,550	–858,210
Japan	3,660	3,660	7,320	–	–	–	253,360	–	98,460	–	366,460	360,480
Other Asia	–	–	–	–	–	6,260	105,590	5,980	–	–	117,830	12,830
Oceania	–	–	–	–	–	–	–	–	–	–	–	–
Communist countries	–	–	–	–	–	–	–	–	–	–	–	–
Total exports	24,980	40,240	252,050	820	370,000	125,000	861,760	5,980	105,000	–	1,785,830	

Table VI–*Continued*

[a] These net trade figures differ somewhat from those in Table V both because of (1) partial differences in sources, and (2) because of the fact that the figures in the present table relate exclusively to data calculated from points of origin, while those in Table V relate to an area's export and import statistics.

Source: The 1965 data were primarily adapted from release dated May 1966, U. S. Department of the Interior, Office of Oil and Gas, covering crude oil and refined petroleum products combined. This source expressed data in barrels per day; these were converted into metric tons per year by multiplying by 50. Data on exports by destination and imports by origin for Japan, Other Asia, and Oceania were not separately shown in the cited source and had to be constructed from three additional sources: (1) Crude oil flows for all three areas from United Nations, *World Energy Supplies,* Series J, No. 10; (2) Japanese refined product flows from OECD, *Oil Statistics, 1965* (Paris, 1966); and (3) refined product flow of Other Asia area from U.S. Department of the Interior, Office of Oil and Gas, *1965 Petroleum Supply and Demand in the Free World* (June 1967). Oceania's insignificant refined product flows were roughly estimated. Some geographic reclassification (done by rough estimation) was necessary to make the May 1966 Office of Oil and Gas release consistent with the scheme used in our other tables. Principally, this involved moving Puerto Rico and Jamaica from "Other Latin America" into the "Caribbean Area." Also, Cuba was put into the Caribbean region and estimated roughly on the basis of figures in *World Energy Supplies.*

The projected data are the result of the following steps: The net trade projections of Table V were translated into the net trade projections shown in the last column of the present table by taking into account the base-year differences–generally slight–between the two series, referred to in footnote a, above. The net trade projections in the last column were then used to calculate *gross* imports (next-to-last column), relying primarily upon the base-year net-gross import relationship. Gross imports were then distributed, by point of origin, according to the percentages shown in Table VII. (Some successive approximation was inherent in this procedure since the now-completed matrix of 1975 and 1980 interregional flows might yield regional net trade balances somewhat different from those initially stipulated.)

Table VII

Interregional Imports of Liquid Fuels: Percentage Distribution by Area of Origin, 1965, 1975, 1980

Importing area	United States	Canada	Caribbean	Western Europe	North Africa	Other Africa	Middle East	Japan	Other Asia	Communist countries	Total imports
	Exporting area										
						1965					
United States	–	12.8	65.3	0.1	2.9	0.7	15.5	–	2.6	–	100.0
Canada	4.2	–	73.2	0.7	–	–	21.3	–	0.5	–	100.0
Caribbean	13.9	–	–	0.3	–	–	22.2	–	–	63.6	100.0
Other Latin America	–	–	43.5	0.7	3.3	3.3	30.5	–	–	18.8	100.0
Western Europe	0.6	–	10.2	–	22.9	4.3	54.3	–	0.4	7.4	100.0
North Africa	0.5	–	5.1	12.8	–	2.6	63.8	–	–	15.3	100.0
Other Africa	0.8	–	20.8	4.2	2.0	–	72.1	–	–	–	100.0
Middle East	2.6	–	10.5	21.1	–	26.3	–	–	31.6	7.9	100.0
Japan	2.7	–	2.3	–	0.1	–	83.4	–	9.3	2.2	100.0
Other Asia	–	–	–	–	–	–	95.7	3.6	–	0.6	100.0
Oceania	–	–	–	–	–	–	72.7	–	27.3	–	100.0
Communist countries	–	–	–	45.4	54.6	–	–	–	–	–	100.0
Total exports	1.1	2.2	22.0	0.5	12.5	2.5	50.3	0.1	2.6	6.1	100.0
						1975					
United States	–	16.0	65.8	–	3.4	2.5	9.8	–	2.5	–	100.0
Canada	5.0	–	64.6	–	2.9	5.0	22.5	–	–	–	100.0
Caribbean	24.3	–	–	–	9.6	18.3	47.7	–	–	–	100.0
Other Latin America	–	–	49.8	–	4.9	6.2	39.1	–	–	–	100.0
Western Europe	1.0	–	5.0	–	38.4	8.5	47.2	–	–	–	100.0
North Africa	–	–	4.9	9.9	–	12.4	72.8	–	–	–	100.0
Other Africa	1.0	–	20.3	2.0	5.0	–	71.7	–	–	–	100.0
Middle East	1.9	–	9.4	14.3	14.3	29.3	–	–	30.8	–	100.0
Japan	1.4	1.0	2.6	–	–	–	73.6	–	21.4	–	100.0
Other Asia	–	–	–	–	–	6.3	88.7	5.0	–	–	100.0
Oceania	–	–	–	–	–	–	–	–	–	–	–
Communist countries	–	–	–	–	–	–	–	–	–	–	–
Total exports	1.4	2.3	15.9	0.1	21.1	6.0	48.8	0.3	4.0	–	100.0
						1980					
United States	–	17.0	65.5	–	4.1	2.6	8.3	–	2.5	–	100.0
Canada	5.0	–	64.1	–	3.6	4.7	22.6	–	–	–	100.0
Caribbean	24.9	–	–	–	15.2	20.9	39.0	–	–	–	100.0
Other Latin America	–	–	55.2	–	5.1	5.2	34.4	–	–	–	100.0
Western Europe	1.0	–	4.5	–	38.4	10.9	45.2	–	–	–	100.0
North Africa	–	–	5.1	5.1	–	10.6	79.3	–	–	–	100.0
Other Africa	–	–	20.3	–	8.3	–	71.5	–	–	–	100.0
Middle East	2.0	–	9.9	14.6	14.9	25.9	–	–	32.7	–	100.0
Japan	1.0	1.0	2.0	–	–	–	69.1	–	26.9	–	100.0
Other Asia	–	–	–	–	–	5.3	89.6	5.1	–	–	100.0
Oceania	–	–	–	–	–	–	–	–	–	–	–
Communist countries	–	–	–	–	–	–	–	–	–	–	–
Total exports	1.4	2.3	14.1	0.0	20.7	7.0	48.3	0.3	5.9	–	100.0

Source: Data for 1965 are percentages of absolutes given in Table VI. Projections are described in the accompanying text.

Table VIII

Summary: Liquid Fuels Consumption and Net Trade by Major Consuming Areas, 1965, 1975, 1980

(Million Metric Tons)

	Consumption	Net imports[a]	Principal supplier or export market(–)[a]
United States			
1965 . . .	494.4	119.4	Canada, 15.1; Caribbean, 81.2; Middle East, 19.7; Africa, 4.4
1975 . . .	692.3	167.3	Canada, 27.8; Caribbean, 116.2; Middle East, 18.6; Africa, 9.9
1980 . . .	799.5	190.3	Canada, 34.2; Caribbean, 131.0; Middle East, 18.2; Africa, 14.0
Canada			
1965 . . .	50.0	11.4	Caribbean, 20.2; Middle East, 5.9; United States, –15.1
1975 . . .	79.4	7.6	Caribbean, 25.7; Africa, 3.2; Middle East, 8.9; United States, –27.8
1980 . . .	102.9	6.2	Caribbean, 29.8; Africa, 3.8; Middle East, 10.5; United States, –34.2
Non-Caribbean Latin America			
1965 . . .	41.2	23.0	Caribbean, 10.0; Africa, 1.5; Middle East, 7.0; Communist countries, 4.3
1975 . . .	69.4	38.4	Caribbean, 19.1; Africa, 4.3; Middle East, 15.0
1980 . . .	88.9	44.0	Caribbean, 24.3; Africa, 4.6; Middle East, 15.2
Western Europe			
1965 . . .	343.0	368.7	Caribbean, 38.0; North Africa, 84.0; Other Africa, 14.8; Middle East, 201.8; Communist countries, 27.0
1975 . . .	670.2	716.6	Caribbean, 35.8; North Africa, 274.8; Other Africa, 60.2; Middle East, 338.6
1980 . . .	837.3	902.1	Caribbean, 40.3; North Africa, 346.8; Other Africa, 98.7; Middle East, 407.4
Japan			
1965 . . .	76.7	80.5	Middle East, 67.8; Other Asia, 6.8
1975 . . .	217.9	227.8	Middle East, 170.7; Other Asia, 45.5
1980 . . .	346.2	360.5	Middle East, 253.4; Other Asia, 92.5
Other Asia			
1965 . . .	40.2	11.3	Not available[b]
1975 . . .	86.9	25.3	Middle East, 70.4; Japan, –45.5; United States, –4.7
1980 . . .	129.1	12.8	Middle East, 104.4; Japan, –92.5; United States, –5.4
Oceania			
1965 . . .	19.6	20.0	Middle East, 14.6; Other Asia, 5.5
1975 . . .	33.5	–	
1980 . . .	41.0	–	

[a] Trade balance with a particular area may exceed overall trade balance.

[b] As derived from Table VI (similarly to the other net import figures in the present table), net imports of "Other Asia" amounted to 3.1 million metric tons in 1965. This figure seems suspect when compared to the 11.3 million ton figure and the longer historical series appearing in Table V. The 11.3 million ton net import figure for 1965 (from Table V) is therefore used above, even though it could not be allocated by origin and destination. The 3.1 million ton geographically allocable (if less reliable) estimate from Table V shows, among principal suppliers or export markets(–): Middle East, 20.1; Japan, –6.8; Oceania, –5.5; United States, –3.3.

Source: Consumption from Table V, and net imports and regional trade balances from Table VI (except for "Other Asia" in 1965, concerning which see footnote b, above). Note that the net import definition in the present table necessarily conforms to that used in Table VI. See footnote a to Table VI.

Table IX

Solid Fuels: Consumption[a] and Net Trade,[b] 1955–80

(1) = Consumption (1,000 metric tons HCE);
(2) = Net imports or net exports(–) (1,000 metric tons HCE);
(3) = Net imports or net exports(–) as percent of consumption

	1955	1957	1960	1963	1965	1975	1980
Canada							
(1) . . .	29,310	27,990	20,080	19,564	23,305	30,180	34,170
(2) . . .	17,500	18,600	11,050	12,500	14,120	15,220	15,710
(3) . . .	59.7%	66.5%	55.0%	63.9%	60.6%	50.4%	46.0%
United States							
(1) . . .	403,680	390,820	359,190	382,988	425,203	549,000	626,000
(2) . . .	–49,160	–73,550	–34,450	–45,600	–46,800	–67,240	–80,910
(3) . . .	–12.2%	–18.8%	–9.6%	–11.9%	–11.0%	–12.2%	–12.9%
Latin America							
(1) . . .	10,600	10,881	9,740	10,237	10,595	17,670	21,680
(2) . . .	2,370	2,610	3,010	2,170	2,640	4,420	5,420
(3) . . .	22.4%	24.0%	30.9%	21.2%	24.9%	25.0%	25.0%
Western Europe							
(1) . . .	568,080	585,750	540,250	556,780	515,414	370,500	340,000
(2) . . .	38,020	58,700	31,850	45,800	43,900	53,000	59,970
(3) . . .	6.7%	10.0%	5.9%	8.2%	8.5%	14.3%	17.6%
Far East							
(1) . . .	91,180	116,990	125,740	145,857	150,358	247,580	337,660
(2) . . .	4,070	7,730	9,190	11,710	17,800	30,890	36,050
(3) . . .	4.5%	6.6%	7.3%	8.0%	11.8%	12.5%	10.7%
of which: Japan							
(1) . . .	46,110	55,330	62,240	64,791	64,534	64,600	69,600
(2) . . .	2,900	6,480	8,200	11,050	17,290	29,600	34,600
(3) . . .	6.3%	11.7%	13.2%	17.1%	26.8%	45.8%	49.7%
Oceania							
(1) . . .	24,370	24,510	27,860	30,283	34,287	40,990	47,490
(2) . . .	–30	–560	–1,390	–3,080	–6,980	–11,390	–13,500
(3) . . .	–0.1%	–2.3%	–5.0%	–10.2%	–20.4%	–27.8%	–28.4%
Middle East and Africa							
(1) . . .	42,400	46,090	48,070	51,160	58,685	98,210	124,760
(2) . . .	390	–90	–10	–540	–320	360	210
(3) . . .	0.9%	–0.2%	–0.0%	–1.1%	–0.5%	0.4%	0.2%
Communist countries							
(1) . . .	646,950	740,520	1,073,430	976,832	1,043,438	1,633,230	2,083,550
(2) . . .	–13,550	–14,350	–19,300	–23,150	–24,660	–25,250	–22,950
(3) . . .	–2.1%	–1.9%	–1.8%	–2.4%	–2.4%	–1.5%	–1.1%

[a] Consumption refers to inland consumption, thus excluding negligible quantities of coal used as bunker fuel.

[b] Net trade, as shown here, refers to shipments measured at points of origin. This is done for consistency with Tables X and XI. The United Nations source cited below records also somewhat different trade figures (not shown in this study), calculated from the importing-country side.

Source: Consumption as in Table IV. Historical net imports from United Nations, *World Energy Supplies*, Series J, various numbers. On projections, see accompanying text. Note, however, that only net *import* consumption shares were explicitly projected. Net exports were derived following the allocation of imports by geographic sources (in Table X). When projected net exports–so derived–appeared to require additional adjustment, this computational procedure was repeated one or more times.

Table IX.A

Solid Fuels: Consumption[a] and Net Trade,[b] 1955–80

(1) - Consumption (1,000 metric tons oil equivalent);
(2) = Net imports or net exports(–) (1,000 metric tons oil equivalent);
(3) = Net imports or net exports(–) as percent of consumption

	1955	1957	1960	1963	1965	1975	1980
Canada							
(1) . . .	19,541	18,661	13,387	13,043	15,537	20,120	22,780
(2) . . .	11,667	18,600	7,367	8,333	9,413	10,150	10,470
(3) . . .	59.7%	66.5%	55.0%	63.9%	60.6%	50.4%	46.0%
United States							
(1) . . .	269,133	260,560	239,472	255,338	283,483	366,020	417,350
(2) . . .	–32,773	–49,033	–22,967	–30,400	–31,200	–44,830	–53,940
(3) . . .	–12.2%	–18.8%	–9.6%	–11.9%	–11.0%	–12.2%	–12.9%
Latin America							
(1) . . .	7,067	7,254	6,494	6,825	7,064	11,780	14,450
(2) . . .	1,580	1,740	2,007	1,447	1,760	2,947	3,614
(3) . . .	22.4%	24.0%	30.9%	21.2%	24.9%	25.0%	25.0%
Western Europe							
(1) . . .	378,739	390,520	360,185	371,205	343,627	247,000	226,670
(2) . . .	25,347	39,133	21,233	30,533	29,267	35,330	39,980
(3) . . .	6.7%	10.0%	5.9%	8.2%	8.5%	14.3%	17.6%
Far East							
(1) . . .	60,790	77,997	83,831	97,243	100,244	165,060	225,110
(2) . . .	2,713	5,153	6,127	7,807	11,867	20,590	24,030
(3) . . .	4.5%	6.6%	7.3%	8.0%	11.8%	12.5%	10.7%
of which: Japan							
(1) . . .	30,742	36,889	41,495	43,196	43,025	43,070	46,400
(2) . . .	1,933	4,320	5,467	7,367	11,527	19,730	23,070
(3) . . .	6.3%	11.7%	13.2%	17.1%	26.8%	45.8%	49.7%
Oceania							
(1) . . .	16,247	16,341	18,574	20,190	22,859	27,330	31,660
(2) . . .	–20	–373	–927	–2,053	–4,653	–7,590	–9,000
(3) . . .	–0.1%	–2.3%	–5.0%	–10.2%	–20.4%	–27.8%	–28.4%
Middle East and Africa							
(1) . . .	28,268	30,728	32,048	34,108	39,125	65,470	83,180
(2) . . .	260	–60	–7	–360	–213	240	140
(3) . . .	0.9%	–0.2%	–0.0%	–1.1%	–0.5%	0.4%	0.2%
Communist countries							
(1) . . .	431,322	493,705	715,656	651,254	695,660	1,088,840	1,389,040
(2) . . .	–9,033	–9,567	–12,867	–15,433	–16,440	–16,830	–15,300
(3) . . .	–2.1%	–1.9%	–1.8%	–2.4%	–2.4%	–1.5%	–1.1%

[a] Consumption refers to inland consumption, thus excluding negligible quantities of coal used as bunker fuel.

[b] Net trade, as shown here, refers to shipments measured at points of origin. This is done for consistency with Tables X and XI. United Nations, *World Energy Supplies,* Series J, records also somewhat different trade figures (not shown in this study), calculated from the importing-country side.

Source: Table IX coal equivalent figures divided by 1.5 in order to approximate oil equivalents.

Table X

Interregional Shipments of Solid Fuels by Origin and Destination, Selected Years, 1955–80

(Thousand Metric Tons HCE)

Importing area	United States	Canada	Western Europe	Africa	Oceania	Communist Countries	Total imports	Net imports or net exports(–)
	Exporting area							
	1955							
United States	–	310	–	–	–	–	310	–49,160
Canada	17,800	–	240	–	–	–	18,040	17,500
Latin America	1,300	–	190	10	–	870	2,370	2,370
Western Europe	26,560	220	–	220	–	12,290	39,290	38,020
Africa and Middle East	20	–	420	–	–	–	750	390
Far East	3,410	–	–	90	30	540	4,070	4,070
Japan	2,500	–	–	20	10	370	2,900	2,900
Oceania	–		–	–	–	–	–	–30
Communist countries	–	–	340	–	–	–	340	–13,550
Total exports	49,470	540	1,270	360	30	13,890	65,560	
	1960							
United States	–	300	–	–	–	–	350	–34,450
Canada	11,900	–	60	–	–	–	11,950	11,050
Latin America	2,120	–	160	–	–	650	3,010	3,010
Western Europe	15,500	–	–	180	–	17,400	33,100	31,850
Africa and Middle East	65	–	370	–	–	200	635	–10
Far East	5,230	550	50	400	1,460	1,500	9,190	9,190
Japan	5,200	550	10	20	1,400	1,020	8,200	8,200
Oceania	–	–	60	6	–	–	70	–1,390
Communist countries	–	–	540	–	–	–	500	–19,300
Total exports	34,800	900	1,250	645	1,460	19,800	58,860	
	1965							
United States	–	240	10	–	–	–	250	–46,800
Canada	15,290	–	5	–	–	–	15,300	14,120
Latin America	2,090	–	280	–	–	270	2,640	2,640
Western Europe	22,600	–	–	540	–	21,800	44,950	43,900
Africa and Middle East	–	–	95	–	–	500	590	–320
Far East	6,840	920	55	340	6,990	2,650	17,800	17,800
Japan	6,800	920	45	190	6,860	2,270	17,290	17,290
Oceania	5	–	20	15	–	–	40	–6,980
Communist countries	210	–	450	–	–	–	670	–24,660
Total exports	47,050	1,180	1,050	910	7,020	25,330	82,540	

Table X–*Continued*

Importing area	Exporting area: United States	Canada	Western Europe	Africa	Oceania	Communist Countries	Total imports	Net imports or net exports(–)
				1975				
United States	–	–	–	–	–	–	–	–67,240
Canada	19,620	–	–	–	–	–	19,620	15,220
Latin America	3,760	–	130	–	–	530	4,420	4,420
Western Europe	33,300	–	–	–	–	20,350	53,650	53,000
Africa and Middle East	30	–	170	–	–	340	550	360
Far East	10,530	4,400	–	190	11,390	4,380	30,890	30,890
Japan	10,400	4,400	–	–	11,000	3,800	29,600	29,600
Oceania	–	–	–	–	–	–	–	–11,390
Communist countries	–	–	350	–	–	–	350	–25,250
Total exports	67,240	4,400	650	190	11,390	25,600	109,470	
				1980				
United States	–	–	–	–	–	–	–	–80,910
Canada	22,210	–	–	–	–	–	22,210	15,710
Latin America	4,880	–	–	–	–	540	5,420	5,420
Western Europe	42,100	–	–	–	–	18,300	60,400	59,970
Africa and Middle East	40	–	130	–	–	220	390	210
Far East	11,680	6,500	–	180	13,500	4,190	36,050	36,050
Japan	11,500	6,500	–	–	13,000	3,600	34,600	35,600
Oceania	–	–	–	–	–	–	–	–13,500
Communist countries	–	–	300	–	–	–	300	–22,950
Total exports	80,910	6,500	430	180	13,500	23,250	124,770	

Source: Historical data from United Nations, *World Energy Supplies*, Series J, various numbers. Projections calculated by applying regional percentages of Table XI to total projected solid fuels imports shown in the next-to-last column of the present table. Total imports (that is, *gross* imports) were, in turn, derived from projected *net* imports (shown in the last column above and taken from Table IX) on the basis of the historical trends in the gross-net import relationship. (Some successive approximation was inherent in this procedure since the now-completed matrix of 1975 and 1980 interregional flows might yield net trade balances somewhat different from those initially stipulated.)

Note: Figures may not add to row or column totals, both because of rounding and due to geographically unallocable shipments. (E.g., African and Middle East imports in 1955 and United States imports in 1960 are not fully allocable.)

Table X.A

Interregional Shipments of Solid Fuels by Origin and Destination, Selected Years, 1955–80

(Thousand Metric Tons Oil Equivalent)

Importing Area	Exporting area							
	United States	Canada	Western Europe	Africa	Oceania	Communist countries	Total imports	Net imports or net exports(–)
				1955				
United States	–	205	–	–	–	–	205	–32,775
Canada	11,865	–	160	–	–	–	12,025	11,665
Latin America	865	–	125	5	–	580	1,580	1,580
Western Europe	17,710	145	–	145	–	8,195	26,195	25,350
Africa and Middle East	15	–	280	–	–	–	500	260
Far East	2,280	–	–	60	20	360	2,715	2,715
Japan	1,665	–	–	15	5	245	1,935	1,935
Oceania	–	–	–	–	–	–	–	–20
Communist countries	–	–	225	–	–	–	225	–9,035
Total exports	32,980	360	845	240	20	9,260	43,705	
				1960				
United States	–	200	–	–	–	–	235	–22,965
Canada	7,935	–	40	–	–	–	7,965	7,665
Latin America	1,415	–	105	–	–	435	2,005	2,005
Western Europe	10,335	–	–	120	–	11,600	22,065	19,230
Africa and Middle East	45	–	245	–	–	135	425	–5
Far East	3,485	365	35	265	975	1,000	6,125	6,125
Japan	3,465	365	5	15	935	680	5,465	5,465
Oceania	–	–	40	5	–	–	45	–930
Communist countries	–	–	360	–	–	–	335	–12,865
Total exports	23,200	600	835	430	975	13,200	39,240	
				1965				
United States	–	160	5	–	–	–	165	–31,205
Canada	10,195	–	5	–	–	–	10,200	–9,415
Latin America	1,395	–	185	–	–	180	1,760	1,760
Western Europe	15,065	–	–	360	–	14,535	29,965	29,265
Africa and Middle East	–	–	65	–	–	335	395	–210
Far East	4,560	615	35	225	4,660	1,765	11,865	11,865
Japan	4,535	615	30	125	4,575	1,515	11,525	11,525
Oceania	5	–	15	10	–	–	25	–4,655
Communist countries	140	–	300	–	–	–	445	–16,440
Total exports	31,370	787	700	605	4,680	16,885	55,025	

Table X. A–*Continued*

Importing Area	United States	Canada	Western Europe	Africa	Oceania	Communist countries	Total imports	Net imports or net exports(–)
	Exporting area							
	1975							
United States	–	–	–	–	–	–	–	–44,830
Canada	13,080	–	–	–	–	–	13,080	10,150
Latin America	2,510	–	90	–	–	350	2,950	2,950
Western Europe	22,200	–	–	–	–	13,570	35,770	35,340
Africa and Middle East	20	–	110	–	–	230	370	240
Far East	7,020	2,930	–	130	7,590	2,920	20,590	20,590
Japan	6,930	2,930	–	–	7,330	2,530	19,730	19,730
Oceania	–	–	–	–	–	–	–	–7,590
Communist countries	–	–	230	–	–	–	230	–16,840
Total exports	44,830	2,930	430	130	7,590	17,070	72,980	
	1980							
United States	–	–	–	–	–	–	–	–53,940
Canada	14,810	–	–	–	–	–	14,810	10,480
Latin America	3,250	–	–	–	–	360	3,610	3,610
Western Europe	28,070	–	–	–	–	12,200	40,270	39,980
Africa and Middle East	30	–	90	–	–	150	260	140
Far East	7,790	4,330	–	120	9,000	2,790	24,030	24,030
Japan	7,670	4,330	–	–	8,670	2,400	23,070	23,070
Oceania	–	–	–	–	–	–	–	–9,000
Communist countries	–	–	200	–	–	–	200	–15,300
Total exports	53,940	4,330	290	120	9,000	15,500	83,180	

Source: Table X coal equivalent figures divided by 1.5 in order to approximate oil equivalents. Note to Table X applies here as well.

Table XI

Interregional Imports of Solid Fuels: Percentage Distribution by Area of Origin, Selected Years, 1955–80

Importing area	Exporting area						
	United States	Canada	Western Europe	Africa	Oceania	Communist countries	Total imports
				1955			
United States	–	100.0	–	–	–	–	100.0
Canada	98.7	–	1.3	–	–	–	100.0
Latin America	54.9	–	8.0	0.4	–	36.7	100.0
Western Europe	67.6	0.6	–	0.6	–	31.3	100.0
Africa and Middle East	2.7	–	56.0	–	–	–	100.0
Far East	83.8	–	–	2.2	0.7	13.3	100.0
Japan	86.2	–	–	0.7	0.3	12.8	100.0
Oceania	–	–	–	–	–	–	–
Communist countries	–	–	100.0	–	–	–	100.0
Total exports	75.4	0.8	1.9	0.5	0.0	21.2	100.0
				1960			
United States	–	85.7	–	–	–	–	100.0
Canada	99.6	–	0.5	–	–	–	100.0
Latin America	70.4	–	5.3	–	–	21.6	100.0
Western Europe	46.8	–	–	0.5	–	52.6	100.0
Africa and Middle East	10.2	–	58.3	–	–	31.5	100.0
Far East	56.9	6.0	0.5	4.4	15.9	16.3	100.0
Japan	63.4	6.7	0.1	0.2	17.1	12.4	100.0
Oceania	–	–	85.7	8.6	–	–	100.0
Communist countries	–	–	100.0	–		–	100.0
Total exports	59.1	1.5	2.1	1.1	2.5	33.6	100.0
				1965			
United States	–	96.0	4.0	–	–	–	100.0
Canada	99.9	–	0.0	–		–	100.0
Latin America	79.2	–	10.6	–	–	10.2	100.0
Western Europe	50.3	–	–	1.2	–	48.5	100.0
Africa and Middle East	–	–	16.1	–	–	84.7	100.0
Far East	38.4	5.2	0.3	1.9	39.3	14.9	100.0
Japan	39.3	5.3	0.3	1.1	39.7	13.1	100.0
Oceania	12.5	–	50.0	37.5	–	–	100.0
Communist countries	31.3	–	67.2	–	–	–	100.0
Total exports	57.0	1.4	1.3	1.1	8.5	30.7	100.0

Table XI–*Continued*

Importing area	Exporting area						
	United States	Canada	Western Europe	Africa	Oceania	Communist countries	Total imports
				1975			
United States	–	–	–	–	–	–	–
Canada	100.0	–	–	–	–	–	100.0
Latin America	85.1	–	2.9	–	–	12.0	100.0
Western Europe	62.1	–	–	–	–	37.9	100.0
Africa and Middle East	5.5	–	30.9	–	–	61.8	100.0
Far East	34.1	14.2	–	0.6	36.9	14.2	100.0
Japan	35.1	14.9	–	–	37.2	12.8	100.0
Oceania	–	–	–	–	–	–	–
Communist countries	–	–	100.0	–	–	–	100.0
Total exports	61.4	4.0	0.6	0.2	10.4	23.4	100.0
				1980			
United States	–	–	–	–	–	–	–
Canada	100.0	–	–	–	–	–	100.0
Latin America	90.0	–	–	–	–	10.0	100.0
Western Europe	69.7	–	–	–	–	30.3	100.0
Africa and Middle East	10.3	–	33.3	–	–	56.4	100.0
Far East	32.9	18.0	–	0.5	37.4	11.6	100.0
Japan	33.2	18.8	–	–	37.6	10.4	100.0
Oceania	–	–	–	–	–	–	–
Communist countries	–	–	100.0	–	–	–	100.0
Total exports	64.8	5.2	0.3	0.1	10.8	18.6	100.0

Source: Historical data represent percentage distributions of data in Table X; projections are described in accompanying text.

Note: Figures may not add to row or column totals, both because of rounding and due to geographically unallocable shipments. (E.g., African and Middle East imports in 1955 and United States imports in 1960 are not fully allocable.)

Table XII

Summary: Solid Fuels Balances, Selected Areas,[a] 1955–80

(Million Metric Tons HCE)

	Production	Consumption	Net imports or net exports(–)[b]	Principal suppliers or export markets(–)[b]
Canada				
1955 ..	12.0	29.3	17.5	United States, 17.5
1960 ..	8.7	20.1	11.1	United States, 11.6
1965 ..	9.3	23.3	14.1	United States, 15.0
1975 ..	15.0	30.2	15.2	United States, 19.6
1980 ..	18.5	34.2	15.7	United States, 22.2
United States				
1955 ..	443.4	403.7	–49.2	Canada, –17.5; Western Europe,–26.6; Japan,–2.5
1960 ..	392.4	359.2	–34.4	Canada, –11.6; Western Europe,–15.5; Japan,–5.2
1965 ..	476.2	425.2	–46.8	Canada, –15.0; Western Europe,–22.6; Japan,–6.8
1975 ..	616.2	549.0	–67.2	Canada, –19.6; Western Europe,–33.3; Japan,–10.4
1980 ..	706.9	626.0	–80.9	Canada, –22.2; Western Europe,–42.1; Japan,–11.5
Western Europe				
1955 ..	529.3	568.1	38.0	United States, 26.6; Communist countries, 12.0
1960 ..	497.8	540.2	31.8	United States, 15.5; Communist countries, 16.9
1965 ..	481.1	515.4	43.9	United States, 22.6; Communist countries, 21.6
1975 ..	317.5	370.5	53.0	United States, 33.3; Communist countries, 20.0
1980 ..	280.0	340.0	60.0	United States, 42.1; Communist countries, 18.0
Japan				
1955 ..	42.9	46.1	2.9	United States, 2.5
1960 ..	51.5	62.2	8.2	United States, 5.2; Canada, 0.6; Oceania, 1.4; Communist countries,1.0
1965 ..	49.7	64.5	17.3	United States, 6.8; Canada, 0.9; Oceania, 6.9; Communist countries, 2.3
1975 ..	35.0	64.6	29.6	United States, 10.4; Canada, 4.4; Oceania, 11.0; Communist countries, 3.8
1980 ..	35.0	69.6	34.6	United States, 11.5; Canada, 6.5; Oceania, 13.0; Communist countries, 3.6
Oceania				
1955 ..	24.4	24.4	–	
1960 ..	29.5	27.9	–1.4	Japan, –1.4
1965 ..	40.9	34.3	–7.0	Japan, –6.9
1975 ..	52.4	41.0	–11.4	Japan, –11.0
1980 ..	61.0	47.5	–13.5	Japan, –13.0
Communist countries				
1955 ..	661.2	647.0	–13.6	Western Europe, –12.0
1960 ..	1,093.1	1,073.4	–19.3	Western Europe, –16.9; Japan, –1.0
1965 ..	1,068.7	1,043.4	–24.7	Western Europe, –21.6; Japan, –2.3
1975 ..	1,658.4	1,633.2	–25.2	Western Europe, –20.0; Japan, –3.8
1980 ..	2,106.6	2,083.6	–23.0	Western Europe, –18.0; Japan, –3.6

[a] Latin America and Middle East and Africa are excluded from the tabulations.
[b] Trade balance with a particular area may exceed overall trade balance.

Source: Historical data from United Nations, *World Energy Supplies,* Series J, various numbers.
Projections from our Tables IV and X, except production, which represents consumption less net imports. (See accompanying text remarks.)

Note: For historical years, net trade may not precisely equal the difference between (inland) consumption and production owing (1) to stock changes and (to a negligible extent) bunkers (both factors being ignored in projected years); and (2) to the fact that net imports (as pointed out in the note to Table IX) are calculated exclusively from the exporting-country side.

Table XII.A

Summary: Solid Fuels Balances, Selected Areas,[a] 1955–80

(Million Metric Tons Oil Equivalent)

	Production	Consumption	Net imports or net exports(–)[b]	Principal suppliers or export markets(–)[b]
Canada				
1955 ...	8.0	19.5	11.7	United States, 11.7
1960 ...	5.8	13.4	7.4	United States, 7.7
1965 ...	6.2	15.5	9.4	United States, 10.0
1975 ...	10.0	20.1	10.1	United States, 13.1
1980 ...	12.3	22.8	10.5	United States, 14.8
United States				
1955 ...	295.6	269.1	–32.8	Canada, –11.7; Western Europe, –17.7; Japan, –1.7
1960 ...	261.6	239.5	–22.9	Canada, –7.7; Western Europe, –10.3; Japan, –3.5
1965 ...	317.5	283.5	–31.2	Canada, –10.0; Western Europe, –15.1; Japan, –4.5
1975 ...	410.8	366.0	–44.8	Canada, –13.1; Western Europe, –22.2; Japan, –6.9
1980 ...	471.3	417.4	–53.9	Canada, –14.8; Western Europe, –28.1; Japan, –7.7
Western Europe				
1955 ...	352.9	378.8	25.3	United States, 17.7; Communist countries, 8.0
1960 ...	331.9	360.2	21.2	United States, 10.3; Communist countries, 11.3
1965 ...	320.7	343.6	29.3	United States, 15.1; Communist countries, 14.4
1975 ...	211.7	247.0	35.3	United States, 22.2; Communist countries, 13.3
1980 ...	186.7	226.7	40.0	United States, 28.1; Communist countries, 12.0
Japan				
1955 ...	28.6	30.7	1.9	United States, 1.7
1960 ...	34.3	41.5	5.5	United States, 3.5; Canada, 0.4; Oceania, 0.9; Communist countries, 0.7
1965 ...	33.1	43.0	11.5	United States, 4.5; Canada, 0.6; Oceania, 4.6; Communist countries, 1.5
1975 ...	23.3	43.1	19.7	United States, 6.9; Canada, 2.9; Oceania, 7.3; Communist countries, 2.5
1980 ...	23.3	46.4	23.1	United States, 7.7; Canada, 4.3; Oceania, 8.7; Communist countries, 2.4
Oceania				
1955 ...	16.3	16.3	–	
1960 ...	19.7	18.6	–0.9	Japan, –0.9
1965 ...	27.3	22.9	–4.7	Japan, –4.6
1975 ...	34.9	27.3	–7.6	Japan, –7.3
1980 ...	40.7	31.7	–9.0	Japan, –8.7
Communist countries				
1955 ...	440.8	431.4	–9.1	Western Europe, –8.0
1960 ...	728.8	715.6	–12.9	Western Europe, –11.3; Japan, –0.7
1965 ...	712.5	695.6	–16.5	Western Europe, –14.4; Japan, –1.5
1975 ...	1,105.6	1,088.8	–16.8	Western Europe, –13.3; Japan, –2.5
1980 ...	1,404.3	1,389.0	–15.3	Western Europe, –12.0; Japan, –2.4

[a] Latin America and Middle East and Africa are excluded from the tabulations.
[b] Trade balance with a particular area may exceed overall trade balance.

Source: Table XII coal equivalents divided by 1.5 in order to approximate oil equivalents. Note to Table XII applies here as well.

Author Index

Numbers in parentheses indicate the numbers of the references when these are cited in the text without the names of the authors.

Numbers set in *italics* designate the page numbers on which the complete literature citations are given.

Subject Index

Selected Rand *Books*

Becker, Abraham S. *Soviet National Income 1958-1964.* University of California Press, Berkeley and Los Angeles. 1969.

Bergson, A. *The Real National Income of Soviet Russia Since 1928.* Harvard University Press, Cambridge, Massachusetts. 1961.

Bergson, Abram, and Hans Heymann, Jr. *Soviet National Income and Product,* 1940-48. Columbia University Press, New York. 1954.

Chapman, Janet G. *Real Wages in Soviet Russia Since 1928.* Harvard University Press, Cambridge, Massachusetts. 1963.

Downs, Anthony. *Inside Bureaucracy.* Little Brown and Company, Boston, Massachusetts. 1967.

Gurtov, Melvin. *Southeast Asia Tomorrow: Problems and Prospects For U.S. Policy.* Johns Hopkins Press, Baltimore, Maryland. 1970.

Hirshleifer, Jack, James C. DeHaven, and Jerome W. Milliman. *Water Supply: Economics, Technology, and Policy.* The University of Chicago, Chicago, Illinois. 1960.

Hitch, Charles J., and Roland McKean. *The Economics of Defense in the Nuclear Age.* Harvard University Press, Cambridge, Massachusetts. 1960.

Hoeffding, Oleg. *Soviet National Income and Product in 1928.* Columbia University Press, New York. 1954.

Johnson, William A. *The Steel Industry of India.* Harvard University Press, Cambridge, Massachusetts. 1966.

Johnstone, William C. *Burma's Foreign Policy: A Study in Neutralism.* Harvard University Press, Cambridge, Massachusetts. 1963.

Kershaw, Joseph A., and Roland N. McKean. *Teacher Shortages and Salary Schedules.* McGraw-Hill Book Company. Inc., New York. 1962.

Leites, Nathan, and C. Wolf. *Rebellion and Authority.* Markham Publishing Company, Chicago, Illinois. 1970.

Liu, Ta-Chung, and Kung-Chia Yeh. *The Economy of the Chinese Mainland: National Income and Economic Development, 1933-1959.* Princeton University Press, Princeton, New Jersey. 1965.

Lubell, Harold. *Middle East Oil Crises and Western Europe's Energy Supplies.* The Johns Hopkins Press, Baltimore, Maryland. 1963.

McKean, Roland N. *Efficiency in Government through Systems Analysis with Emphasis on Water Resource Development.* John Wiley and Sons, Inc., New York. 1958.

Marschak, Thomas, Thomas K. Glennan Jr., and Robert Summers. *Strategy for R & D,* Springer-Verlag, Inc., Berlin, Heidelberg, New York, 1967.

Moorsteen, Richard. *Prices and Production of Machinery in the Soviet Union, 1928-1958.* Harvard University Press, Cambridge, Massachusetts. 1962.

Nelson, Richard R., Merton J. Peck, and Edward D. Kalachek. *Technology, Economic Growth and Public Policy.* The Brookings Institution, Washington, D. C. 1967.

Pascal, Anthony. *Thinking About Cities: New Perspectives on Urban Problems.* Dickenson Publishing Company, Belmont, California. 1970.

Pincus, John A. *Economic Aid and International Cost Sharing.* The Johns Hopkins Press, Baltimore, Maryland. 1965.

Rosen, George. *Democracy and Economic Change in India.* University of California Press, Berkeley and Los Angeles, California. 1966.

Wolf, Charles, Jr. *Foreign Aid: Theory and Practice in Southern Asia.* Princeton University Press, Princeton, New Jersey. 1960.

Selected Resources for the Future *Books*

Grunwald, Joseph, and Philip Musgrove. *Natural Resources in Latin American Development.* 1970. 512 pp. $20.00

Mikesell, Raymond F., and Associates. *Foreign Investment in the Petroleum and Mineral Industries: Case Studies of Investor-Host Country Relations.* 1970. 460 pp. $15.00.

Herfindahl, Orris C. *Natural Resource Information for Economic Development.* 1969. 232 pp. $7.00.

Christy, Francis T., Jr., and Anthony Scott. *The Common Wealth in Ocean Fisheries: Some Problems of Growth and Economic Allocation.* 1966. 296 pp. $6.00.

Kneese, Allen V., and Blair T. Bower. *Managing Water Quality: Economics, Technology, Institutions.* 1968. 338 pp. $8.95.

Landsberg, Hans H., Leonard L. Fischman, and Joseph L. Fisher. *Resources in America's Future: Patterns of Requirements and Availabilities, 1960-2000.* 1963. 1,040 pp. $15.00.

Barnett, Harold J., and Chandler Morse. *Scarcity and Growth: The Economics of Natural Resource Availability.* 1963. 304 pp. $9.00. Paper 1969, $2.25.

Crosson, Pierre R. *Agricultural Development and Productivity: Lessons from the Chilean Experience*. 1970. 208 pp. $7.00.

Brubaker, Sterling. *Trends in the World Aluminum Industry.* 1967. 276 pp. $6.95.

Campbell, Robert W. *The Economics of Soviet Oil and Gas.* 1968. 294 pp. $6.95.

Clawson, Marion, ed. *Natural Resources and International Development: Essays Based on the RFF Forum Lectures of 1963.* 1964. 474 pp. $10.00.

Arrow, Kenneth J., and Mordecai Kurz. *Public Investment, the Rate of Return, and Optimal Fiscal Policy.* 1970. 288 pp. $9.00.

Lovejoy, Wallace F., and Paul T. Homan. *Methods of Estimating Reserves of Crude Oil, Natural Gas, and Natural Gas Liquids.* 1965. 180 pp. Paper, $3.00

McDonald, Stephen L. *Petroleum Conservation in the United States: An Economic Analysis.* 1971. In press.

Resources for the Future. *U.S. Energy Policies: An Agenda for Research.* 1968. 164 pp. Paper, $4.00.

Schurr, Sam H., and Bruce C. Netschert, with Vera F. Eliasberg, Joseph Lerner, and Hans H. Landsberg. *Energy in the American Economy, 1850-1975.* 1960. 796 pp. $15.00.

(All titles cited above were published for RFF by The Johns Hopkins Press, Baltimore, Maryland 21218.)